An Introduction to 5G

An Introduction to 5G

The New Radio, 5G Network and Beyond

Christopher Cox
Chris Cox Communications Ltd
Cambridge, UK.

Registered Offices
John Wiley & Sons, Inc., 111 River Street, Hoboken, NJ 07030, USA
John Wiley & Sons Ltd, The Atrium, Southern Gate, Chichester, West Sussex, PO19 8SQ, UK

Editorial Office
The Atrium, Southern Gate, Chichester, West Sussex, PO19 8SQ, UK

For details of our global editorial offices, customer services, and more information about Wiley products visit us at www.wiley.com.

Wiley also publishes its books in a variety of electronic formats and by print-on-demand. Some content that appears in standard print versions of this book may not be available in other formats.

Library of Congress Cataloging-in-Publication Data
Names: Cox, Christopher (Christopher Ian), 1965- author.
Title: An introduction to 5G : the new radio, 5G network and beyond / Christopher Cox.
Description: Hoboken, NJ : Wiley, [2021] | Includes index.
Identifiers: LCCN 2020021380 (print) | LCCN 2020021381 (ebook) | ISBN 9781119602668 (cloth) | ISBN 9781119602699 (adobe pdf) | ISBN 9781119602712 (epub)
Subjects: LCSH: 5G mobile communication systems.
Classification: LCC TK5103.25 .C68 2021 (print) | LCC TK5103.25 (ebook) | DDC 621.3845/6—dc23
LC record available at https://lccn.loc.gov/2020021380
LC ebook record available at https://lccn.loc.gov/2020021381

Cover image: © KTSDESIGN/SCIENCE PHOTO LIBRARY/Getty Images
Cover design by Wiley

Set in 9.5/12.5pt STIXTwoText by SPi Global, Chennai, India

10 9 8 7 6 5 4 3 2 1

To Susie

Contents

Preface

This book is an introduction to fifth-generation (5G) mobile telecommunication systems.

5G builds on earlier generations of mobile telecommunications, but it differs in several ways. It is designed for a wider range of applications than before: not only consumer applications such as voice, video and data, but also industrial applications such as machine-type communications. In turn, those applications have a wider range of requirements than before: some require a peak data rate of several gigabits per second, others require the delivery of data packets with delays as low as a few milliseconds, while others require a battery life of several years. To help meet those requirements, the network uses technologies such as network function virtualization, software-defined networking and network slicing, so as to provide a flexible, underlying resource that can be easily reconfigured to support new requirements as they arise. In pursuit of higher data rates and higher system capacities, the air interface supports higher radio frequencies than previous generations have done, and makes extensive use of multiple antennas.

There are many other books on 5G which contain detailed accounts of the underlying technologies, the likely applications, the system architecture and the air interface. The aim of this book is to complement them by offering an end-to-end, system-level perspective. It contains fewer details about individual topics than many of the others, but it is broader in scope, covering the requirements and architecture of 5G, the principles and implementation of the air interface and the signalling procedures within the network. It is aimed at mobile telecommunication engineers who want an introduction to the architecture and operation of 5G; at those who are experts in one part of the system, but who want to understand what is taking place elsewhere; and as a technical foundation for those in related roles such as management, marketing, intellectual property and consultancy.

The first few chapters cover the foundations. Chapter 1 reviews the history of mobile telecommunications, and introduces the applications, requirements and key technical features of 5G. Chapter 2 covers the architecture of the core network and the signalling protocols that it uses, while Chapter 3 covers similar ground in the case of the radio access network.

The next three chapters address the underlying technologies that are used by the air interface. Chapter 4 reviews the use of radio spectrum, and discusses the coverage issues that appear when operating at high radio frequencies. Chapter 5 reviews the techniques that are used for digital signal processing, for example modulation, demodulation, orthogonal frequency division multiple access and error correction. (Much of that chapter has been condensed from the author's previous work, 'An Introduction to LTE'.) Chapter 6 explains how 5G uses multiple antennas to compensate for the coverage problems that appear at high radio frequencies, and to improve the system capacity when the radio frequency is low.

The next six chapters describe how the air interface is actually implemented. Chapter 7 is an introduction to the air interface, while Chapter 8 describes the acquisition procedure in which a mobile discovers a nearby cell, and Chapter 9 describes the random access procedure in which the mobile makes its initial communication. Chapter 10 covers the procedures for link adaptation, which compensate for changes in the amplitude and phase of the incoming radio signal, and help to control the use of multiple antennas. Chapter 11 addresses the procedures that are used for scheduling and data transmission, while Chapter 12 explains how the air interface's higher-level protocols handle tasks such as scheduling and re-transmission.

We then move on to the end-to-end operation of 5G. Chapter 13 covers the procedures that a mobile runs when it switches on, to discover a nearby cell and register with its chosen network. Chapter 14 addresses the security procedures in 5G, while Chapter 15 covers the procedures which connect a mobile to an external data network, and which allow an external application server to monitor and influence its behaviour.

The next four chapters address different aspects of mobility management, in which the network keeps track of the mobile's location and controls the cells with which it is communicating. Chapter 16 covers the procedures in a state known as RRC_CONNECTED, in which the mobile is carrying out normal communications with the network. Chapter 17 addresses the state of RRC_IDLE, in which the mobile is on standby, while Chapter 18 deals with the new state of RRC_INACTIVE, which is an optimization for low data rate communications. Chapter 19 covers the procedures used for inter-operation between the core networks of 4G and 5G. As the final topic, Chapter 20 addresses the enhancements that have been made to 5G in the most recent release of the specifications, and reviews some of the plans for future releases.

Like any technical subject, 5G uses a large number of acronyms and abbreviations. To help make the text readable, I have tried to strike a balance between the uses of abbreviations and text, for example by using 'downlink' in place of DL, but AMF in place of 'access and mobility management function'. New terms are highlighted in italics throughout the text, and there is a full list of abbreviations at the start of the book.

Unavoidably, some of the topics are mathematical ones. Some previous knowledge of Fourier transforms would help the reader understand the discussion of orthogonal frequency division multiple access in Chapter 5. Similarly, some previous knowledge of matrices would help with the discussion of multiple antennas in Chapter 6, while familiarity with complex numbers would be helpful for both. However, I have attempted to make the maths as lightweight as possible, and to explain the important concepts using English instead. The Bibliography contains some references for further reading.

Acknowledgements

My first acknowledgement is to the team at Wiley, for taking on the publication of this book and for guiding me through the writing and production process. I am particularly grateful for the support received from my Commissioning Editor, Sandra Grayson; from my Project Editor, Steve Fassioms; from my Production Editor, Juliet Booker; and from the whole of the production team. I would also like to extend my appreciation to the directors and staff of Wray Castle Limited. My first exposure to 5G was as part of a contract to write training material on their behalf, and I am grateful for the opportunities that the work provided to extend my knowledge and understanding of the subject.

Particular thanks are due to Paul Mason, Jeff Cartwright and Tony Wakefield, for many valuable discussions about telecommunications in general and 5G in particular; to Ignacio Rodriguez Larrad, for providing the data on penetration losses underlying Figure 4.5; and to David Herbert, for his guidance about the signalling protocols that are used by the internet. I would also like to offer a big thank you to the delegates in my training classes, whose probing questions have so often proved the ideal trigger for learning new aspects of the technology. Nevertheless, the responsibility for any errors and omissions in the text, and for any lack of clarity in the explanations, is entirely my own.

Several diagrams in this book have been reproduced from the technical specifications for 5G, with permission from the European Telecommunications Standards Institute, © 2019. 3GPP™ TSs and TRs are the property of ARIB, ATIS, CCSA, ETSI, TSDSI, TTA and TTC who jointly own the copyright in them. They are subject to further modifications and are therefore provided to you 'as is' for information purposes only. Further use is strictly prohibited.

On a lighter note, I would like to thank the creators of the internet blocking application Cold Turkey, for removing innumerable opportunities for procrastination and keeping me focussed on the task. I would also like to thank the spirits of J. S. Bach and Frédéric Chopin, for providing me with some more valuable diversions, and for reminding me that there are more things in life than mobile telecommunications.

My last and greatest thank you is to my beloved wife, Susie. In a piece of timing that the author could perhaps have planned more effectively, the earlier parts of writing this book coincided with our engagement, while the later parts coincided with the first few months of our marriage. Her kindness, patience and understanding have been extraordinary throughout, and I dedicate this book to her.

List of Abbreviations

1G	First generation
16-QAM	16-quadrature amplitude modulation
2G	Second generation
256-QAM	256-quadrature amplitude modulation
3G	Third generation
3GPP	Third Generation Partnership Project
3GPP2	Third Generation Partnership Project 2
4G	Fourth generation
5G	Fifth generation
5G AKA	5G authentication and key agreement
5G-AN	5G access network
5GC	5G core network
5G-EIR	5G equipment identity register
5G-GUTI	5G globally unique temporary identity
5GMM	5G mobility management
5GS	5G system
5GSM	5G session management
5G-S-TMSI	5G S temporary mobile subscriber identity
5G-TMSI	5G temporary mobile subscriber identity
5QI	5G QoS identifier
64-QAM	64-quadrature amplitude modulation
A/D	Analogue to digital
AAA	Authentication, authorization and accounting
AAS	Active antenna system
ABMF	Account balance management function
AES	Advanced Encryption Standard
AF	Application function
AKA	Authentication and key agreement
AM	Acknowledged mode
AMBR	Aggregate maximum bit rate
AMF	Access and mobility management function
AMFI	AMF identifier

AMPS	Advanced Mobile Phone System
AN	Access network
API	Application programming interface
APN	Access point name
ARIB	Association of Radio Industries and Businesses
ARP	Allocation and retention priority
ARPF	Authentication credential repository and processing function
ARQ	Automatic repeat request
AS	Access stratum
ATIS	Alliance for Telecommunications Industry Solutions
AuC	Authentication centre
AUSF	Authentication server function
AUTN	Authentication token
BAP	Backhaul adaptation protocol
BCCH	Broadcast control channel
BCH	Broadcast channel
BD	Billing domain
BLER	Block error ratio
BPSK	Binary phase shift keying
BSF	Binding support function
BSR	Buffer status report
BTS	Base transceiver station
BWP	Bandwidth part
CA	Carrier aggregation
CBG	Code block group
CCCH	Common control channel
CCE	Control channel element
CCSA	China Communications Standards Association
CDM	Code division multiplexing
CDMA	Code division multiple access
CDR	Charging data record
CE	Control element
CGF	Charging gateway function
CHF	Charging function
CIoT	Cellular internet of things
cIPX	Consumer's IPX
CI-RNTI	Cancellation indication RNTI
CLI	Cross-link interference
CM	Connection management
CMAS	Commercial mobile alert system
CN	Core network
cNF	Consumer's network function
CORESET	Control resource set

CP	Control plane *or* Cyclic prefix
CPRI	Common Public Radio Interface
CQI	Channel quality indicator
C-RAN	Cloud radio access network
CRB	Common resource block
CRC	Cyclic redundancy check
CRI	CSI-RS resource indicator
CriC	Critical communication
C-RNTI	Cell RNTI
CRUD	Create, read, update and delete
CS	Circuit switched
cSEPP	Consumer's SEPP
CSFB	Circuit-switched fallback
CSI	Channel state information
CSI-IM	CSI interference measurement
CSI-RS	CSI reference signal
CS-RNTI	Configured scheduling RNTI
CTF	Charging trigger function
CU	Central unit
D/A	Digital to analogue
D2D	Device-to-device
dB	Decibel
dBi	Decibels relative to an isotropic antenna
dBm	Decibels relative to 1 milliwatt
DC	Dual connectivity
DCCH	Dedicated control channel
DCI	Downlink control information
DFT	Discrete Fourier transform
DFT-s-OFDMA	Discrete Fourier transform spread OFDMA
DHCP	Dynamic host configuration protocol
DiffServ	Differentiated services
DL	Downlink
DL-SCH	Downlink shared channel
DM-RS	Demodulation reference signal
DN	Data network
DNN	Data network name
DNS	Domain name server
DRB	Data radio bearer
DRX	Discontinuous reception
DSCP	Differentiated services code point
DSRC	Dedicated short-range communication
DTCH	Dedicated traffic channel
DTLS	Datagram transport layer security
DU	Distributed unit

E1-AP	E1 application protocol
EAP	Extensible authentication protocol
ECGI	EUTRA cell global identification
EC-GSM	Extended coverage GSM
ECI	EUTRA cell identity
ECIES	Elliptic curve integrated encryption scheme
ECM	EPS connection management
eCPRI	Evolved Common Public Radio Interface
EDGE	Enhanced Data Rates for GSM Evolution
eDRX	Extended discontinuous reception
EHF	Extremely high frequency
EIR	Equipment identity register
eMBB	Enhanced mobile broadband
EMM	EPS mobility management
eMTC	Enhanced machine-type communications
eNB	Evolved Node B
EN-DC	EUTRA–NR dual connectivity
EPC	Evolved packet core
EPS	Evolved packet system
ESM	EPS session management
ESP	Encapsulating security payload
ETSI	European Telecommunications Standards Institute
ETWS	Earthquake and tsunami warning system
EUTRA	Evolved UMTS terrestrial radio access
E-UTRAN	Evolved UMTS terrestrial radio access network
F1-AP	F1 application protocol
FDD	Frequency division duplex
FDMA	Frequency division multiple access
FFT	Fast Fourier transform
FR1	Frequency range 1
FR2	Frequency range 2
GBR	Guaranteed bit rate
GERAN	GSM EDGE radio access network
GFBR	Guaranteed flow bit rate
GMLC	Gateway mobile location centre
gNB	Next-generation Node B
gNB-CU	gNB central unit
gNB-CU-CP	gNB central unit control plane
gNB-CU-UP	gNB central unit user plane
gNB-DU	gNB distributed unit
GNSS	Global navigation satellite system
GP	Guard period
GPRS	General Packet Radio Service

GPS	Global Positioning System
GSCN	Global synchronization channel number
GSM	Global System for Mobile Communications
GTP	GPRS tunnelling protocol
GTP-C	GPRS tunnelling protocol control part
GTP-U	GPRS tunnelling protocol user part
GUAMI	Globally unique AMF identifier
GUTI	Globally unique temporary identity
HARQ	Hybrid ARQ
HARQ-ACK	Hybrid ARQ acknowledgement
HF	High frequency
HLR	Home location register
HRES	Hashed response
HSDPA	High-speed downlink packet access
HSPA	High-speed packet access
HSS	Home subscriber server
HSUPA	High-speed uplink packet access
HTML	Hypertext Markup Language
HTTP	Hypertext Transfer Protocol
HXRES	Hashed expected response
IAB	Integrated access and backhaul
IEEE	Institute of Electrical and Electronics Engineers
IETF	Internet Engineering Task Force
IF	Intermediate frequency
IKE	Internet Key Exchange
IMEI	International mobile equipment identity
IMEISV	IMEI and software version number
IMS	IP multimedia subsystem
IMSI	International mobile subscriber identity
IMT	International Mobile Telecommunications
INT-RNTI	Interruption RNTI
IoT	Internet of things
IP	Internet Protocol
IPSec	IP security
IPv4	Internet Protocol version 4
IPv6	Internet Protocol version 6
IPX	IP packet exchange
I-RNTI	Inactive RNTI
ISI	Inter-symbol interference
ITS	Intelligent transport system
ITU	International Telecommunication Union
I-UPF	Intermediate UPF

JOSE	JSON object signing and encryption
JSON	JavaScript Object Notation
JWE	JSON web encryption
JWS	JSON web signature
L1-RSRP	Layer 1 reference signal received power
LAA	Licence-assisted access
LBT	Listen before talk
LCID	Logical channel identity
LCS	Location service
LDPC	Low-density parity check
LI	Layer indicator
LMF	Location management function
LoRaWAN	Long-range Wide-area Network
LOS	Line of sight
LPP	LTE positioning protocol
LSB	Least significant bit
LTE	Long-term Evolution
LWA	LTE WLAN aggregation
MAC	Medium access control *or* Message authentication code
MC	Mission critical
MCC	Mission-critical communication *or* Mobile country code
MCG	Master cell group
MCPTT	Mission-critical push to talk
MCS	Modulation and coding scheme
MCS-C-RNTI	Modulation and coding scheme cell RNTI
ME	Mobile equipment
MEC	Mobile edge computing *or* Multi-access edge computing
MeNB	Master eNB
MFBR	Maximum flow bit rate
MgNB	Master gNB
MIB	Master information block
MICO	Mobile-initiated connection only
MIMO	Multiple input multiple output
mIoT	Massive internet of things
MME	Mobility management entity
mMIMO	Massive MIMO
MMS	Multimedia Messaging Service
MMSE	Minimum mean square error
mMTC	Massive machine-type communications
MN	Master node
MNC	Mobile network code
MPLS	Multi-protocol label switching
MR	Maximum ratio

MR-DC	Multi-radio dual connectivity
MSB	Most significant bit
MT	Mobile termination
MTC	Machine-type communications
MU-MIMO	Multiple-user MIMO
MVNO	Mobile virtual network operator
N3IWF	Non-3GPP interworking function
NAI	Network access identifier
NAS	Non-access stratum
NAT	Network address translation
NB-IoT	Narrowband internet of things
NCC	Next-hop chaining counter
NCGI	New Radio cell global identity
NCI	New Radio cell identity
NDS	Network domain security
NE-DC	NR–EUTRA dual connectivity
NEA	Encryption algorithm for 5G
NEF	Network exposure function
NEO	Network operation
NF	Network function
NFV	Network function virtualization
NG-AP	Next-generation application protocol
ng-eNB	Next-generation evolved Node B
NGEN-DC	NG-RAN–EUTRA–NR dual connectivity
ngKSI	Key set identifier for 5G
NG-RAN	Next-generation radio access network
NH	Next hop
NIA	Integrity algorithm for 5G
NLOS	Non-line of sight
NMT	Nordic Mobile Telephone
NOMA	Non-orthogonal multiple access
NPN	Non-public network
NR	New Radio
NR-ARFCN	New Radio absolute radio frequency channel number
NR-DC	NR–NR dual connectivity
NRF	Network repository function
NRPPa	New Radio positioning protocol A
NSA	Non-standalone
NSI-ID	Network slice instance identifier
NSSAI	Network slice selection assistance information
NSSF	Network slice selection function
NTN	Non-terrestrial network
NWDAF	Network data analytics function
NZP	Non-zero power

OFDM	Orthogonal frequency division multiplexing
OFDMA	Orthogonal frequency division multiple access
OSA	Open service access
OSI	Open Systems Interconnection
PBCH	Physical broadcast channel
PCC	Policy and charging control
PCCH	Paging control channel
PCell	Primary cell
PCF	Policy control function
PCH	Paging channel
PCI	Physical cell identity
PCRF	Policy and charging rules function
P-CSCF	Proxy call session control function
PDB	Packet delay budget
PDCCH	Physical downlink control channel
PDCP	Packet data convergence protocol
PDF	Portable document format
PDN	Packet data network
PDN-GW	Packet data network gateway
PDP	Packet data protocol
PDR	Packet detection rule
PDSCH	Physical downlink shared channel
PDU	Protocol data unit
PEI	Permanent equipment identifier
PER	Packet error rate
PF	Paging frame
PFCP	Packet forwarding control protocol
PFD	Packet flow description
PGW	Packet data network gateway
PGW-C	Packet data network gateway control plane
PGW-U	Packet data network gateway user plane
PHR	Power headroom report
PHY	Physical layer
pIPX	Producer's IPX
PLMN	Public land mobile network
PLMN-ID	Public land mobile network identity
PMI	Pre-coding matrix indicator
pNF	Producer's network function
PO	Paging occasion
PQI	PC5 5G QoS identifier
PRACH	Physical random access channel
PRB	Physical resource block
P-RNTI	Paging RNTI
ProSe	Proximity services

PRS	Positioning reference signal
PS	Packet switched
PSCell	Primary SCG cell
pSEPP	Producer's SEPP
PSS	Primary synchronization signal
PSTN	Public-switched telephone network
PT-RS	Phase-tracking reference signal
PTT	Push-to-talk
PUCCH	Physical uplink control channel
PUSCH	Physical uplink shared channel
QAM	Quadrature amplitude modulation
QCI	QoS class identifier
QCL	Quasi co-location
QFI	QoS flow identifier
QNC	QoS notification control
QoS	Quality of service
QPSK	Quadrature phase shift keying
RACH	Random access channel
RAN	Radio access network
RAND	Random number
RAPID	Random access preamble identifier
RA-RNTI	Random access RNTI
RAT	Radio access technology
RB	Resource block
RBG	Resource block group
RE	Resource element
REG	Resource element group
RES	Response
REST	Representational state transfer
RF	Radio frequency *or* Rating function
RFSP	RAT/frequency selection priority
RI	Rank indication
RIM-RS	Remote interference management reference signal
RLC	Radio link control
RM	Registration management
RNA	RAN-based notification area
RNTI	Radio network temporary identifier
ROHC	Robust header compression
RQA	Reflective QoS attribute
RRC	Radio resource control
RRH	Remote radio head
RSRP	Reference signal received power
RSRQ	Reference signal received quality

RSU	Roadside unit
RTP	Real-time transport protocol
RV	Redundancy version
S1-AP	S1 application protocol
SA	Standalone
SCEF	Service capability exposure function
SCell	Secondary cell
SC-FDMA	Single-carrier frequency division multiple access
SCG	Secondary cell group
SCH	Shared channel
SCP	Service communication proxy
SCTP	Stream control transmission protocol
SD	Slice differentiator
SDAP	Service data adaptation protocol
SDF	Service data flow
SDL	Supplementary downlink
SDN	Software-defined networking
SDU	Service data unit
SEAF	Security anchor function
SEG	Security gateway
SeNB	Secondary eNB
SEPP	Security edge protection proxy
SFI	Slot format indication
SFI-RNTI	Slot format indication RNTI
SFN	System frame number
SgNB	Secondary gNB
SGW	Serving gateway
SHF	Super high frequency
SI	Segmentation information
SIB	System information block
SIDF	Subscription identifier de-concealing function
SIM	Subscriber identity module
SINR	Signal-to-interference plus noise ratio
SIR	Signal-to-interference ratio
SI-RNTI	System information RNTI
SL	Sidelink
SLIV	Start and length indicator value
SMF	Session management function
SMS	Short Message Service
SMSF	SMS function
SMS-GMSC	SMS gateway mobile switching centre
SMS-IWMSC	SMS interworking mobile switching centre
SMS-SC	SMS service centre
SN	Secondary node *or* Sequence number

SNR	Signal-to-noise ratio
S-NSSAI	Single network slice selection assistance information
SO	Segment offset
SON	Self-optimizing network *or* Self-organizing network
SpCell	Special cell
SP-CSI-RNTI	Semi-persistent CSI RNTI
SPS	Semi-persistent scheduling
SR	Scheduling request
SRB	Signalling radio bearer
SRS	Sounding reference signal
SRVCC	Single radio voice call continuity
SS	Synchronization signal
SSB	SS/PBCH block
SSBRI	SS/PBCH block resource indicator
SSC	Session and service continuity
SS-RSRP	Synchronization signal reference signal received power
SS-RSRQ	Synchronization signal reference signal received quality
SSS	Secondary synchronization signal
SST	Slice/service type
SUCI	Subscription concealed identifier
SUL	Supplementary uplink
SU-MIMO	Single-user MIMO
SUPI	Subscription permanent identifier
SVD	Singular value decomposition
TA	Timing advance *or* Tracking area
TAC	Tracking area code
TACS	Total Access Communication System
TAG	Timing advance group
TAI	Tracking area identity
TCI	Transmission configuration indicator
TCP	Transmission Control Protocol
TC-RNTI	Temporary cell RNTI
TDD	Time division duplex
TDMA	Time division multiple access
TD-SCDMA	Time division synchronous code division multiple access
TE	Terminal equipment
TEID	Tunnel endpoint identifier
TLS	Transport Layer Security
TM	Transparent mode
TMSI	Temporary mobile subscriber identity
TPC	Transmit power control
TPC-PUCCH-RNTI	Transmit power control PUCCH RNTI
TPC-PUSCH-RNTI	Transmit power control PUSCH RNTI
TPC-SRS-RNTI	Transmit power control SRS RNTI

TR	Technical report
TRS	Tracking reference signal
TS	Technical specification
TSDSI	Telecommunications Standards Development Society, India
TTA	Telecommunications Technology Association
TTI	Telecommunication Technology Committee
UCI	Uplink control information
UDM	Unified data management
UDP	User Datagram Protocol
UDR	Unified data repository
UDSF	Unstructured data storage function
UE	User equipment
UHF	Ultra high frequency
UICC	Universal integrated circuit card
UL	Uplink
UL CL	Uplink classifier
ULA	Uniform linear array
UL-SCH	Uplink shared channel
UM	Unacknowledged mode
UMB	Ultra Mobile Broadband
UMTS	Universal Mobile Telecommunication System
UP	User plane
UPF	User plane function
URI	Uniform resource identifier
URL	Uniform resource locator
URLLC	Ultra-reliable low-latency communication
USIM	Universal subscriber identity module
UTC	Coordinated universal time
UTRAN	UMTS terrestrial radio access network
UUID	Universally unique identifier
V2I	Vehicle to infrastructure
V2N	Vehicle to network
V2P	Vehicle to pedestrian
V2V	Vehicle to vehicle
V2X	Vehicle to everything
VHF	Very high frequency
VNF	Virtualized network function
VoIP	Voice over IP
VoLTE	Voice over LTE
VRB	Virtual resource block
WCDMA	Wideband code division multiple access
WEA	Wireless emergency alert

WiMAX	Worldwide Interoperability for Microwave Access
WLAN	Wireless local area network
WRC	World Radiocommunication Conference
X2-AP	X2 application protocol
XMAC	Expected message authentication code
Xn-AP	Xn application protocol
XRES	Expected response
ZF	Zero forcing
ZP	Zero power
ZUC	Zu Chongzhi

1

Introduction

This book is about *fifth-generation* (5G) mobile telecommunications. Our first chapter puts 5G into its historical context, and lays out its requirements and its most important technical features. We will begin by setting out the architecture of a mobile telecommunication system, and by reviewing the history of mobile telecommunications and the current state of the market. We will then discuss the likely applications for 5G, the resulting technical requirements and the underlying technologies that 5G uses. The chapter closes with a review of the standardization process and an introduction to the architecture of the 5G system.

1.1 Architecture of a Mobile Telecommunication System

1.1.1 High-level Architecture

A mobile telecommunication system is run by a *network operator* such as Vodafone, AT&T or China Mobile, and is officially known as a *public land mobile network* (PLMN). As shown in Figure 1.1, it has four main components, namely the core network (CN), the radio access network (RAN), the management system and the user's device. The last of these is known colloquially as the *mobile*, and more formally as the *user equipment* (UE).

The core network transports traffic between the mobile and one or more external networks, such as the *public switched telephone network* (PSTN) or the internet. The core network also controls the mobile's communications with those external networks, and stores information about the network operator's subscribers.

The radio access network handles the network's radio communications with the mobile. It communicates with the core network over an interface known as the *backhaul*, and with the mobile over the *air interface*, also known as the radio interface. On that interface, the direction from network to mobile is known as the *downlink* (DL) or forward link, and the direction from mobile to network is known as the *uplink* (UL) or reverse link.

The network is controlled by a separate *management system*. Its tasks include configuring the various components of the core and radio access networks, monitoring their performance, reporting any faults to the network operator and billing the user.

A mobile can communicate outside the coverage area of its network operator by using the resources from two PLMNs, namely the *visited network*, where the mobile is located, and the operator's *home network*. That situation is known as *roaming*.

An Introduction to 5G: The New Radio, 5G Network and Beyond, First Edition. Christopher Cox.

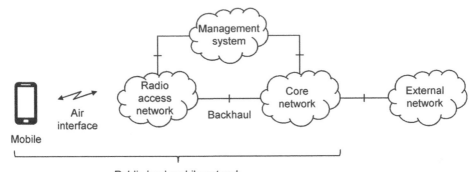

Figure 1.1 Architecture of a mobile telecommunication system.

1.1.2 Internal Architecture of the Mobile

Internally, the mobile has two main components. The actual communication device is known as the *mobile equipment* (ME). In the case of a voice mobile or a smartphone, that is just a single device. However, the ME can itself have two parts, namely the *mobile termination* (MT), which handles all the communication functions, and the *terminal equipment* (TE), which terminates the data streams. The MT might be a plug-in 5G card for a laptop, for example, in which case the TE would be the laptop itself.

The *universal integrated circuit card* (UICC) is a smart card, colloquially known as the SIM card. It runs an application known as the *universal subscriber identity module* (USIM), which stores user-specific data such as the home network identity and carries out security-related calculations using secure keys that the smart card stores.

1.1.3 Architecture of the Radio Access Network

The most important component of the radio access network is the *base station*. In a medium-sized country like the United Kingdom, a typical network might contain several thousand of these. Figure 1.2 shows a group of nearby base stations, as viewed from above.

On the air interface, a base station transmits and receives using one or more radio frequencies, each of which is known as a *carrier frequency*. Around each carrier, the radio signal occupies a certain amount of radio spectrum, which is known as the *bandwidth*. For example, a 5G base station might use a carrier frequency of 3500 MHz and a bandwidth of 40 MHz, so that its transmissions span the range from 3480 to 3520 MHz.

Usually, the base station operates in *licensed spectrum*, in which the network operator has purchased an exclusive license to carry out radio communications from the corresponding national regulator. The alternative is *unlicensed spectrum*, in which there are no licenses, but the transmitters have to use a low power so that they do not cause undue interference to other receivers in the same band.

The base station controls one or more *cells*, each of which is a radio transmission with a particular carrier frequency and bandwidth, which spans a particular coverage area. On any one radio frequency, a base station can control multiple cells, known as *sectors*, by transmitting in different directions. There are typically three sectors per base station, with each one

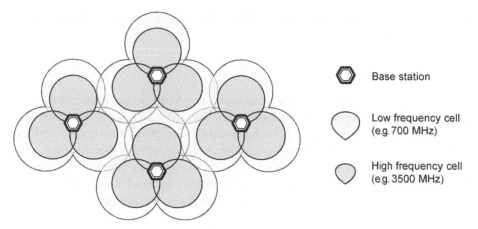

Figure 1.2 Example of base stations with two carrier frequencies and three sectors.

spanning an arc of 120°. The base station can also control multiple cells in the same direction, by using different radio frequencies to ensure that the radio signals do not interfere.

1.1.4 Coverage and Capacity

There are two main limits on a cell's performance. Firstly, each cell has a coverage limit, which is the maximum distance at which the receiver can successfully hear the transmitter. A cell's coverage is greatest if the transmissions are at a high power and if the data rate is low, because both conditions imply that the energy per bit at the receiver is high. In a simple communication system, in which the base station and mobile both have a single antenna, the cell's coverage is also greatest if the carrier frequency is low, for example 700 MHz rather than 3500 MHz.

Secondly, each cell has a capacity limit, which is the maximum combined data rate of all the mobiles that are communicating through the cell. A cell's capacity is greatest if the transmissions occupy a wide bandwidth and are received with a high *signal-to-interference plus noise ratio* (SINR), which is the power at the receiver from the desired signal divided by the power from noise and interference. More precisely, a cell's capacity is constrained by the *Shannon limit* [1], which in a single-antenna communication system is as follows:

$$C = B\log_2\left(1 + \text{SINR}\right) \tag{1.1}$$

where B is the bandwidth in Hz, and C is the cell's capacity in bits s^{-1}. Improvements in the radio communication technology allow the cell's capacity to approach the Shannon limit, but not to exceed it.

Bearing those limits in mind, cells can be grouped into a number of different classes. *Macrocells* are a few kilometres across and provide wide-area coverage in rural areas and suburbs, with the signals travelling over the building roofs. *Microcells* are a few hundred metres across and are used in urban areas, with the signals confined to individual streets by the buildings that surround them. *Picocells* are a few tens of metres across and are used for indoor communications in shopping centres and offices. *Femtocells* are a few metres across and are installed by consumers within the home. The same area can be spanned by one macrocell or by a large number of smaller cells, so the collective capacity of those small cells can be very large.

1.1.5 Architecture of the Core Network

In older systems, the core network contains two domains, which transport different types of traffic using different networking techniques. The *circuit-switched* (CS) domain transports fixed-rate traffic such as voice, so that the user can make phone calls with other devices that are in the public switched telephone network or the circuit-switched domains of other network operators. It does that using a technique known as *circuit switching*, which sets aside a dedicated two-way connection for each individual phone call. The CS domain transports voice traffic with a constant data rate and minimal delay, but it is inappropriate for services in which the data rate can vary.

The *packet-switched* (PS) domain transports variable-rate data traffic, such as web pages and emails, between the user and external data networks such as the internet. It does that using a different technique, known as *packet switching*, in which a data stream is divided into packets, each of which is labelled with the address of the required destination device. Within the network, *routers* read the destination addresses of the incoming data packets, and forward them towards those destinations by following the instructions in internal *routing tables*. The network's resources are shared amongst all the users, so the technique is more efficient than circuit switching. However, delays can result if too many devices try to transmit at the same time.

More recently, the widespread use of smartphones has led to a situation in which mobile network traffic is dominated by data rather than voice. In response, designers have abandoned the use of circuit switching, and have introduced mobile communication systems that use packet switching alone. That simplifies the design and allows the system to be optimized for the delivery of data traffic, but it implies that voice calls have to be handled in other ways. We will discuss those in Section 1.2.4.

1.1.6 Communication Protocols

In any communication network, information is transported using hardware and software functions known as *protocols*. The best examples are the protocols used for packet-switched data transport over the internet, because those protocols are also used by the core and radio access networks of 5G. The internet's protocols are designed by the *Internet Engineering Task Force* (IETF) and are grouped into various numbered *layers*, each of which handles one aspect of the transmission and reception process. The usual grouping follows a seven-layer model known as the *Open Systems Interconnection* (OSI) model.

As an example (Figure 1.3), let us suppose that a web server is sending information to a user's browser. In the first step, an *application layer* protocol – in this case, the *Hypertext Transfer Protocol* (HTTP) – receives information from the server's application software, and passes it to the next layers down by representing it in a way that the user's application layer will eventually be able to understand.

The *transport layer* manages the end-to-end data transmission. There are two main alternatives. The *Transmission Control Protocol* (TCP) re-transmits a packet from end to end if

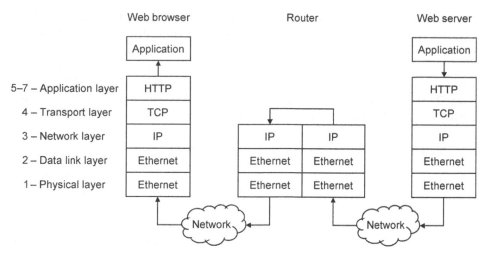

Figure 1.3 Examples of the communication protocols used by the internet.

it does not arrive correctly, and is suitable for data such as web pages and emails that have to be received reliably. The *User Datagram Protocol* (UDP) sends the packet without any re-transmission, and is suitable for information such as real-time voice or video for which timely arrival is more important.

In the *network layer*, the *Internet Protocol* (IP) sends packets on the correct route from source to destination, using the *IP address* of the destination device. The process is handled by the intervening routers, which inspect the destination IP addresses by implementing just the lowest three layers of the protocol stack. The *data link layer* manages the transmission of packets from one device to the next, for example by re-transmitting a packet across a single interface if it does not arrive correctly. Finally, the *physical layer* deals with the actual transmission details, for example by setting the voltage of the transmitted signal. The internet can use any suitable protocols for the data link and physical layers, such as *Ethernet*.

At each level of the transmitter's stack, a protocol receives a data packet from the protocol above in the form of a *service data unit* (SDU). It processes the packet, adds a header to describe the processing it has carried out, and outputs the result as a *protocol data unit* (PDU). This immediately becomes the incoming SDU of the next protocol down. The process continues until the packet reaches the bottom of the protocol stack, at which point it is transmitted. The receiver reverses the process, using the headers to help it undo the effect of the transmitter's processing.

1.2 History of Mobile Telecommunications

1.2.1 Introduction

In the time since they were first introduced in the early 1980s, a new *generation* of mobile communication systems has appeared every 10 years or so. The term generation emerges by industry consensus: it does not have any formal definition, but it is typically associated with a new air interface and with a new architecture for the network.

The *first-generation* (1G) systems used analogue communication techniques which were similar to those in a traditional analogue radio. The individual cells were large and the systems did not use the available radio spectrum efficiently, so their capacity was by today's standards very small. The mobile devices were large and expensive, and were marketed almost exclusively to business users. Some popular systems included *Nordic Mobile Telephone* (NMT) in Scandinavia, the *Advanced Mobile Phone System* (AMPS) in the USA and the *Total Access Communication System* (TACS) in the UK.

1.2.2 Global System for Mobile Communications (GSM)

Mobile telecommunications took off as a consumer product with the introduction of *second-generation* (2G) systems in the early 1990s. Those systems were the first to use digital technology, which permitted a more efficient use of radio spectrum and the introduction of smaller, cheaper devices. The most popular 2G system was the *Global System for Mobile Communications* (GSM), which was originally specified by the *European Telecommunications Standards Institute* (ETSI).

GSM was originally designed just for voice, but was later enhanced to support instant messaging by means of the *Short Message Service* (SMS), and the delivery of packet data to a mobile device by means of the *General Packet Radio Service* (GPRS). Another enhancement was *Enhanced Data Rates for GSM Evolution* (EDGE), which modified the air interface so as to increase the available data rate.

In GSM, the core network has the traditional two domains: the circuit-switched domain handles voice calls, while the packet-switched domain handles data. The RAN is known as the *GSM EDGE radio access network* (GERAN), while the base station is known as the *base transceiver station* (BTS). Figure 1.4 shows the high-level architecture of GSM and its descendants, and highlights the most important interfaces.

1.2.3 Universal Mobile Telecommunication System (UMTS)

Buoyed by the success of 2G, the *International Telecommunication Union* (ITU) published a set of requirements in 1997 for a *third-generation* (3G) mobile communication system, under the name *International Mobile Telecommunications* (IMT) *2000* (IMT-2000) [2]. A number of standards bodies responded by submitting specifications, which the ITU eventually accepted as meeting the requirements of IMT-2000. The most popular was the *Universal Mobile Telecommunication System* (UMTS), which was designed by the *Third Generation Partnership Project* (3GPP) [3], a specially formed collaboration that later took responsibility for the specifications for GSM.

UMTS uses the same core network as GSM, but has a new RAN known as the *UMTS terrestrial radio access network* (UTRAN) and a new base station known as the *Node B*. The UMTS air interface is more efficient than GSM's, and has two implementations: *wideband code division multiple access* (WCDMA) is used throughout most of the world, while *time division synchronous code division multiple access* (TD-SCDMA) is derived from WCDMA and is mainly used in China. The most important later enhancement is known as *high-speed packet access* (HSPA), which improves the performance of data applications by increasing the air interface's average data rate, at the expense of short-term variations in the data rate for an individual user.

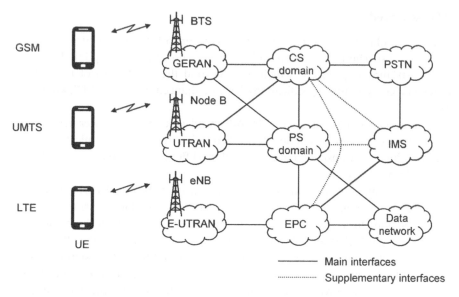

Figure 1.4 Architecture of GSM, UMTS and LTE.

1.2.4 Long-term Evolution (LTE)

The introduction of HSPA coincided with the introduction of smartphones, notably the Apple iPhone in 2007, followed by devices based on Google's Android operating system from 2008. Together, those factors led to an explosion in the use of mobile data applications and in the amount of mobile data traffic.

In response, *fourth-generation* (4G) systems are optimized for the delivery of data. By far the most popular is the 3GPP system known as *Long-term Evolution* (LTE). In LTE, the base station is known as an *evolved Node B* (eNB), the RAN is the *evolved UMTS terrestrial radio access network* (E-UTRAN) and the *evolved packet core* (EPC) is a direct replacement for the packet-switched domain of UMTS and GSM. Officially, the whole system is known as the *evolved packet system* (EPS), although in practice that term is used only occasionally.

LTE does not have a circuit-switched domain of its own, but there are still various ways for an LTE user to make a voice call. The first one specified was *circuit-switched fallback* (CSFB), in which the network transfers the mobile to a legacy 2G or 3G cell, through which the mobile contacts the CS domain of the 2G or 3G core. Alternatively, the mobile can make a *voice over IP* (VoIP) call by means of the EPC, using a third-party application such as Skype or WhatsApp. Finally, network operators can provide their own IP voice services by means of an external 3GPP network known as the *IP multimedia subsystem* (IMS), in a technology known as *voice over LTE* (VoLTE). Although rather complex, the IMS brings several advantages: for example, it can handle services such as emergency calls that are not supported by third-party applications, it maintains the operators' voice revenue and it frees network operators from reliance on nearby 2G or 3G cells. It also supports a technique known as *single radio voice call continuity* (SRVCC), in which the network can hand a mobile over to a 2G or 3G cell if its leaves the coverage area of LTE, and can convert an ongoing VoIP call to a circuit-switched call that is handled by the 2G or 3G core.

1.2.5 LTE-Advanced

In 2008, the ITU published requirements for a successor to IMT-2000, which is known as *IMT-Advanced* [4]. The requirements called for peak downlink and uplink data rates of 1000 and 500 Mbps respectively, which were beyond the capabilities of LTE. 3GPP therefore responded by writing specifications for an enhanced version of LTE, known as *LTE-Advanced*.

The most important enhancement in LTE-Advanced is known as *carrier aggregation* (CA). This technique increases a mobile's data rate by allowing it to communicate through multiple cells, which are transmitted and received on different radio frequencies, but are still controlled by a single eNB.

A later enhancement, known as *dual connectivity* (DC), allows those cells to be controlled by two separate base stations rather than by one. The original base station carries traffic for the user and signalling messages that control the mobile, and is known as the *master eNB* (MeNB). Typically, the MeNB controls one or more macrocells, which provide the mobile with wide-area coverage and reliable communications. The additional base station mainly carries traffic and is known as the *secondary eNB* (SeNB). Typically, the SeNB controls one or more microcells, which provide mobiles that can receive them with a much higher data rate. Dual connectivity was not widely implemented at first, but it is an important step towards the introduction of 5G.

1.2.6 LTE-Advanced Pro

Later enhancements to LTE are known by the name *LTE-Advanced Pro*. The enhancements broaden the capabilities that LTE offers and the services that it supports, and act as a precursor to 5G.

LTE was originally designed for use by people, but there is also a need for wide-area *machine-type communications* (MTC) for devices such as vehicle trackers and smart utility meters. LTE supports those devices using two enhancements. *Enhanced machine-type communications* (eMTC) is backwards compatible with LTE, in the sense that an eMTC base station can communicate with both legacy mobiles and machine-type devices at the same time. The *narrowband internet of things* (NB-IoT) has better performance characteristics, but is defined as a separate air interface technology that is not backwards compatible with LTE: an NB-IoT base station cannot communicate with legacy mobiles, and vice versa.

The LTE specifications have always allowed a mobile to access the evolved packet core by means of a non-3GPP access network. The most common choice is a *wireless local area network* (WLAN), also known as WiFi, which is specified by the *Institute of Electrical and Electronics Engineers* (IEEE) in a set of standards denoted IEEE 802.11. Later enhancements build on that capability in various ways. *Licence assisted access* (LAA) is based on carrier aggregation, but it introduces additional LTE cells into the unlicensed spectrum that WiFi would normally use. *LTE WLAN aggregation* (LWA) is based on dual connectivity, except that the mobile communicates with a master eNB and a WiFi access point, rather than a master and secondary eNB.

Device-to-device (D2D) communications, also known as *proximity services* (ProSe), allow two mobiles to communicate directly with one another over a new air interface that is known as the *sidelink* (SL). The initial application is in public safety networks such as those used by the police and other emergency services, in which such communication is important if the base station has been taken down by an emergency situation or is otherwise out of range. D2D communication is also a useful way to offload traffic from the base station, for applications such as file exchange between nearby devices.

The sidelink is also a valuable part of vehicle-related communications, in other words the exchange of information to and from a road vehicle. Although full support for vehicle communications requires 5G, LTE can address some of the less stringent, non-safety-critical aspects. Enhancements include support for ETSI's architecture for an *intelligent transport system* (ITS), and improvements to the sidelink so that vehicles can communicate even if their relative speed is high.

1.2.7 Other Mobile Communication Systems

Although 3GPP systems have come to dominate the mobile communication market, two other technology families have been important. *IS-95* was a technology designed by Qualcomm, which became the dominant 2G system in the United States. It was succeeded by a 3G system known as *cdma2000*, which was specified by a similar collaboration to 3GPP known as the *Third Generation Partnership Project 2* (3GPP2), and which met the requirements for IMT-2000. Qualcomm intended to develop a 4G successor to cdma2000 under the name *Ultra Mobile Broadband* (UMB), but no network operator announced plans to adopt the technology, and the project was dropped in 2008.

Worldwide Interoperability for Microwave Access (WiMAX) is specified in the IEEE 802.16 standards. The original specification was for a point-to-point microwave system, but that gradually evolved into *Mobile WiMAX 1.0* (IEEE 802.16e), which had all the characteristics of a mobile cellular network and met the requirements for IMT-2000. The IEEE later specified an enhancement known as *Mobile WiMAX 2.0* (IEEE 802.16m), which met the requirements of IMT-Advanced. However, LTE had far more support amongst network operators and equipment manufacturers, and WiMAX no longer has a significant market share.

1.3 The Mobile Telecommunication Market

1.3.1 Traffic Levels

For many years, mobile telecommunication networks mainly handled voice calls. The amount of mobile data grew slowly at first, but it overtook the amount of voice traffic in 2010. Mobile networks are now dominated by data, especially by video downloads, and the traffic levels are forecast to grow still further.

To illustrate that issue, Figure 1.5 shows the total traffic being handled by mobile telecommunication networks throughout the world, in units of exabytes (10^{18} bytes, i.e. one billion gigabytes) per month. The figures are by Ericsson [5, 6], and are based on measurements up to the third quarter of 2019 and forecasts thereafter.

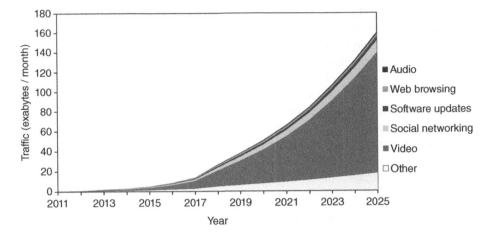

Figure 1.5 Monthly traffic due to different mobile telecommunication applications, with measurements up to 2019 and forecasts thereafter. *Source:* Data from Ericsson 2019.

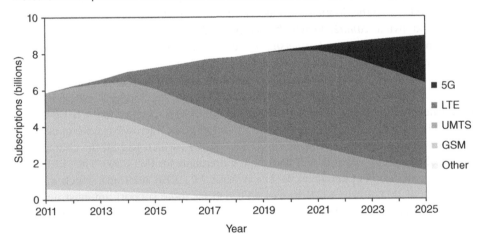

Figure 1.6 Numbers of subscriptions to different mobile telecommunication technologies, with measurements up to 2019 and forecasts thereafter. *Source:* Data from Ericsson 2019.

1.3.2 Numbers of Subscriptions

Although traffic levels are continuing to grow, developed markets have become saturated with smartphones, and sales have begun to stall. To illustrate that, Figure 1.6 shows Ericsson's measurements and forecasts of the numbers of mobile subscriptions worldwide. The total number of subscriptions overtook the world's population in 2018, because some subscriptions are inactive, and because some users own more than one device. The actual number of subscribers in the third quarter of 2019 was about 5.8 billion.

1.3.3 Operator Revenue

To make matters worse, increasing levels of competition have reduced network operators' revenue per bit, while subscribers have also been making greater use of free WiFi access and

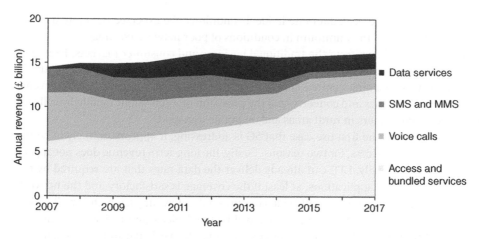

Figure 1.7 Annual revenue earned by mobile network operators in the UK. *Source:* Data from Ofcom 2019.

of third-party services for voice and messaging. Together, these issues have caused network operators' income to stall. To illustrate that point, Figure 1.7 shows estimates of the annual income earned by network operators in the UK up to the end of 2017, which have been published by the UK regulator Ofcom [7, 8]. (Ofcom's data for 2018 and 2019 use a different accounting standard, so they are not directly comparable.)

Taking all of these issues into account, network operators have tended to conclude that the traditional consumer market is not a promising source of future revenue. Their response has been to look elsewhere.

1.4 Use Cases and Markets for 5G

1.4.1 5G Research Projects

In 2012, the ITU started a programme of work to define a successor to IMT-Advanced, which it denoted *IMT-2020* [9–12]. That triggered a surge of national and industrial research projects into 5G, which are summarized in Reference [13]. It also triggered a 3GPP study item that began in 2015.

The 3GPP study [14] identified several potential markets and a large number of individual use cases for 5G, and grouped those use cases into five families that share common requirements and characteristics. Those families, and their associated markets, are discussed in Sections 1.4.2 through 1.4.6. The first three were also identified by the ITU [15] and are generally considered to be the main use case families for 5G, while the last two have been separated out from the others by 3GPP.

1.4.2 Enhanced Mobile Broadband

Enhanced mobile broadband (eMBB) addresses similar markets to LTE, but with improved capabilities [16]. In particular, the eMBB use case calls for a higher data rate than LTE can provide, in terms of the peak data rate in ideal conditions, the expected data rate in more typical conditions, and the minimum in conditions of poor network coverage.

Applications are drawn from the traditional business and consumer markets. Examples include data downloads and real-time video in an indoor office environment, traffic hotspots such as shopping centres and other dense urban environments, public events such as football matches and concerts, and the provision of a consistent mobile broadband experience for consumers in rural areas and on public transport.

Although eMBB is the first use case that 5G is addressing, it appears unlikely to be the key to its long-term success, for two reasons. Firstly, the long-term revenue does not appear to be sufficient. Secondly, LTE can already deliver the data rates that are required by the most common mobile applications, at least if the coverage is satisfactory and the network is uncongested. (For example, mobile video requires only a few Mbps, which is well within the capabilities of LTE, and is ultimately capped by the device's physical size and display resolution [17].) Nevertheless, 5G does offer some benefits, particular in terms of a higher network capacity and a lower consumption of electrical power, which are discussed later in this chapter.

1.4.3 Massive Machine-type Communications

The second use case is *machine-type communications* (MTC), also known as the *internet of things* (IoT) [18]. These terms both refer to wireless communications between autonomous machine-type devices without any direct human interaction. However, the first term is sometimes restricted to point-to-point communications, whereas the second always implies that the devices communicate over the internet. In turn, the terms *massive machine-type communications* (mMTC) and *massive internet of things* (MIoT) both imply the aggregation and analysis of data from large numbers of connected devices.

The applications for MTC are wide-ranging [19]. One example is wireless tracking devices, for example to report the physical location of goods in a warehouse, vehicles in a delivery system or cats in a local neighbourhood. Another is wireless sensors, such as motion-triggered security cameras, environmental sensors, and strain gauges for bridges, buildings and other civil engineering structures. A third application is electronic health, including the reporting of medical information such as body temperature, blood pressure and glucose levels, to help permit independent living for the elderly and for those with long-term medical conditions.

The market for MTC has been widely expected to grow, with some early commentators speaking of 50 billion IoT devices by the year 2020 [20]. Those forecasts have not been ful-filled, at least in part because of the difficulty of extrapolation, but the numbers are still promising. In 2019, there were 10.8 billion IoT devices worldwide, of which the majority used short-range technologies such as WiFi, but 1.4 billion made use of cellular networks. As shown in Figure 1.8, which contains additional measurements and forecasts by Ericsson, those numbers are expected to grow further.

MTC has different performance requirements from mobile broadband. The devices must be cheap, and they require a low power consumption to ensure a long battery life. The devices must also communicate successfully if the received signal is very weak, to support devices such as smart meters that might be installed deep inside a building. The data rate

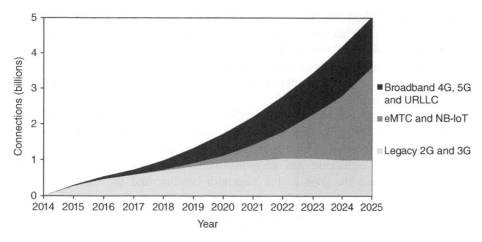

Figure 1.8 Numbers of cellular IoT devices using different mobile telecommunication technologies, with measurements up to 2019 and forecasts thereafter. *Source:* Data from Ericsson 2019.

of an individual machine-type device is usually low, but the collective data rate can still be high. Furthermore, the associated signalling messages can overwhelm the network if they are not carefully controlled, for example if the devices try to re-establish communications following a network outage.

There are already several technologies for wide-area MTC. GPRS offers a basic set of capabilities, which was subsequently enhanced by a technology known as *extended coverage GSM* (EC-GSM). LTE was originally designed for use by smartphones, but was later enhanced to support machine-type communications by means of eMTC and NB-IoT. There are also two popular wide-area technologies that operate in unlicensed spectrum, namely *Long Range Wide Area Network* (LoRaWAN) and *Sigfox*.

At least in the short to medium term, MTC does not appear to be a core use case for 5G, but the reasons are different from those for mobile broadband. Firstly, eMTC and NB-IoT can handle most of the requirements, at least in the short to medium term. Secondly, the growth of MTC appears to be boosted more by improvements to network coverage, to ensure that devices can operate successfully wherever they are installed with minimal human intervention. That issue is better addressed by improving the roll-out of existing air interfaces rather than by introducing a new one.

1.4.4 Ultra-reliable Low-latency Communication

Ultra-reliable low-latency communication (URLLC), also known as *critical communication* (CriC), is a use case characterized by the need for very low latency, often in conjunction with very high reliability [21].

Latency is the time required to deliver packets between the mobile and an external server. LTE supports latencies of a few tens of milliseconds for non-roaming mobiles, which are low enough for conversational VoIP and video over IP, but are still too high for other applications. For example, *tactile networks* are responsive over the timescales of human touch and require feedback within a few milliseconds to give an impression of immediate response.

Virtual reality headsets find consumer applications in immersive entertainment and online games, and require similar latencies to avoid feelings of nausea on the part of the user.

LTE networks can already deliver traffic with very high reliability, but only if the latency is high enough (tens or hundreds of milliseconds) for packets to be re-transmitted if they arrive incorrectly. It is far more difficult to deliver packets reliably with a lower latency, and a new air interface is required.

Low latency and high reliability come together in the concept of wireless industrial automation [22, 23]. This is sometimes depicted under the buzzword *Industry 4.0* as a fourth industrial revolution, which it is hoped will succeed the original industrial revolution of the late eighteenth and early nineteenth centuries, the introduction of mass-production around the start of the twentieth century, and the introduction of computers from the 1950s. For example, a factory or larger industrial site might contain robots and other industrial machinery, which are equipped with large numbers of sensors and actuators, and are governed in a closed-loop control system by a central server. In such a factory, wireless control brings several advantages over a network of fixed cables, because it is cheaper and easier to install the system, to add new machinery and especially to reconfigure the factory in response to changes in customer demand. Although such a wireless control system might operate using unlicensed spectrum, a licensed system such as 5G can operate over longer ranges and with lower levels of interference, and hence with higher reliability.

A still more stringent set of requirements appears in wireless remote surgery, in other words the tactile control of robotic tools by a surgeon at a remote location. Although the application has been demonstrated successfully, its safety-critical nature may make widespread adoption more of a long-term aspiration for 5G.

URLLC is the use case in which 5G can offer the greatest benefits over LTE, both because of its technical requirements and because of the opportunity it offers network operators to go beyond their traditional consumer markets. 5G supports it through a number of enhancements to the air interface and the network architecture [24], which will be an ongoing theme throughout this book.

1.4.5 Vehicle-to-everything Communication

Vehicle-to-everything (V2X) communication refers to the exchange of information between a road vehicle and the mobile telecommunication network, and with other vehicles and pedestrians that are nearby. It is often considered as an aspect of URLLC, but is separated out into its own use case by 3GPP.

Vehicle communications can be carried out using unlicensed spectrum, for example using a variant of WiFi known as *dedicated short-range communication* (DSRC) that is defined in the standard IEEE 802.11p. As in the case of URLLC, however, cellular networks offer several advantages, in particular a higher transmit power and a lower level of interference, which lead to longer range and higher reliability [25]. LTE was enhanced to offer basic support for vehicle communications as part of LTE-Advanced Pro, but that was limited by LTE's capabilities for data rate, latency, and reliability. Such support is greatly improved by 5G, but is only implemented in later releases of the specifications. We will address it as part of Chapter 20.

1.4.6 Network Operation

In the light of this discussion, it is evident that the 5G network has to address a far wider variety of use cases than LTE. To help it do so, 3GPP defined a final set of internal requirements for *network operation* (NEO) [26]. These include, as examples, requirements to optimize and re-route the traffic path in the case of low-latency applications, to interact with a client's application servers and to reconfigure a network to support new use cases as they arise. Those requirements are addressed by a number of enabling technologies for the 5G system, which we will discuss later in this chapter.

1.5 Technical Performance Requirements

Based on its earlier studies, the ITU prepared a set of performance requirements for IMT-2020, which it published in November 2017 [27]. In parallel, 3GPP published its own requirements for the 5G air interface in October 2016 and its service requirements for the 5G system in March 2017 [28, 29]. The 3GPP system is designed to be compliant with IMT-2020, so the two sets of requirements are very similar; they are summarized in Table 1.1. The table also states the earlier requirements for LTE-Advanced, incorporating those for IMT-Advanced and, where relevant, NB-IoT [4, 30, 31].

Looking first at the peak data rate, the design targets for LTE-Advanced were 1000 Mbps on the downlink and 500 Mbps on the uplink. The design targets for 5G are far greater:

Table 1.1 Technical performance requirements for 5G.

Parameter	LTE-Advanced	5G	Main 5G use case
Peak data rate	500 Mbps (UL)	10 Gbps (UL)	eMBB
	1000 Mbps (DL)	20 Gbps (DL)	
Typical spectral efficiency[a]	0.7–2.25 bits s^{-1} Hz^{-1} (UL)	1.6–6.75 bits s^{-1} Hz^{-1} (UL)	eMBB
	1.1–3 bits s^{-1} Hz^{-1} (DL)	3.3–9 bits s^{-1} Hz^{-1} (DL)	
Energy efficiency	n/a	Maximize	eMBB
Maximum UE speed	350 km h^{-1}	500 km h^{-1}	eMBB
User plane latency	5 ms	0.5 ms	URLLC
User plane reliability[b]	n/a	99.999%	URLLC
Maximum coupling loss[c]	n/a	143 dB	eMBB
	164 dB	164 dB	mMTC[e]
Battery life[d]	10 yr	10–15 yr	mMTC[e]
Connection density	60 000 km^{-2}	1 000 000 km^{-2}	mMTC

a) Targets in various ITU performance environments.
b) Delivery of a 32 byte packet within 1 ms.
c) Data rate of 1 Mbps DL/30 kbps UL for eMBB, and 160 bps for mMTC.
d) Delivery of 200 uplink bytes per day, with a coupling loss of 164 dB, and a 5 Wh battery.
e) Addressed in Release 16 by integrating eMTC and NB-IoT into the 5G system.

20 Gbps on the downlink and 10 Gbps on the uplink. The increase is mainly achieved by increasing the maximum bandwidth of each individual cell, from 20 MHz in the cases of LTE and LTE-Advanced, to 400 MHz in the case of 5G. (We will discuss the peak data rates that the two systems can actually handle as part of Chapter 11.)

It cannot be stressed too strongly, however, that these peak data rates can only be reached in ideal conditions, for example if the bandwidth is very large and the received signal is very strong. A better measure is the typical *spectral efficiency*, which expresses the typical capacity of one cell per unit bandwidth, in units of bits s^{-1} Hz^{-1}. 5G supports a spectral efficiency that is about three times greater than the original design target for LTE-Advanced. (Once again, we will discuss the figures that are actually achieved as part of Chapter 11.)

In LTE networks, around 15% of the operating expenditure is for electrical power alone, with around 80% of that consumed by the base stations [32]. There is a risk of that situation worsening as the data rates increase, so there is a qualitative requirement to maximize the 5G network's *energy efficiency*, which expresses the capacity of one cell per unit of electrical power, in units of bits s^{-1} W^{-1}.

The air interface's latency is the time required for a data packet to travel between the mobile and the fixed network, provided that the air interface is not congested. (Additional delays can arise within the fixed network itself, or in cases of congestion.) To handle URLLC applications, 5G requires an air interface latency of 0.5 ms, and also requires the air interface to deliver packets within that time with a reliability of 99.999%.

The remaining requirements are mainly applicable to machine-type communications. At present, there is no attempt to address those requirements by means of the 5G air interface itself: instead, the requirements are met by integrating the earlier technologies of eMTC and NB-IoT into the rest of the 5G system. (The coupling loss is closely related to the range of the communication system, and will be defined in Chapter 4.)

5G does not have to meet all of these requirements at the same time: for example, the requirements for low latency and high reliability are only relevant to the URLLC use case, while those for enhanced coverage and long battery life are only relevant for MTC. Nevertheless, the requirements for 5G are far more challenging than those for LTE, and are far more diverse. In Section 1.6, we will discuss the technologies through which those requirements are addressed.

1.6 Technologies for 5G

1.6.1 Network Function Virtualization

The core network of a mobile telecommunication system contains a large number of network elements, for example routers, signalling functions and databases. Traditionally, these have been implemented using special-purpose hardware, which is procured from an equipment vendor such as Ericsson, Nokia, or Huawei. That approach is perfectly satisfactory, but has several limitations: the hardware is manufactured in low volumes, so it is expensive; the hardware and software are coupled together, with the operator locked to a single vendor for both; and it is difficult, slow and expensive to upgrade the hardware as new requirements arise, an issue that becomes ever more problematic as development lifecycles shorten.

Network function virtualization (NFV) is a more recent approach, in which the network elements are implemented using software in the form of *virtualized network functions* (VNFs) [33–35]. Those functions then run on general-purpose, commercial off-the-shelf hardware. NFV brings several benefits: the hardware is generic and cheaper; the software is easier, faster and cheaper to write and upgrade; the software and hardware can be uncoupled from one another, with the two procured from different vendors; and it is easier for new vendors to enter the market. Network operators were already adopting these techniques in their existing networks [36]. However, 5G has been designed with NFV in mind from the beginning, for example in the design of a management system which can easily set up, modify and tear down the individual network functions, and which can administer them independently of the underlying hardware [37].

The concept can also be applied to the radio access network. A *centralized RAN* is a network in which the higher-level functions of the base station are implemented at a more central hub, which is some distance from the local cell site. By itself, RAN centralization offers improved security and a smaller footprint at the cell site, and allows the higher-level functions of nearby cells to communicate more effectively. If those functions are then implemented using software, the result is a *virtualized RAN* or *cloud RAN* (C-RAN). As well as bringing the generic benefits of NFV, a cloud RAN allows for load balancing, in which heavily loaded base stations can acquire more resources from the underlying hardware than their lightly loaded neighbours.

1.6.2 Software-defined Networking

At the highest level, each element in a traditional communication network handles two main tasks. Functions in the *user plane* (UP) forward traffic from one network element to another, for example by the use of routing tables, while those in the *control plane* (CP) carry out higher-level tasks such as configuring the routing tables and managing the network's resources. Once again, that approach is perfectly satisfactory, but has several limitations: the decision-making processes are distributed over the entire network, which makes the control plane unnecessarily complex, while those complexities limit how far the operator can distribute the functions in the user plane.

The central feature of *software-defined networking* (SDN) is a clear separation between the control and user planes [38]. In a software-defined network, the control plane functions can be centralized, which allows for simpler, more integrated control over the network. Once centralized, those functions can easily communicate with authorized third-party application servers over a so-called *northbound application programming interface* (API), through which those servers can influence or control the network's operation. The third-party servers do not know any details of the network's physical implementation, as those are confined to the *southbound API*, through which the control plane functions configure the ones in the user plane. In turn, the user plane functions can be distributed to whichever physical locations are desired, without the constraints that the control plane previously imposed.

NFV and SDN are complementary technologies: the first separates the hardware and software in a communication network, while the second separates its control and user planes. Once again, 5G has been designed with SDN in mind from the beginning.

SDN can be applied to the radio access network as well as the core. *Multi-access edge computing* (MEC), also known as mobile edge computing, is a technology in which the user

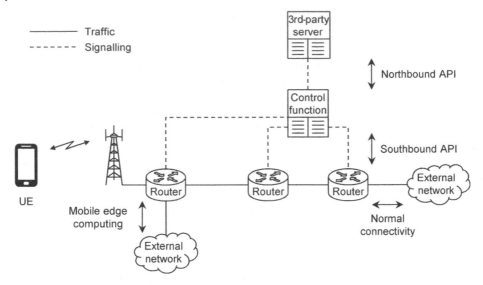

Figure 1.9 Architecture of a software-defined mobile telecommunication network.

plane connection to an external network is implemented close to the cell site, or even at the cell site itself [39–42]. By shortening the communication path, the architecture provides the mobile with low-latency access to external servers, to help support the most demanding forms of low-latency application. Furthermore, if the user plane function is implemented in software by means of NFV, then it can be relocated dynamically, for example if the mobile starts running a low-latency service or if it subsequently changes cell. Figure 1.9 is a high-level view of the resulting architecture for a mobile telecommunication network that is using the two technologies.

1.6.3 Network Slicing

NFV and SDN provide a network operator with a flexible underlying resource, which is easily reconfigured to handle changes in network traffic. Taking those concepts further, *network slices* are virtual logical sub-networks which run on shared, underlying physical hardware, and which are realized by means of NFV and SDN [43, 44]. A network operator can set up, reconfigure and tear down individual slices, as the need for them arises.

Each network slice is optimized for the traffic that it carries. For example, a network operator might use different slices for each of the main 5G use cases, with the eMBB, mMTC and URLLC slices respectively optimized for a high data rate, a high connection density, and low latency and high reliability. An operator might also use different slices to subdivide those use cases: for example, the eMBB use case might be realized using different slices for internet traffic, IMS voice and video, and public safety. Different slices might also be used for an operator's individual customers, such as *mobile virtual network operators* (MVNOs) that do not own any infrastructure of their own, or industrial clients.

An operator's network slices are isolated from each other, most critically in respect of data transport and security. If, for example, there is congestion in a slice that is handling internet traffic, then there should be no additional delays in slices that are handling time-sensitive applications such as IMS voice, public safety or URLLC. Similarly, any intrusion into a slice that is handling MTC should not increase the risk of security issues in any of the other network slices.

LTE networks already implement some of the ideas underlying network slicing: for example, LTE transport networks often isolate IMS voice from internet data to minimize congestion problems. However, network slicing can only be fully realized in a 5G network by means of NFV and SDN. With those capabilities in place, a 5G network slice extends not only through the core and radio access networks but also into the air interface itself, which can be understood as another resource that is easily shared amongst network slices as the need arises.

1.6.4 Technologies for the Air Interface

The 5G air interface uses two important technologies that we will address in detail in later chapters. Firstly, 5G can operate over a wider range of radio frequencies than previous systems have done, up to a maximum of about 50 GHz. At the highest frequencies, the radio waves have a wavelength of millimetres or tens of millimetres, so they are often known as *millimetre waves*. When using millimetre waves, the 5G air interface can occupy a far wider bandwidth than it does at low radio frequencies, so its capacity can be far greater. However, the use of millimetre waves also leads to a number of coverage issues, which we will explore as part of Chapter 4.

Secondly, 5G makes greater use of multiple antennas than previous systems have done [45], typically using just a few antennas at the mobile, but using tens or even a few hundred at the base station. It applies them in two distinct ways: to improve the coverage of the air interface at high radio frequencies by means of *beamforming*, and to improve its capacity at low radio frequencies by means of *multiple-input multiple-output* (MIMO) antennas. We will examine both of these techniques as part of Chapter 6.

1.7 The 3GPP Specifications for 5G

The 5G specifications are written by 3GPP in the same way as the earlier specifications for LTE, UMTS and GSM. They are organized into *releases* [46], each of which contains a stable and clearly defined set of features. Within each release, the specifications progress through a number of different versions. New features can be added to successive versions until the release is frozen, after which the only changes involve refinement of the technical details, corrections, and clarifications. Table 1.2 lists the releases that 3GPP has used since the introduction of UMTS, together with their most important features. (The numbering scheme was changed after Release 99, with later releases numbered from 4 onwards.)

5G is specified over two initial releases. Release 15 focusses mainly on eMBB, but also includes support for low-latency and high-reliability communications. Within that release, the specifications were frozen in three separate drops, to help equipment manufacturers introduce the most important capabilities early on without waiting for later capabilities to

Table 1.2 3GPP specification releases for UMTS, LTE and 5G.

Release	End date	New features
Rel-99	December 1999	UMTS, WCDMA
Rel-4	June 2001	TD-SCDMA
Rel-5	September 2002	HSDPA, IP multimedia subsystem
Rel-6	September 2005	HSUPA
Rel-7	March 2008	Enhancements to HSPA
Rel-8	March 2009	LTE
Rel-9	March 2010	Location services for LTE
Rel-10	June 2011	LTE-Advanced, carrier aggregation
Rel-11	March 2013	Co-ordinated multi-point transmission
Rel-12	March 2015	Dual connectivity, LTE sidelink
Rel-13	March 2016	LTE-Advanced Pro, eMTC, NB-IoT
Rel-14	June 2017	V2X services for LTE
Rel-15	June 2019	5G phase 1
Rel-16	June 2020	5G phase 2
Rel-17	December 2021 (target)	Enhancements to 5G

be specified in full [47]. Release 16 supports all the use cases and requirements for 5G, and is a candidate technology for IMT-2020.

The specifications are also organized into several *series*, each of which covers a particular part of the system. Table 1.3 summarizes the contents of series 21–38, which contain all the specifications for UMTS, LTE, and 5G, as well as specifications that are common with GSM. (Some other series numbers are used exclusively for GSM.) Although most series contain specifications for each of the 3GPP technologies, the 38 series is devoted to the 5G air interface and RAN, so it is an important source of information for this book.

When written in full, an example specification number is TS 23.501 v. 16.2.0. Here, TS stands for *Technical Specification*, which is the output from the 3GPP *work item* in which the specification was created. The series number is 23, 501 is the specification number within that series, 16 is the release number, 2 is the technical version number within that release and the final 0 is an editorial version number that is occasionally used for non-technical changes. 3GPP also produces *Technical Reports* (TR), which are the output from preliminary 3GPP *study items*. They are purely informative and have three-digit specification numbers beginning with 9, 8 or occasionally 7.

In a final division, each specification belongs to one of three *stages*, following a scheme that is defined by the ITU [48]. Stage 1 specifications define the service from the user's point of view, and lie exclusively in the 22 series. Stage 2 specifications define the system's high-level architecture and operation, and lie mainly (but not exclusively) in the 23 series. Finally, stage 3 specifications define all the functional details.

Table 1.3 3GPP specification series used by UMTS, LTE and 5G.

Series	Scope
21	High-level requirements
22	Stage 1 service specifications
23	Stage 2 service and architecture specifications
24	Non-access stratum protocols
25	UMTS air interface and radio access network
26	Codecs
27	Data terminal equipment
28	Network management, including for 5G
29	Core network protocols
30	Programme management
31	UICC and USIM
32	Network management and charging
33	Security
34	UE test specifications
35	Security algorithms
36	LTE air interface and radio access network
37	Multiple radio access technologies
38	5G air interface and radio access network

The stage 2 specifications are especially useful for acquiring a high-level understanding of the system. The most useful ones for 5G are TS 23.501, TS 23.502, TS 37.340 and TS 38.300 [49–52], which respectively cover the system architecture, the high-level signalling procedures, dual connectivity with LTE, and the 5G air interface. There is, however, an important note of caution: those specifications are superseded later on and cannot be relied upon for complete accuracy. Instead, the details should be checked in the relevant stage 3 specifications.

The individual specifications can be downloaded from 3GPP's specification numbering page or from its FTP server [53, 54]. The 3GPP website has summaries of the features in early releases [55], and later releases are summarized as technical reports [56–58].

1.8 Architecture of 5G

1.8.1 High-level Architecture

Architecturally, 5G builds on LTE by adding a number of new components that are shown in Figure 1.10. The *5G core network* (5GC) contains new network functions that support NFV, SDN and network slicing. The *next-generation radio access network* (NG-RAN)

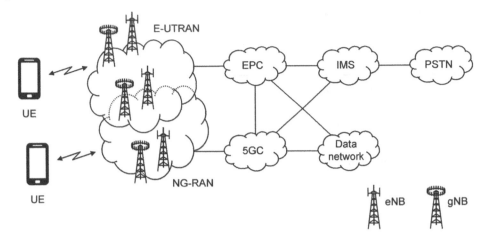

Figure 1.10 High-level architecture of 5G.

introduces new interfaces and signalling procedures for the backhaul and for communications between nearby base stations, and supports RAN centralization and virtualization, edge computing and network slicing. Together, the two networks form the *5G system* (5GS).

The 5G base station is known as a *next-generation Node B* (gNB). The gNB communicates with the mobile over an air interface known as the *New Radio* (NR), which supports millimetre-wave and multiple antenna communications.

Generically, an individual eNB or gNB is known as a *node*. As shown in Figure 1.10, an individual node can lie in either the E-UTRAN, the NG-RAN or both. Specifically, a node lies in the E-UTRAN if it supports the legacy LTE backhaul and is connected to the evolved packet core, and in the NG-RAN if it supports the new 5G backhaul and is connected to the 5G core.

In Release 15, the 5G core network handles voice calls using VoIP alone, with the calls controlled by either a third-party application or the IP multimedia subsystem. Release 16 introduces support for SRVCC, in which the network can convert an IMS voice call to a circuit-switched call, and can hand the mobile over to a 3G cell [59]. However, that is the only direct interaction between the 5G core network and the core networks of 2G and 3G. If, for example, an SRVCC handover takes place, then the mobile's communications with any external packet data networks are all torn down.

1.8.2 Architectural Options

As implied by Figure 1.10, one of the important features of 5G is its close inter-operation with LTE. To help enumerate the different possibilities, the 5G specifications support a number of different architectural options, which were identified and numbered as part of the early 3GPP studies into 5G [60]. We will discuss these options more fully as part of Chapter 3, but they are summarized in Figure 1.11. An individual network operator can support one, some or all of these options at the same time.

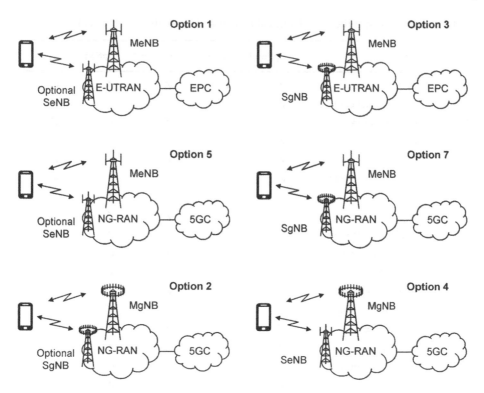

Figure 1.11 Architectural options for 5G.

Option 1 is a legacy LTE system that supports dual connectivity. Option 3 replaces the secondary eNB with a secondary gNB, and is the one most commonly used by early roll-outs of 5G. The next two options introduce the 5G core network and the NG-RAN, using a master eNB and an optional secondary eNB in option 5, and a secondary gNB in option 7. Each of these options is a *non-standalone* (NSA) implementation of 5G, because the gNB can only operate in conjunction with a master eNB, and cannot operate alone.

In the last two options, the gNB takes on the role of the master, operating either alone or with a secondary gNB in option 2, and with a secondary eNB in option 4. These options are known as *standalone* (SA) implementations of 5G. Standalone operation is only supported when using the 5G core network, not when using the evolved packet core.

With the exception of option 1, each of these options uses aspects of 5G. Even so, parts of the book are only relevant to certain architectural options, and not to others. In an attempt to resolve any confusion, Table 1.4 shows how the material in this book maps onto the options listed above, with ticks denoting topics that are relevant, and brackets denoting those that are relevant with restrictions or minor modifications. As an example, the material on dual connectivity is written from the viewpoint of the NG-RAN, but can also be applied to the E-UTRAN with only a few changes to the signalling procedures.

Table 1.4 Mapping between the 5G architectural options and the contents of the book.

Section	Topic	Option 3	Option 5	Option 7	Option 2	Option 4
1	Introduction	✓	✓	✓	✓	✓
2.1	Evolved packet core	✓				
2.2–2.7	5G core network		✓	✓	✓	✓
3.1	E-UTRAN	✓				
3.2	gNB	✓		✓	✓	✓
3.3	Architectural options	✓	✓	✓	✓	✓
3.4–3.6	NG-RAN		✓	✓	✓	✓
4	Radio communications	✓	(✓)	✓	✓	✓
5	Digital signal processing	✓	(✓)	✓	✓	✓
6	Multiple antennas	✓	(✓)	✓	✓	✓
7	5G New Radio	✓		✓	✓	✓
8.1–8.3	Initial access	✓		✓	✓	✓
8.4	System information				✓	✓
9	Random access	✓		✓	✓	✓
10	Link adaptation	✓		✓	✓	✓
11	Data transmission	✓		✓	✓	✓
12.1–12.2	MAC, RLC	✓		✓	✓	✓
12.3	PDCP	✓	✓	✓	✓	✓
12.4	SDAP		✓	✓	✓	✓
13.1	Power-on procedures	✓	✓	✓	✓	✓
13.2	Network and cell selection	✓	✓	✓	✓	✓
13.3	RRC connection setup		(✓)	(✓)	✓	✓
13.4	Registration		✓	✓	✓	✓
13.5	Deregistration		✓	✓	✓	✓
14	Security		✓	✓	✓	✓
15	Session management		✓	✓	✓	✓
16.1	RRC_CONNECTED	✓	✓	✓	✓	✓
16.2	Measurements	✓		✓	✓	✓
16.3	Handover		✓	✓	✓	✓
16.4	Dual connectivity	(✓)	✓	✓	✓	✓
16.5	RRC connection release		✓	✓	✓	✓
17	RRC_IDLE		(✓)	(✓)	✓	✓
18	RRC_INACTIVE		(✓)	(✓)	✓	✓
19	Inter-operation with EPC	✓	✓	✓	✓	✓
20	Releases 16 and 17	(✓)	✓	✓	✓	✓

MAC: Medium access control; PDCP: packet data convergence protocol; RLC: radio link control; RRC: radio resource control; SDAP: service data adaptation protocol.

References

1 Shannon, C.E. (1948). A mathematical theory of communication. *The Bell System Technical Journal* 27: 379–428. and 623–656.

2 ITU-R M.1034-1 (1997) Requirements for the radio interface(s) for International Mobile Telecommunications 2000 (IMT-2000).

3 3GPP (2020) 3GPP: The mobile broadband standard. www.3gpp.org (accessed 18 January 2020).

4 ITU-R M.2134 (2008) Requirements related to technical performance for IMT-Advanced radio interface(s).

5 Ericsson (2019) Ericsson mobility report, November 2019. https://www.ericsson.com/en/mobility-report (accessed 18 January 2020).

6 Ericsson (2019) Ericsson mobility visualizer. https://www.ericsson.com/en/mobility-report/mobility-visualizer (accessed 18 January 2020).

7 Ofcom (2019) Telecommunications market data update Q2 2019. www.ofcom.org.uk/research-and-data/telecoms-research/data-updates (accessed 18 January 2020).

8 Ofcom (2019) Telecommunications data revenues, volumes and market share update Q2 2019. www.ofcom.org.uk/__data/assets/file/0027/175581/q2-2019-telecoms-data-update.csv (accessed 18 January 2020).

9 ITU-R M.2290-0 (2013) Future spectrum requirements estimate for terrestrial IMT.

10 ITU-R M.2320-0 (2014) Future technology trends of terrestrial IMT systems.

11 ITU-R M.2370-0 (2015) IMT traffic estimates for the years 2020 to 2030.

12 ITU-R M.2376-0 (2015) Technical feasibility of IMT in bands above 6 GHz.

13 5G Americas (2016) Global organizations forge new frontier of 5G: 5G Americas 5G global update.

14 3GPP TR 22.891 (2016) Feasibility study on new services and markets technology enablers; Stage 1 (Release 14), September 2016.

15 ITU-R M.2083-0 (2015) IMT Vision – Framework and overall objectives of the future development of IMT for 2020 and beyond.

16 3GPP TR 22.863 (2016) Feasibility study on new services and markets technology enablers – Enhanced mobile broadband; Stage 1 (Release 14), September 2016.

17 Webb, W. (2018). *The 5G Myth*, 3e. De G Press, Chapter 3.

18 3GPP TR 22.861 (2016) Feasibility study on new services and markets technology enablers for massive internet of things; Stage 1 (Release 14), September 2016.

19 5G Americas (2019) 5G – The future of IoT.

20 Nordrum, A. (2016) Popular internet of things forecast of 50 billion devices by 2020 is outdated. https://spectrum.ieee.org/tech-talk/telecom/internet/popular-internet-of-things-forecast-of-50-billion-devices-by-2020-is-outdated (accessed 18 January 2020).

21 3GPP TR 22.862 (2016) Feasibility study on new services and markets technology enablers for critical communications; Stage 1 (Release 14), September 2016.

22 Ericsson (2018) Industrial automation enabled by robotics, machine intelligence and 5G. Ericsson Technology Review, February 2018.

23 5G Americas (2018) 5G communications for automation in vertical domains.

24 5G Americas (2018) New services & applications with 5g ultra-reliable low latency communications.

25 5G Americas (2018) Cellular V2X communications towards 5G.

26 3GPP TR 22.864 (2016) Feasibility study on new services and markets technology enablers – Network operation; Stage 1 (Release 14), September 2016.

27 ITU-R M.2410-0 (2017) Minimum requirements related to technical performance for IMT-2020 radio interface(s).

28 3GPP TR 38.913 (2018) Study on scenarios and requirements for next generation access technologies (Release 15), June 2018.

29 3GPP TS 22.261 (2019) Service requirements for the 5G system; Stage 1 (Release 15), September 2019.

30 3GPP TR 36.913 (2018) Requirements for further advancements for Evolved Universal Terrestrial Radio Access (E-UTRA) (LTE-Advanced) (Release 15), June 2018.

31 3GPP TR 45.820 (2015) Cellular system support for ultra-low complexity and low throughput internet of things (CIoT) (Release 13), November 2015.

32 Nokia (2016) 5G network energy efficiency: Massive capacity boost with flat energy consumption. https://gsacom.com/paper/5g-network-energy-efficiency-nokia-white-paper (accessed 18 January 2020).

33 Chayapathi, R., Hassan, S.F., and Shah, P. (2016). *Network Functions Virtualization (NFV) with a Touch of SDN*. Addison Wesley.

34 ETSI (2012) Network functions virtualisation: An introduction, benefits, enablers, challenges & call for action. https://portal.etsi.org/NFV/NFV_White_Paper.pdf (accessed 18 January 2020).

35 ETSI GS NFV 002 (2014) Network functions virtualisation (NFV): Architectural framework, December 2014.

36 Nguyen, V.-G., Brunstrom, A., Grinnemo, K.-J. et al. (2017). SDN/NFV-based mobile packet core network architectures: a survey. *IEEE Communications Surveys and Tutorials* 19 (3): 1567–1602.

37 3GPP TS 28.500 (2018) Management concept, architecture and requirements for mobile networks that include virtualized network functions (Release 15), June 2018.

38 Goransson, P., Black, C., and Culver, T. (2016). *Software Defined Networks: A Comprehensive Approach*, 2e. Morgan Kaufmann.

39 ETSI (2014) Mobile-edge computing – Introductory technical white paper. https://portal.etsi.org/Portals/0/TBpages/MEC/Docs/Mobile-edge_Computing_-_Introductory_Technical_White_Paper_V1%2018-09-14.pdf (accessed 18 January 2020).

40 ETSI GS MEC 003 (2016) Mobile edge computing (MEC); Framework and reference architecture, January 2019.

41 ETSI (2018) MEC in 5G networks. https://www.etsi.org/images/files/ETSIWhitePapers/etsi_wp28_mec_in_5G_FINAL.pdf (accessed 18 January 2020).

42 Taleb, T., Samdanis, K., Mada, B. et al. (2017). On multi-access edge computing: a survey of the emerging 5G network edge cloud architecture and orchestration. *IEEE Communications Surveys and Tutorials* 19 (3): 1657–1681.

43 Kalixylos, A. (2018). A survey and an analysis of network slicing in 5G networks. *IEEE Communications Standards Magazine* 2 (1): 60–65.

44 Afolabi, I., Taleb, T., Samdanis, K. et al. (2018). Network slicing and softwarization: a survey on principles, enabling technologies, and solutions. *IEEE Communications Surveys and Tutorials* 20 (3): 2429–2453.

45 5G Americas (2019) Advanced antenna systems for 5G.

46 3GPP (2020) Releases. https://www.3gpp.org/specifications/releases (accessed 18 January 2020).

47 3GPP (2019) Release 15. https://www.3gpp.org/release-15 (accessed 18 January 2020).

48 ITU-T I.130 (1988) Method for the characterization of telecommunication services supported by an ISDN and network capabilities of an ISDN.

49 3GPP TS 23.501 (2019) System architecture for the 5G System (5GS); Stage 2 (Release 15), December 2019.

50 3GPP TS 23.502 (2019) Procedures for the 5G System (5GS); Stage 2 (Release 15), December 2019.

51 3GPP TS 37.340 (2019) Evolved Universal Terrestrial Radio Access (E-UTRA) and NR; Multi-connectivity; Stage 2 (Release 15), December 2019.

52 3GPP TS 38.300 (2019) NR; NR and NG-RAN overall description; Stage 2 (Release 15), December 2019.

53 3GPP (2020) Specification numbering. https://www.3gpp.org/specifications/specification-numbering (accessed 18 January 2020).

54 3GPP (2020) ftp://ftp.3gpp.org (accessed 18 January 2020).

55 3GPP (2020) ftp://ftp.3gpp.org/Information/WORK_PLAN/Description_Releases (accessed 18 January 2020).

56 3GPP TR 21.914 (2018) Release 14 description; Summary of Rel-14 work items (Release 14), May 2018.

57 3GPP TR 21.915 (2019) Release 15 description; Summary of Rel-15 work items (Release 15), September 2019.

58 3GPP TR 21.916 (2019) Release 16 description; Summary of Rel-16 work items (Release 16), December 2019.

59 3GPP TR 23.756 (2018) Study for single radio voice continuity from 5GS to 3G (Release 16), September 2018.

60 3GPP TR 38.801 (2017) Study on new radio access technology: Radio access architecture and interfaces (Release 14), March 2017, Section 7.2.

2

Architecture of the Core Network

This is the first of two chapters that lay out the system architecture of 5G. We will begin by reviewing the architecture of the LTE evolved packet core, partly because it is still used by early 5G roll-outs, and partly because it will act as an introduction to the 5G core. We will then move onto the 5G core network itself, covering its architecture and the roles of its most important network functions.

The main signalling protocol in the 5G core network is version 2 of the Hypertext Transfer Protocol (HTTP/2). That protocol is very different from the ones used by older mobile telecommunication systems, so we will review the protocol and provide some examples of the resulting signalling procedures. Several specifications are relevant to this chapter, but the most important is the stage 2 description of the 5G core network, Technical Specification (TS) 23.501 [1].

2.1   The Evolved Packet Core

2.1.1   Release 8 Architecture

Figure 2.1 shows the most important elements in the LTE evolved packet core (EPC) [2]. The different icons represent network elements that are used for data transport, signalling and information storage. An individual network might contain several instances of each one, both for geographical organization and for redundancy.

The interfaces between network elements are more accurately known as *reference points*. We will stick to that terminology within this book to avoid confusion with the separate service-based interfaces discussed later on. Solid lines denote reference points in the user plane which carry traffic, while dashed lines denote reference points in the control plane which carry signalling. Despite appearances, the reference points are not usually implemented as point-to-point connections: instead, they run over an underlying IP-based transport network, in which routers use the network elements' IP addresses to deliver traffic and signalling messages between them.

Using the evolved packet core, the mobile connects with one or more external *packet data networks* (PDNs), with each connection made through a point of contact known as a *packet data network gateway* (PDN-GW or PGW). The PDN gateway's role in the user plane is delivering incoming and outgoing traffic to and from the mobile. Its roles in the control

An Introduction to 5G: The New Radio, 5G Network and Beyond, First Edition. Christopher Cox.
© 2021 John Wiley & Sons Ltd. Published 2021 by John Wiley & Sons Ltd.

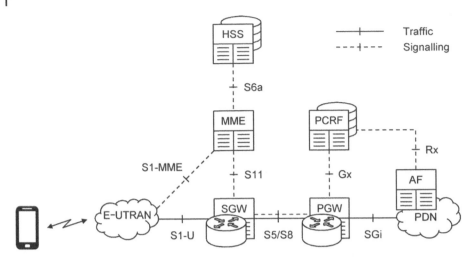

Figure 2.1 Architecture of the evolved packet core. *Source:* Adapted from 3GPP TS 23.401.

plane are managing the mobile's communications with the data network and informing a separate charging system about the amount of traffic that the mobile has transferred.

A mobile is assigned a single *serving gateway* (SGW), which forwards the mobile's traffic between the radio access network and each of its PDN gateways. The mobile is also assigned a *mobility management entity* (MME), which is known as its *serving MME*. That device controls the high-level operation of the mobile using two types of signalling message: *mobility management* signalling, which controls its communications with the evolved packet core, and *session management* signalling, which controls its communications with the packet data networks.

The *home subscriber server* (HSS) has two parts: a database that contains information about the network operator's subscribers, and a front end that presents the information to the other elements in the network. Finally, the *policy and charging rules function* (PCRF) determines sets of parameters that constrain the mobile's interactions with the network, which are known as *network policies*. These include the maximum data rate and end-to-end delay that a mobile will receive for a particular data stream, and the amount that the user will be charged. As part of that role, the PCRF can accept requests from external servers, known as *application functions* (AFs), to set up streams of traffic in the evolved packet core. In LTE, such requests are generally restricted to application functions that are controlled by the network operator itself, most often those in the operator's IP multimedia subsystem (IMS).

2.1.2 Control and User Plane Separation

In the original architecture for the evolved packet core, the PDN gateway carried out tasks in both the control and user planes. Although the serving gateway was mainly a user plane device, it also handled a few control plane tasks such as its own interactions with the charging system.

The control and user planes were split apart in a Release 14 work item known as *control and user plane separation*. In the resulting architecture [3], a *PDN gateway control plane*

(PGW-C) handles the signalling aspects of the PDN gateway and controls one or more *PDN gateway user planes* (PGW-Us) that handle traffic. The serving gateway is divided in the same way. The architecture allows network operators to reap some of the benefits of software-defined networking, anticipates the architecture of the 5G core network and eases subsequent interworking between the two.

2.2 The 5G Core Network

2.2.1 Representation Using Reference Points

Figure 2.2 shows the most important network functions in the 5G core network (5GC) [4]. Although the icons are the same as the ones used in Figure 2.1, the intention is to implement them using software, in the form of virtualized network functions that are running on general-purpose hardware. As before, an individual network might contain several instances of each one.

Using the 5G core network, the mobile can connect to one or more packet data networks (abbreviated in 5G as DN), for example the Internet, the IMS, or a low-latency network for a private industrial client. Each connection is associated with a *session management function* (SMF). The SMF controls the mobile's interactions with the data network, communicates with the mobile by means of session management signalling and sends information about the mobile's data usage to a separate charging system. The SMF also controls one or more *user plane functions* (UPFs), which handle the user's traffic.

Each mobile is assigned an *access and mobility management function* (AMF), which is known as its *serving AMF*. The AMF controls the mobile's interactions with the 5G core by means of mobility management signalling, and relays session management signalling between the mobile and its respective SMFs. Both sets of messages are relayed by the

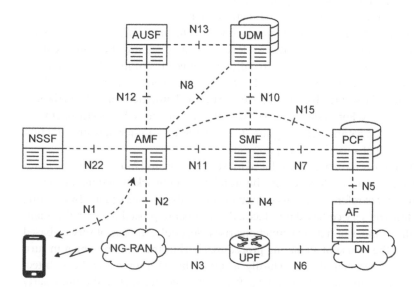

Figure 2.2 Representation of the 5G core network using reference points. *Source:* Adapted from 3GPP TS 23.501.

next-generation radio access network (NG-RAN) over a reference point denoted N1, while the AMF communicates with the NG-RAN itself over a reference point denoted N2.

Unified data management (UDM) combines a subscriber database with a front end that presents the information to the other functions in the network. Finally, *policy control functions* (PCFs) supply a mobile's SMFs with network policies that relate to session management, for example the maximum data rate and the amount that the user will be charged, and supply its AMF with network policies that relate to mobility management, such as any restrictions on the mobile's movement. Interactions between the PCFs and external application functions are richer than in LTE, and are discussed in Chapter 15.

A comparison with LTE shows that the UDM, AMF and PCF have similar roles to the HSS, MME and PCRF, while the SMF and UPF are similar to the control and user planes of the serving and PDN gateways. The main distinction is that the mobile's session management signalling is handled by the SMF, not the AMF.

The remaining network functions are new to 5G. The *network slice selection function* (NSSF) determines the network slices with which the mobile will be registered and assigns a serving AMF to the mobile that supports those slices. The *authentication server function* (AUSF) helps the AMF during the security procedure of authentication.

2.2.2 Representation Using Service-based Interfaces

Although the use of reference points is familiar from previous generations, the 5G core network is usually represented by means of *service-based interfaces* in the architecture shown in Figure 2.3 [5, 6]. In this representation, almost every network function is associated with a service-based interface, on which it offers services to other network functions over a clearly defined application programming interface (API). Most of the signalling procedures in the 5G core are defined by means of service-based interfaces: in contrast, the specifications hardly use the corresponding reference points at all.

On a service-based interface, the network function offering the service is known as the *network function producer*, while the one using the service is the *network function consumer*. The producer's services are organized into two levels: a coarse organization into *network function services*, and a finer organization into individual *service operations*. Each operation is named using the scheme `Nnfname_ServiceName_ServiceOperation`, in which the three parts identify the service-based interface, the network function service and the service operation respectively. For example, the UDM's interface includes the service operation `Nudm_SubscriberDataManagement_Get`, through which another network function can retrieve some or all of a user's subscription data.

Service-based interfaces have been introduced for several reasons [7–9]. Firstly, the number of service-based interfaces depends on the number of network functions in the 5G core, while the number of reference points potentially depends upon its square. By defining signalling procedures on service-based interfaces, the number of procedures remains manageable, even if the number of network functions rises. Secondly, the use of service-based interfaces promotes the re-use of signalling procedures within the system, which in turn contributes to a cleaner system design. Thirdly, a service-based interface is the architecture expected for communications between a software-defined network and a third-party application server over a northbound API. By following that architecture for its external

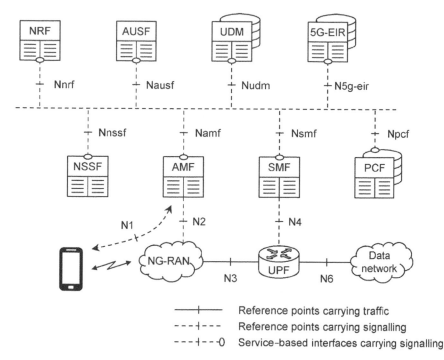

Figure 2.3 Representation of the 5G core network using service-based interfaces. *Source:* Adapted from 3GPP TS 23.501.

signalling and using the same architecture internally, the Third Generation Partnership Project (3GPP) has adopted a consistent design throughout the 5G core.

Nevertheless, the use of service-based interfaces is still incomplete. In particular, there are no service-based interfaces for the radio access network and for the UPF: instead, those network functions continue to use custom signalling procedures that are defined on traditional reference points.

Figure 2.3 also introduces two new network functions. An individual network function can register itself with a *network repository function* (NRF), so that it can be discovered by other network functions within the 5G core. The *5G equipment identity register* (5G-EIR) acts as a database of lost or stolen mobiles, and handles the same tasks as the EIR from previous generations.

2.2.3 Data Transport

The connection between a mobile and an external data network is known as a *protocol data unit session* (PDU session) [10]. Figure 2.4 shows some example PDU sessions for a mobile that is connected to the Internet and the IMS.

Each PDU session is associated with one or more UPFs, which can take on various roles. A *PDU session anchor* is a point of contact between the data network and the 5G core. There is usually one anchor for each PDU session, but there may be more. There may also be *intermediate UPFs* (I-UPFs), which forward traffic between the anchors and the radio access network.

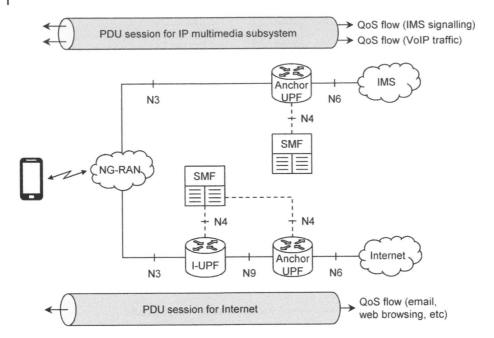

Figure 2.4 PDU sessions and QoS flows.

Adding some more detail, the backhaul for each PDU session terminates in a single UPF, which might be acting as a PDU session anchor, an intermediate UPF or both. If the user is not roaming, then each PDU session is associated with a single SMF in Release 15, but can be associated with more than one SMF from Release 16.

A PDU session contains one or more *quality of service flows* (QoS flows), which have similar roles to the default and dedicated bearers from LTE. Each QoS flow is a set of data packets that have a defined set of QoS targets, for parameters such as data rate, error rate and end-to-end delay. One QoS flow is established when a PDU session is set up, for example a QoS flow carrying Internet traffic, or a QoS flow carrying high-level signalling messages between the mobile and the IMS. Others, with different QoS targets, can be established later, for example a QoS flow carrying IMS voice traffic.

LTE only supported IP-based data networks, such as the Internet. In contrast, 5G also supports data networks that use different communication protocols, for example Ethernet. That helps the system support machine-type and low-latency applications, in which the individual packets can be small and the impact of delivering large IP packet headers can be high.

2.2.4 Roaming Architectures

If the user is roaming, then the mobile, NG-RAN and AMF are always in the visited network, while the AUSF and UDM are always in the home network. There are then two ways to route the traffic in a particular PDU session.

Data communications with the Internet generally use *home routed traffic* (Figure 2.5). In that architecture, the PDU session anchor is in the user's home network and is controlled by

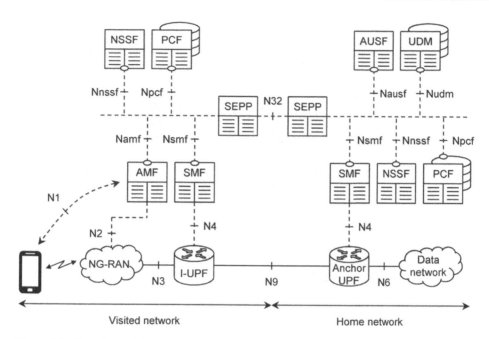

Figure 2.5 Roaming architecture using home routed traffic. *Source:* Adapted from 3GPP TS 23.501.

a home SMF. The visited network contains an intermediate UPF, which shields the anchor from the need to know about the visited network's base stations, and is controlled by a visited SMF. The two networks communicate over an interface that is outside their control, for example an inter-operator backbone known as the *IP packet exchange* (IPX) [11]. The signalling messages on that interface are secured by another network function, which is known as the *security edge protection proxy* (SEPP).

When using home routed traffic, the home and visited networks both contain an NSSF, because the network slice's user plane lies within both networks. Similarly, both networks contain a PCF, so that they can both supply the relevant network policies for the mobile.

Local breakout (Figure 2.6) is valuable for communications with the IMS and essential for low-latency roaming applications. In that architecture, the PDU session anchor is in the visited network, which allows a roaming user to place a local IP voice call without the need for the traffic to travel back to the home network. There is only one NSSF, which is in the visited network because the network slice's user plane only lies within that network. However, there is still a PCF in both networks, for the same reason as before.

2.2.5 Data Storage Architectures

In the architectures described so far in this chapter, the UDM and PCF both had two components: databases containing *structured data*, in other words data whose structure is defined by the 3GPP specifications, and front ends. However, some of the information in those databases is shared. As shown in Figure 2.7, the databases can optionally be combined within a single network function, known as the *unified data repository* (UDR). The UDR can also be used by another new network function, the *network exposure function* (NEF),

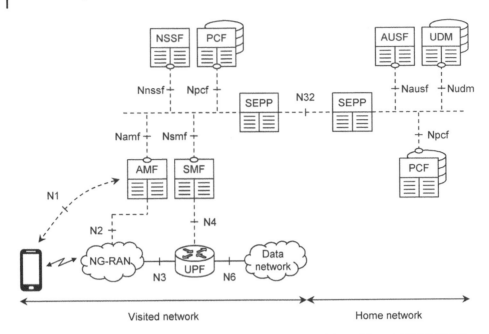

Figure 2.6 Roaming architecture using local breakout. *Source:* Adapted from 3GPP TS 23.501.

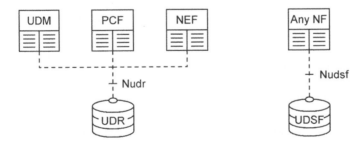

Figure 2.7 Data storage architectures. *Source:* Adapted from 3GPP TS 23.501.

which acts as an additional point of contact between external third-party application functions and the 5G core.

Network functions also store information about the mobiles that they are controlling. If that information is defined by the specifications, then it is known as a *context*. As an example, the AMF stores a *user equipment (UE) context* for each of its mobiles, which contains information such as the subscriber data that it has retrieved from the UDM, and the details of the mobiles' states. The network functions can also store information that is not defined by the specifications, such as operator-specific configuration parameters.

Optionally, the network functions can store this information in a separate database known as an *unstructured data storage function* (UDSF). Depending on its implementation, one UDSF can store data from a single instance of a network function, from several instances of the same type of function or from network functions of different types.

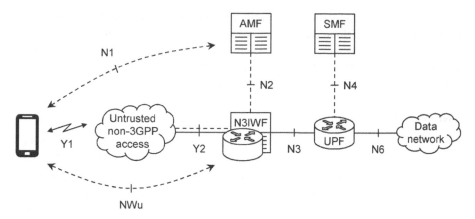

Figure 2.8 Architecture for non-3GPP access to the 5G core network. *Source:* Adapted from 3GPP TS 23.501.

2.2.6 Non-3GPP Access to the 5G Core

Mobiles can access the 5G core network over a non-3GPP access network, such as IEEE 802.11 WiFi. Release 15 supports access over an untrusted wireless network, in which the 5G core network assumes that the access network is insecure. Release 16 extends that support to trusted wireless networks and also to wireline access [12].

Non-3GPP access has been supported by previous generations, but the 5G architecture is more closely aligned with the one for 3GPP access than before. Figure 2.8 shows the most important network functions in Release 15: the other functions from previous figures in this chapter are all retained.

In this architecture, the term *5G access network* (5G-AN) covers both 3GPP access networks, in other words the NG-RAN, and non-3GPP access networks such as WiFi. In both cases, the mobile is controlled by a serving AMF, and is authenticated by the AMF and AUSF. The AMF communicates with the mobile over the N1 reference point and communicates with the access network over N2, using the same signalling procedures as it does for 3GPP access.

There is one new network function. The *non-3GPP interworking function* (N3IWF) terminates the N2 and N3 reference points, so as to act as an interface between the access network and the 5G core. It also secures the mobile's communications over the access network by establishing a secure tunnel to the mobile using signalling messages over a reference point denoted NWu. In Release 15, an individual PDU session traverses either a non-3GPP access network or the NG-RAN, but not both. More flexibility is introduced in Release 16 [13].

2.3   Network Areas, Slices and Identities

2.3.1   Network Identities

The 5G core network uses several different types of identity [14]. Some of these are inherited from previous generations, while others are new.

An operator's network is identified by a *public land mobile network identity* (PLMN-ID). The PLMN-ID comprises a three-digit *mobile country code* (MCC), and a two or three-digit *mobile network code* (MNC) that identifies the network within that country [15]. An external data network is identified by a *data network name* (DNN). That is identical to the *access point name* (APN) from previous generations: the only change is the name.

2.3.2 Network Slices

In 5G, a public land mobile network comprises a number of network slices, each of which is a virtual logical sub-network running on shared physical hardware [16, 17]. An individual network function does not have to support all of the network's slices: instead, it declares which slices it supports when registering with the network repository function.

When a mobile registers with the network, it does so with one or more of the network's slices. The mobile's AMF serves all the slices with which a mobile is registered, and the radio access network does so as well. In contrast, an individual PDU session lies in a single slice, so a mobile's SMFs and UPFs serve one slice each. Even so, a mobile can still connect to a particular data network through multiple slices by means of multiple PDU sessions.

Each type of network slice is identified by a field known as *single network slice selection assistance information* (S-NSSAI). That has two components. The first is the 8-bit *slice/service type* (SST), which defines the high-level behaviour that the slice should deliver. Three values were standardized in Release 15 for the three main use cases of enhanced mobile broadband (eMBB), ultra-reliable low-latency communication (URLLC) and massive internet of things, (mIoT), while a fourth was added in Release 16 for vehicle-to-everything (V2X) communications. Network operators can also define other values of their own. The second component is an optional 24-bit *slice differentiator* (SD), which distinguishes different types of slice that share the same SST. There are no standardized values, but network operators might use them to distinguish slices for different services, for example mobile broadband access to the Internet and to the IMS, or for different clients or devices.

In roaming scenarios, the home and visited networks can use the standardized values for the S-NSSAI without the need for any further action. The two networks can also set up a conversion between any non-standardized values as part of a service-level agreement, which is implemented in the visited network's NSSF.

Network slice selection assistance information (NSSAI) is a list of S-NSSAIs. There are a few varieties of NSSAI, but only three will be important for us. Firstly, the UDM stores a *subscribed NSSAI* for each user, which lists the types of slice that the user has subscribed to. Secondly, the mobile sends a *requested NSSAI* to the network, which lists the types of slice that it would like to register with. Thirdly, the AMF provides the mobile with an *allowed NSSAI*, which lists the types of slice that the mobile is actually registered with, and which it can use to establish a PDU session.

An individual network slice is known as a *network slice instance* and is identified using an operator-specific character string known as the *network slice instance identifier* (NSI-ID). A network operator can deploy multiple network slice instances that share the same S-NSSAI but are distinguished using parameters such as their coverage area or the maximum number of mobiles that they support. The operator can set up, modify and tear down individual network slice instances using the network management system [18].

2.3.3 AMF Areas and Identities

AMFs are grouped together in two levels [19]. An *AMF set* is a group of AMFs which span the same geographical area and support the same set of network slices. In turn, an *AMF region* comprises one or more AMF sets, each of which might span a different geographical area and/or support a different set of network slices. AMF sets can overlap, in that one AMF can belong to multiple sets that have different geographical boundaries and/or different support for network slicing. Similarly, AMF regions can overlap, in that one AMF set can belong to multiple regions.

These areas are the basis of AMF identities. The 8-bit *AMF region ID* uniquely identifies an AMF region within the network, while the 10-bit *AMF set ID* uniquely identifies an AMF set within the region. In simple implementations, the 6-bit *AMF pointer* identifies a single AMF within an AMF set. In more complex implementations, multiple AMFs can share the same AMF pointer if they store their data in the same unstructured data storage function, such that the mobile's exact choice of AMF is immaterial.

The *AMF identifier* (AMFI) comprises the AMF region ID, AMF set ID and AMF pointer, and identifies one or more AMFs within the network. Similarly, the *globally unique AMF identifier* (GUAMI) comprises the MCC, MNC and AMFI.

2.3.4 UE Identities

As in previous 3GPP systems, a 5G mobile has two components with different identities [20, 21]. The mobile equipment is identified by a *permanent equipment identifier* (PEI). There are two slightly different forms of PEI, namely the 15-bit *international mobile equipment identity* (IMEI) and the 16-bit *IMEI and software version number* (IMEISV). The universal subscriber identity module is identified by the *subscription permanent identifier* (SUPI), which again has two forms: either the *international mobile subscriber identity* (IMSI) that was used for 3GPP access in previous generations, or the *network access identifier* (NAI) that was used for non-3GPP access.

Networks avoid transmitting a mobile's SUPI over the air interface to prevent intruders from discovering it. Instead, a serving AMF identifies a mobile by assigning it a temporary identity. There are various types. The 32-bit *5G temporary mobile subscriber identity* (5G-TMSI) uniquely identifies a mobile within a single AMF or within the group of AMFs that share the same AMF pointer. The 5G-S-TMSI comprises the AMF set ID, AMF pointer and 5G-TMSI, and is used for procedures such as paging in which the mobile's AMF region is known. The *5G globally unique temporary identity* (5G-GUTI) comprises the GUAMI and the 5G-TMSI, and is used for all other procedures.

In LTE, a mobile still had to transmit its IMSI or NAI occasionally, for example on the first use of the universal subscriber identity module (USIM) or after a database failure. That problem has been overcome in 5G by defining the *subscription concealed identifier* (SUCI), which contains the same information as the SUPI but is secured by means of public key encryption.

2.3.5 UE Registration Areas

A mobile's *registration area* is a geographical area through which the mobile is allowed to move, without the need to change its serving AMF or its allowed NSSAI, or, when on standby, without the need to inform the 5G core [22]. Registration areas are implemented by means of geographical regions known as tracking areas, which we will discuss further in Chapter 3. In the case of 3GPP access, each registration area is a list of tracking area identities. In the case of non-3GPP access, the registration area is a single reserved tracking area identity which spans the whole of the non 3GPP access network.

2.4 State Diagrams

2.4.1 Registration Management

The mobile's behaviour is described using state diagrams. Each diagram has a number of different states, in which there are different objectives and different sets of procedures. The mobile and the network both have a copy of each state diagram, so that they can agree on the procedures that the mobile should use. The terminology is somewhat different in the stage 2 [23] and stage 3 [24] specifications for the 5G core network; we will stick to the simpler terminology from stage 2.

The first diagram (Figure 2.9) is the one for *registration management*. There are two states. In the state of RM-REGISTERED, the mobile is in contact with the 5G core network and is registered with a serving AMF. The AMF knows something about the mobile's location and can reach the mobile through the 5G access network.

In RM-DEREGISTERED, the mobile is out of communication with the network, for example because it is switched off or in aeroplane mode. The network knows nothing about the mobile's location and is unable to reach it. However, the mobile and the old serving AMF both store some information about the mobile's previous registration, for example its temporary identities and security keys.

When the mobile switches on, it contacts the 5G core network in a signalling procedure known as *registration*, which moves it to the state of RM-REGISTERED. Similarly, the mobile runs the procedure of *deregistration* before it switches off.

In early releases of LTE, a mobile always connected to a packet data network as part of the registration procedure to provide it with always-on connectivity. That became optional in 3GPP Release 13 to help support machine-type devices that might only communicate using mechanisms such as the short message service. In 5G, a PDU session is no longer established as part of registration: instead, it is always established in a separate signalling procedure. A mobile can therefore be registered with the 5G core without any PDU sessions and without any data network addresses, SMFs or UPFs.

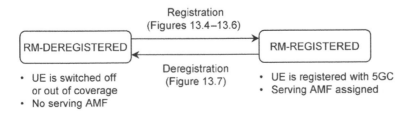

Figure 2.9 Registration management state diagram. *Source:* Adapted from 3GPP TS 23.501.

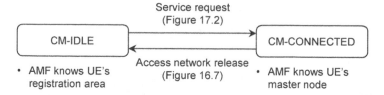

Figure 2.10 Connection management state diagram. *Source:* Adapted from 3GPP TS 23.501.

2.4.2 Connection Management

The second diagram (Figure 2.10) is the one for *connection management*. There are two states which, except in transient situations, are best understood as sub-states of RM-REGISTERED. In the state of CM-CONNECTED, the mobile is active. The serving AMF knows the mobile's master node, and can exchange mobile-specific signalling messages with that node over the N2 reference point and with the mobile itself over N1. If the mobile has any PDU sessions, then it may also be able to exchange data with an external data network.

In CM-IDLE, the mobile is on standby. To reduce the amount of signalling and preserve the mobile's battery life, the AMF does not record the mobile's exact location: instead, the mobile can move throughout its registration area without the need to inform the 5G core. Only limited signalling is possible, and the mobile cannot exchange data with the outside world: if downlink data arrive, then the core network signals to the mobile throughout its registration area in a procedure known as *paging*. If the mobile wishes to contact the core network or to reply to a paging message, then it does so using the *service request* procedure, which takes it to the state of CM-CONNECTED. The reverse procedure is known as *access network release*.

2.4.3 Non-3GPP Access

In cases of non-3GPP access, a mobile has two independent registration management states, one for 3GPP access through the NG-RAN and one for non-3GPP access. It also has two independent connection management states. Thus, for example, the mobile might be in RM-REGISTERED and CM-CONNECTED in respect of 3GPP access, but in RM-REGISTERED and CM-IDLE for non-3GPP.

2.5 Signalling Protocols

2.5.1 Signalling Protocol Architecture

The 5G core network uses a number of signalling protocols, which are illustrated in Figure 2.11. Three of the protocols define signalling procedures between pairs of network functions over the underlying reference points. An SMF controls one or more UPFs by means of the *packet forwarding control protocol* (PFCP) [25], which is inherited from LTE Release 14. Similarly, the mobile communicates with the AMF and SMF using the *5G mobility management* (5GMM) and *5G session management* (5GSM) protocols [26].

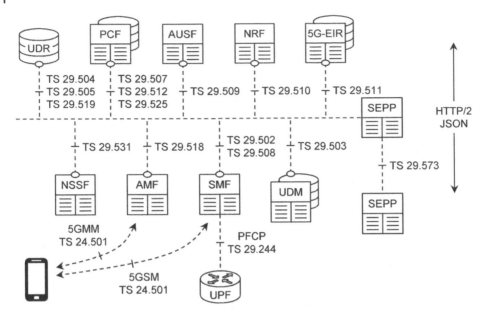

Figure 2.11 Signalling protocols used by the 5G core network.

On the service-based interfaces, the signalling protocol for the unstructured data storage function is not yet defined: any such implementations are proprietary. The other service-based interfaces use version 2 of the Hypertext Transfer Protocol (HTTP/2), supplemented by the use of *JavaScript Object Notation* (JSON). Those protocols are also used on the N32 reference point between security edge protection proxies. The two protocols are unfamiliar in the context of a telecommunication system, so they are discussed in more detail later in this chapter.

The individual service-based interfaces are then defined in a set of stage 3 specifications. The most valuable ones cover the AMF [27], SMF [28, 29] and UDM [30], while others address the UDR [31–33], PCF [34–36], AUSF [37], NRF [38], 5G-EIR [39], NSSF [40] and SEPP [41]. The underlying details are in References [42, 43]. (We will cover a number of other specifications, including some more for the PCF, in Chapter 15.)

The core network's signalling messages are transported using an IP-based protocol stack, the layer 4 protocol being the user datagram protocol (UDP) in the case of PFCP, and the transmission control protocol (TCP) in the case of HTTP/2. On the air interface, the 5GMM and 5GSM messages are transported using lower-level signalling protocols within the radio access network, which we will discuss as part of Chapter 3. Those low-level protocols are said to lie in the *access stratum* (AS), while the 5GMM and 5GSM protocols lie in the *non-access stratrum* (NAS).

2.5.2 Example Signalling Procedures

To illustrate the traditional use of signalling procedures over reference points, Figure 2.12 shows a simple procedure in which a UPF sets up an association with the SMF that will control it [44].

Figure 2.12 PFCP association setup procedure.

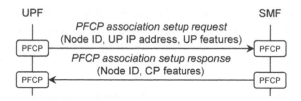

Before the procedure begins, the network management system has instantiated an instance of a UPF and has configured it with the IP addresses of one or more SMFs. (The UPF might also discover the SMFs by contacting the network repository function, as described later in this chapter.) To begin the procedure, the UPF contacts the SMF by sending it a message known as a *PFCP association setup request*. In that message, the UPF states its identity, includes the IP address that it uses for data transport and notes any optional features that it supports. The SMF replies with a *PFCP association setup response*, which includes its own identity and optional features. Once the procedure has completed, the SMF can instruct the UPF to set up, modify or tear down the user planes of individual PDU sessions, by means of further signalling procedures.

2.6 The Hypertext Transfer Protocol

2.6.1 HTTP/1.1 and HTTP/2

HTTP is the protocol used for information retrieval on the World Wide Web, for example to download information from a web server. 3GPP adopted the protocol for signalling on the 5G core network's service-based interfaces for three main reasons [45]. Firstly, the use of a service-based architecture is implicit in the design of HTTP. Secondly, its widespread use outside telecommunications eases deployment and helps with future-proofing. Thirdly, HTTP is widely used for communications between software-defined networks and third-party application functions over the networks' northbound APIs. By using that protocol for both external and internal signalling, 3GPP has again adopted a consistent design throughout the 5G core.

In HTTP, a *resource* is an item of information that can be named. The information could be static, for example an image or text file, or dynamic, such as "Today's BBC news." Its name is known as a *uniform resource identifier* (URI), while a *uniform resource locator* (URL) combines the URI with a means of locating the information. The distinction between the two is a subtle one [46]: consistent with the 3GPP specifications, we will refer to the names as URIs.

In a typical signalling transaction, a client composes an HTTP request for the information that is held at a particular URI. A *domain name server* (DNS) returns the IP address of a server where the information can be found, and the client sends the request there. The server replies with an HTTP response, together with data that represent the information in a suitable format. Examples include a web page that is written using *Hypertext Markup Language* (HTML), or a *Portable Document Format* (PDF) file.

The most familiar version is HTTP/1.1, introduced in 1999 [47–52]. HTTP/1.1 is a text protocol in which there is nothing to distinguish one character from another. To avoid any risk of confusion, individual requests and responses have to be sent in series, which increases the time required for information retrieval. That in turn triggers a risk of *head-of-line blocking*, in which a problem with an early request prevents completion of the ones that follow. Although there are ways to work around these shortcomings [53], none is entirely effective.

HTTP/2, introduced in 2015, has two protocol layers [54, 55]. The upper *data layer* is similar to version 1.1 but has a slightly different syntax. The lower *framing layer* converts the text and any associated data into binary frames, with each frame labelled to distinguish one request and response from another. That allows requests and responses to be sent in parallel, and helps to overcome the shortcomings of version 1.1. Because of these enhancements, the protocol chosen for the 5G core network is HTTP/2.

2.6.2 Representational State Transfer

Representational state transfer (REST) is a set of architectural principles that has been widely adopted in the design and implementation of the World Wide Web [56]. At least in part, 3GPP has followed these principles in the design of the 5G core network [57].

REST uses a client–server architecture, in which a client sends a request to a server, the server returns an immediate response and there is a clear division of responsibilities between the two. The architecture is layered, so that the client knows nothing about the internal details of the server. 3GPP has followed both of these principles in the design of the 5G core network. The roles of client and server are usually filled by the network function consumer and producer respectively, although those roles are reversed in the procedures for subscription and notification that we will discuss at the end of this chapter.

In REST, the client and server interact by manipulating resources that are held by the server. 3GPP has followed this principle wherever possible, a decision that is key to understanding the operation of the 5G core. Most of the 5G network function services are implemented using a set of database-style operations, often abbreviated as CRUD, which *create, read, update and delete* resources that are held by the server. That viewpoint is very different from the traditional use of signalling procedures over reference points, so we will illustrate it in more detail here. However, it has not proved possible to model every network function service in this way, so some have been implemented differently by means of customized operations.

REST is intended to be stateless, such that a server can store details about a client in a separate database but otherwise does not remember anything about a client's requests. Instead, each individual request contains all the information needed to understand it. Although the 5G core network does not follow this principle in full, it has been used to guide certain aspects of system design. If, for example, an AMF stores the details of its mobiles in a separate unstructured data storage function, then the AMF can be made stateless, and the mobile can communicate with any AMF that can retrieve those details.

In its last two principles, REST allows a client to cache responses from the server for later use, and to extend its functionality using code in the form of applets or scripts. The 5G core network occasionally uses caching, but otherwise these principles have not been adopted by 3GPP.

2.6.3   The HTTP/2 Data Layer

Figure 2.13 shows a simple example of an HTTP/2 request and response. An HTTP/2 request begins with four *pseudo-headers*, which replace the start line used by HTTP/1.1. Three of those pseudo-headers contain the different components of the URI. `:scheme` defines a naming scheme for the URI, and is usually the same as the name of the associated signalling protocol. `:authority` contains the domain name or IP address of the server that hosts the URI, together with an optional TCP port number. Finally, `:path` identifies an individual resource within the host.

The fourth pseudo-header, `:method`, identifies the request. Although HTTP/2 supports several different methods, the 5G core network only uses five, for the tasks that are listed in Table 2.1 [58]. Most of these tasks are database-style operations to create, read, update and delete a resource upon a server. POST is also used for custom procedures that cannot easily be modelled using database operations, and for notifications from a network function producer back to a consumer. Various HTTP/2 header fields supply additional details about the request.

The client can send information to the server in two ways. In the case of a GET request, the client can provide more details about the data than are required by extending the URI using *query parameters*. In the cases of PATCH, POST and PUT, the client supplies data of its own, in the same way as for the server's responses.

An HTTP/2 response begins with a single pseudo-header, `:status`, which provides a three-digit code that indicates the nature of the response. The first digit can have five values: 1 indicates a provisional response; 2 indicates success; 3 asks the client to take further action, for example by contacting a different host; while 4 and 5 indicate errors on the client and server respectively. Each individual response code is associated with a text description, for example `200 OK`. Those descriptions are explicitly included in HTTP/1.1 responses, but are omitted from HTTP/2.

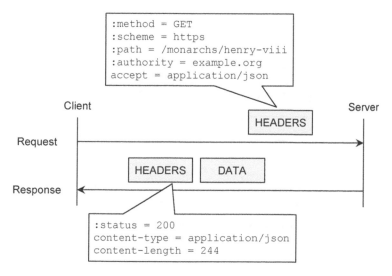

Figure 2.13   Example of an HTTP/2 request and response.

Table 2.1 HTTP/2 methods used by the 5G core network.

Method	Usage in the 5G core network	Additional information
DELETE	Delete a resource.	None
GET	Read a representation of a resource.	URI query parameters
PATCH	Partially update a resource.	JSON modifications
POST	Create a new resource, with the URI chosen by the server. Notify a client in response to a previous subscription. Ask the server to carry out a custom operation.	JSON content
PUT	Create a new resource, with the URI chosen by the client. Replace a resource.	JSON content

In the example in Figure 2.13, the client is requesting information from the server by means of an HTTP/2 GET. The framing layer converts the request to an HTTP/2 frame known as HEADERS, and converts the response in a similar way. The response frame is followed by a second HTTP/2 frame known as DATA, which contains the requested information.

2.6.4 JavaScript Object Notation (JSON)

On the World Wide Web, HTTP transfers data between clients and servers using formats such as HTML web pages and PDF files. In the 5G core network, the data simply comprise parameters that the client and server both understand. The data are most often structured by means of a standard known as *JavaScript Object Notation* (JSON), which includes formatting rules for logical values, numbers, character strings, objects and arrays [59]. Figure 2.14 shows a simple example.

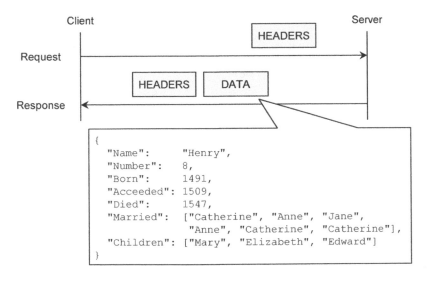

```
{
   "Name":      "Henry",
   "Number":    8,
   "Born":      1491,
   "Acceeded": 1509,
   "Died":      1547,
   "Married":   ["Catherine", "Anne", "Jane",
                "Anne", "Catherine", "Catherine"],
   "Children": ["Mary", "Elizabeth", "Edward"]
}
```

Figure 2.14 Example of a JSON object.

There are two main variations. Firstly, the content for an HTTP/2 PATCH request is written using standards known as *JSON Patch* [60] and *JSON Merge Patch* [61]. These express modifications that the client wishes to apply to a JSON document, the main distinction being that JSON Patch allows modifications to the individual elements in an array, whereas JSON Merge Patch does not. Secondly, network functions sometimes exchange content by means of binary encoded data. For example, the SMF uses binary coding to send 5G session management signalling messages to the AMF, which relays them transparently to the mobile.

2.7 Example Network Function Services

2.7.1 Network Function Service Registration

To illustrate the implementation of network function services, Figure 2.15 shows the procedure in which a newly instantiated network function registers itself with an NRF. The example is for a registering AUSF, but the procedure for another type of network function would be the same.

The specifications define the procedure in two contrasting ways. The stage 2 specification defines it in terms of network function service operations: the AUSF requests the NFRegister operation from the NFManagement service in the NRF, and states its capabilities by supplying a *network function profile* [62]. The stage 3 specification then defines the HTTP implementation: the AUSF sends an HTTP PUT request to the NRF, which asks it to create a new resource at a URI that the AUSF supplied [63].

In common with all the network function services in the 5G core network, the URI has the following structure, in which the quantities in curly brackets are to be replaced with the appropriate text [64]:

{apiRoot}/{apiName}/{apiVersion}/{apiSpecificResourceUriPart}

Here, {apiRoot} contains two of the HTTP/2 pseudo-headers, namely the scheme and the authority. The scheme is either http:// or https://, depending on the choice of security protocol. By default, the authority for an NRF is nrf.5gc.mnc{MNC}.mcc{MCC} .3gppnetwork.org, where {MNC} and {MCC} are the mobile network and mobile country codes respectively [65]. (An operator can set up a different authority for the NRF by means of the network management system.) {apiName} is the name of the network

Figure 2.15 Network function service registration procedure. *Source:* Adapted from 3GPP TS 29.510.

function service's API, in this case `nnrf-nfm`, while `{apiVersion}` is its major version number, currently `v1`. Finally, `{apiSpecificResourceUriPart}` is specific to the network function service operation. In this case, it is `nf-instances/{nfInstanceID}`, where `{nfInstanceID}` is a 128-bit *Universally Unique Identifier* (UUID) version 4 [66], which is supplied by the registering network function, and which should be unique within the network. Bringing all of these components together, the following is an example URI for the network function service registration procedure:

```
https://nrf.5gc.mnc13.mcc901.3gppnetwork.org
/nnrf-nfm/v1/nf-instances/4947a69a-f61b-4bc1-b9da-47c9c5d14b64
```

The AUSF supplies its network function profile as data that are formatted using JSON. The profile includes the unique identifier defined above, the type of network function (in this case, AUSF) and its status (either registered or suspended). It also includes the network identity, the network slices that the AUSF supports, its priority and capacity relative to other network functions of the same type, and its domain name or IP address.

The NRF replies with the response `201 Created`, which confirms that the URI has been created. In its own JSON data, the NRF echoes back the network function profile and supplies a heartbeat timer. Later, the AUSF can update its profile by sending an HTTP PATCH to the same URI that it used earlier. It also contacts the NRF if the heartbeat timer expires, to confirm that it is still available.

2.7.2 Network Function Service Discovery

Once a network function has registered with the NRF, it can be discovered by another network function, for example by an AMF. The procedure is illustrated in Figure 2.16. To avoid clutter, the figure only shows the name of the network function service operation, the HTTP method and response code, the `{apiSpecificResourceUriPart}` field from the URI, and the most important data. That convention is consistent with the stage 3 specifications for 5G, and will be followed throughout the rest of the book.

To ask for the details of any registered AUSFs, the AMF requests the `NFDiscover` operation from the `NFDiscovery` service in the NRF [67]. That operation is implemented as an HTTP GET [68]. The URI has the same structure as before: the `{apiName}` field is `nnrf-disc`, and `{apiSpecificResourceUriPart}` is `nf-instances?<query parameters>`. The query parameters supply further details about the request: they

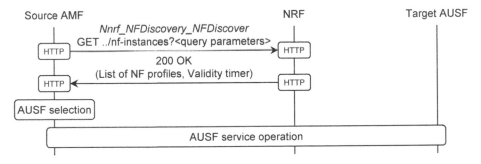

Figure 2.16 Network function service discovery procedure. *Source:* Adapted from 3GPP TS 29.510.

include the types of the two network functions, their two PLMN identities and the network slices that the AUSF should support.

The NRF replies with the HTTP response 200 OK. In its JSON data, the NRF lists the profiles of AUSFs that match the request, and includes a validity period that states how long those profiles can be cached.

Assisted by the priority and capacity fields from their network function profiles, the AMF selects one of the matching AUSFs, sets up TCP and HTTP/2 connections with it, and requests the appropriate service operation. In the corresponding URI, the authority part of {apiRoot} is the domain name or IP address from the network function profile, while the fields {apiName} and {apiSpecificResourceUriPart} are defined in the appropriate stage 3 specification.

To support roaming mobiles, the AMF may have to discover an instance of an AUSF in a different network. To do so, the AMF sends its request to an NRF in its own network, which forwards the request to a second NRF in the destination network. On receiving the response, the AMF can contact the selected AUSF directly.

2.7.3 Network Function Service Subscription and Notification

To supplement the discovery procedure, a network function can ask to be notified if the registration details of another network function change. Figure 2.17 shows the procedure, once again for the example of an AMF and an AUSF.

The AMF begins by requesting the NFStatusSubscribe operation from the NRF's NFManagement service [69]. That operation is implemented as an HTTP POST to the URI whose {apiSpecificResourceUriPart} is ../subscriptions, which asks the NRF to create a child resource of that URI to handle the new subscription [70]. The AMF supplies further details by means of JSON data. These include the type of network function being monitored, the target network identity and a list of events that are of interest (namely registration, deregistration and/or modification). The AMF also supplies a *notification URI*, also known as a *callback URI*, on which it would like to be notified if one of those events takes place.

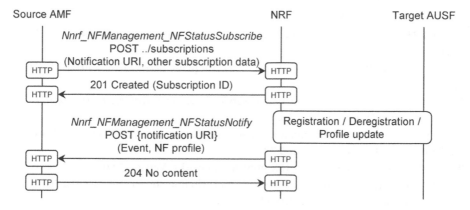

Figure 2.17 Procedures for network function service subscription and notification. Source: Adapted from 3GPP TS 29.510.

The NRF replies with the HTTP response `201 Created`. In its JSON data, the NRF echoes back the original subscription data, and supplies a subscription identity that completes the newly created child resource. (The AMF can delete the subscription whenever it needs to by sending an HTTP DELETE to that resource.)

Later, an AUSF might update its registration data in a way that matches the subscription. The NRF reacts by requesting the `NFStatusNotify` operation from its `NFManagement` service, which is implemented as an HTTP POST to the notification URI that the AMF supplied earlier [71]. In its JSON data, the NRF states the event that triggered the notification and supplies the updated network function profile. The AMF acknowledges with the HTTP response `204 No Content`.

It is worth noting that the 5G core network implements notifications differently from other service operations, in three ways. Firstly, the network function producer (in this case, the NRF) is acting as the client, not as the server. As a result, the service operation forms part of the client's API, not the server's. Finally, the URI is supplied by the network function consumer during the earlier subscription and is not defined by the API itself.

References

1 3GPP TS 23.501 (2019) System architecture for the 5G system (5GS); Stage 2 (Release 15), December 2019.

2 3GPP TS 23.401 (2019) General Packet Radio Service (GPRS) enhancements for Evolved Universal Terrestrial Radio Access Network (E-UTRAN) access (Release 15), December 2019, Sections 4.2, 4.4.

3 3GPP TS 23.214 (2018) Architecture enhancements for control and user plane separation of EPC nodes; Stage 2 (Release 15), December 2018.

4 3GPP TS 23.501 (2019) System architecture for the 5G system (5GS); Stage 2 (Release 15), December 2019, Sections 4.2, 6.2.

5 3GPP TS 23.501 (2019) System architecture for the 5G system (5GS); Stage 2 (Release 15), December 2019, Section 4.2, Annex A.

6 3GPP TS 23.502 (2019) Procedures for the 5G system (5GS); Stage 2 (Release 15), December 2019, Annex A.

7 3GPP TR 23.799 (2016) Study on architecture for next generation system (Release 14), December 2016, Section 8.7.

8 Guttman, E. and Ali, I. (2018). Path to 5G: a control plane perspective. *Journal of ICT Standardization* 6 (1 and 2): 87–100.

9 Mayer, G. (2018). RESTful APIs for the 5G service based architecture. *Journal of ICT Standardization* 6 (1 and 2): 101–116.

10 3GPP TS 23.501 (2019) System architecture for the 5G system (5GS); Stage 2 (Release 15), December 2019, Sections 5.6.1, 5.7.1.

11 GSMA IR.34 (2018) Guidelines for IPX provider networks, Version 14.0.

12 3GPP TR 23.716 (2018) Study on the wireless and wireline convergence for the 5G system architecture (Release 16), December 2018.

13 3GPP TR 23.793 (2018) Study on access traffic steering, switch and splitting support in the 5G system architecture (Release 16), December 2018.

14 3GPP TS 23.501 (2019) System architecture for the 5G system (5GS); Stage 2 (Release 15), December 2019, Section 5.9.

15 Mobile Country Codes (MCC) and Mobile Network Codes (MNC). http://www.mcc-mnc.com (accessed 18 January 2020).

16 3GPP TS 23.501 (2019) System architecture for the 5G system (5GS); Stage 2 (Release 15), December 2019, Section 5.15.

17 3GPP TS 23.003 (2019) Numbering, addressing and identification (Release 15), September 2019, Section 28.4.

18 3GPP TS 28.541 (2019) Management and orchestration; 5G Network Resource Model (NRM); Stage 2 and stage 3 (Release 15), December 2019, Section 6.

19 3GPP TS 23.003 (2019) Numbering, addressing and identification (Release 15), September 2019, Sections 2.10.1, 28.3.2.5.

20 3GPP TS 23.003 (2019) Numbering, addressing and identification (Release 15), September 2019, Sections 2.2A, 2.2B, 2.10, 2.11, 6.4.

21 3GPP TS 33.501 (2019) Security architecture and procedures for 5G system (Release 15), December 2019, Section 6.12.

22 3GPP TS 23.501 (2019) System architecture for the 5G system (5GS); Stage 2 (Release 15), December 2019, Section 5.3.2.3.

23 3GPP TS 23.501 (2019) System architecture for the 5G system (5GS); Stage 2 (Release 15), December 2019, Section 5.3.

24 3GPP TS 24.501 (2019) Non-access-stratum (NAS) protocol for 5G system (5GS); Stage 3 (Release 15), December 2019, Sections 3.1, 5.1.3.

25 3GPP TS 29.244 (2019) Interface between the control plane and the user plane nodes; Stage 3 (Release 15), December 2019.

26 3GPP TS 24.501 (2019) Non-access-stratum (NAS) protocol for 5G system (5GS); Stage 3 (Release 15), December 2019.

27 3GPP TS 29.518 (2019) 5G system; Access and mobility management services; Stage 3 (Release 15), December 2019.

28 3GPP TS 29.502 (2019) 5G system; Session management services; Stage 3 (Release 15), December 2019.

29 3GPP TS 29.508 (2019) 5G system; Session management event exposure service; Stage 3 (Release 15), December 2019.

30 3GPP TS 29.503 (2019) 5G system; Unified data management services; Stage 3 (Release 15), December 2019.

31 3GPP TS 29.504 (2019) 5G system; Unified data repository services; Stage 3 (Release 15), December 2019.

32 3GPP TS 29.505 (2019) 5G system; Usage of the unified data repository services for subscription data; Stage 3 (Release 15), December 2019.

33 3GPP TS 29.519 (2019) 5G system; Usage of the unified data repository service for policy data, application data and structured data for exposure; Stage 3 (Release 15), December 2019.

34 3GPP TS 29.507 (2019) 5G system; Access and mobility policy control service; Stage 3 (Release 15), December 2019.

35 3GPP TS 29.512 (2019) 5G system; Session management policy control service; Stage 3 (Release 15), December 2019.

36 3GPP TS 29.525 (2019) 5G system; UE policy control service; Stage 3 (Release 15), December 2019.

37 3GPP TS 29.509 (2019) 5G system; Authentication server services; Stage 3 (Release 15), October 2019.

38 3GPP TS 29.510 (2019) 5G system; Network function repository services; Stage 3 (Release 15), December 2019.

39 3GPP TS 29.511 (2019) 5G system; Equipment identity register services; Stage 3 (Release 15), December 2019.

40 3GPP TS 29.531 (2019) 5G system; Network slice selection services; Stage 3, (Release 15), September 2019.

41 3GPP TS 29.573 (2019) 5G system; Public land mobile network (PLMN) Interconnection; Stage 3 (Release 15), October 2019.

42 3GPP TS 29.500 (2019) 5G system; Technical realization of service based architecture; Stage 3 (Release 15), September 2019.

43 3GPP TS 29.501 (2019) 5G system; Principles and guidelines for services definition; Stage 3 (Release 15), December 2019.

44 3GPP TS 29.244 (2019) Interface between the control plane and the user plane nodes; Stage 3 (Release 15), December 2019, Sections 6.2.6.3, 7.4.4.1, 7.4.4.2.

45 3GPP TS 29.891 (2017) 5G system – Phase 1; CT WG4 aspects (Release 15), December 2017, Section 11.3.1.2.

46 Miessler, D. (2019) What's the difference between a URI and a URL? https://danielmiessler.com/study/difference-between-uri-url (accessed 18 January 2020).

47 IETF RFC 7230 (2014) Hypertext Transfer Protocol (HTTP/1.1): Message syntax and routing.

48 IETF RFC 7231 (2014) Hypertext Transfer Protocol (HTTP/1.1): Semantics and content.

49 IETF RFC 7232 (2014) Hypertext Transfer Protocol (HTTP/1.1): Conditional requests.

50 IETF RFC 7233 (2014) Hypertext Transfer Protocol (HTTP/1.1): Range requests.

51 IETF RFC 7234 (2014) Hypertext Transfer Protocol (HTTP/1.1): Caching.

52 IETF RFC 7235 (2014) Hypertext Transfer Protocol (HTTP/1.1): Authentication.

53 Stenberg, D. (2019) HTTP2 explained. https://http2-explained.haxx.se (accessed 18 January 2020).

54 IETF RFC 7540 (2015) Hypertext Transfer Protocol Version 2 (HTTP/2).

55 IETF RFC 7541 (2015) HPACK: Header compression for HTTP/2.

56 Fielding, R.T. (2000) Architectural styles and the design of network-based software architectures. Doctoral Thesis, University of California.

57 3GPP TS 29.891 (2017) 5G system – Phase 1; CT WG4 aspects (Release 15), December 2017, Section 6.2.2.4.

58 3GPP TS 29.501 (2019) 5G system; Principles and guidelines for services definition; Stage 3 (Release 15), December 2019, Section 4.6.

59 IETF RFC 8259 (2017) The JavaScript Object Notation (JSON) data interchange format.

60 IETF RFC 6902 (2013) JavaScript Object Notation (JSON) patch.

61 IETF RFC 7396 (2014) JSON merge patch.

62 3GPP TS 23.502 (2019) Procedures for the 5G system (5GS); Stage 2 (Release 15), December 2019, Section 4.17.1.

63 3GPP TS 29.510 (2019) 5G system; Network function repository services; Stage 3 (Release 15), December 2019, Sections 5.2.2.2, 6.1.3.3.3.2.

64 3GPP TS 29.501 (2019) 5G system; Principles and guidelines for services definition; Stage 3 (Release 15), December 2019, Section 4.4.

65 3GPP TS 23.003 (2019) Numbering, addressing and identification (Release 15), September 2019, Section 28.3.2.3.

66 IETF RFC 4122 (2005) A Universally Unique IDentifier (UUID) URN namespace.

67 3GPP TS 23.502 (2019) Procedures for the 5G system (5GS); Stage 2 (Release 15), December 2019, Section 4.17.4.

68 3GPP TS 29.510 (2019) 5G system; Network function repository services; Stage 3 (Release 15), December 2019, Sections 5.3.2.2, 6.2.3.2.3.1.

69 3GPP TS 23.502 (2019) Procedures for the 5G System (5GS); Stage 2 (Release 15), December 2019, Section 4.17.7.

70 3GPP TS 29.510 (2019) 5G system; Network function repository services; Stage 3 (Release 15), December 2019, Sections 5.2.2.5, 6.1.3.4.3.1.

71 3GPP TS 29.510 (2019) 5G system; Network function repository services; Stage 3 (Release 15), December 2019, Sections 5.2.2.6, 6.1.5.2.

3

Architecture of the Radio Access Network

Our introduction to the radio access network (RAN) follows a similar pattern to the material on the core. After reviewing the evolved UMTS terrestrial radio access network, we will introduce the next-generation Node B, and discuss its internal division into central and distributed units. We will then move on to the next-generation radio access network, introduce the concept of multi-radio dual connectivity and detail the architectural options that the radio access network supports.

Several specifications are relevant to this chapter, in the same way as before. The most important are the descriptions of the air interface and the radio access network, TS 38.300 [1] and TS 38.401 [2], and of dual connectivity between 5G and LTE, TS 37.340 [3].

3.1 The Evolved UMTS Terrestrial Radio Access Network

3.1.1 Release 8 Architecture

Figure 3.1 shows the architecture of the evolved UMTS terrestrial radio access network (E-UTRAN) in 3GPP Releases 8 and 9 [4]. There is just one type of network element, namely the base station, which is officially known as the evolved Node B (eNB). The eNB's tasks include radio transmission and reception, digital signal processing and the delivery of signalling messages that control the mobile's radio communications.

In Releases 8 and 9, a mobile communicates with one base station at a time through a single cell, with the two known as the mobile's *serving eNB* and *serving cell* respectively. Between the two devices, the air interface is a reference point that is denoted Uu and is known as *evolved UMTS terrestrial radio access* (EUTRA).

A base station communicates with other network elements over two other reference points. The S1 backhaul carries access stratum (AS) signalling messages through which the core network controls the base station, non-access stratum (NAS) signalling messages which the base station forwards to and from the mobile, and traffic for the user. Optionally, nearby base stations can communicate through another reference point, known as X2. In early releases of LTE, the main purpose of X2 is to convey the signalling messages that hand a mobile over from one base station to another, and to forward undelivered data packets from the old base station to the new one. (Even if there is no X2 reference point between two base stations, they can still hand mobiles over with the help of the evolved

An Introduction to 5G: The New Radio, 5G Network and Beyond, First Edition. Christopher Cox.
© 2021 John Wiley & Sons Ltd. Published 2021 by John Wiley & Sons Ltd.

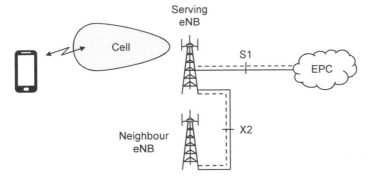

Figure 3.1 Architecture of the evolved UMTS terrestrial radio access network in Releases 8 and 9.

packet core [EPC], using signalling messages that are delivered over two instances of S1.) Both S1 and X2 are implemented over an IP-based transport network.

3.1.2 Carrier Aggregation

Carrier aggregation (CA) is a popular enhancement, which was introduced in LTE Release 10 [5]. When using carrier aggregation, a mobile communicates through multiple cells instead of just one, so as to increase its maximum data rate. The cells are controlled by a single base station and are transmitted on different radio frequencies to ensure that they do not interfere.

As shown in Figure 3.2, a mobile communicates through one *primary cell* (PCell), which carries traffic for the user and signalling messages for the mobile, and which operates in both the uplink and downlink. Optionally, there are one or more *secondary cells* (SCells) which only carry traffic, and which can operate either in the uplink and downlink or in the downlink alone. Usually, a mobile's primary cell is at a lower carrier frequency than its secondaries, to ensure that the primary has the best coverage and the most reliable communications. However, the terms *primary* and *secondary* are simply labels in respect

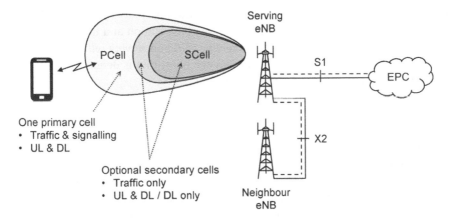

Figure 3.2 Carrier aggregation.

of a particular mobile: a cell can simultaneously act as a primary towards one mobile and as a secondary towards another.

3.1.3 Dual Connectivity

Dual connectivity (DC) was introduced in LTE Release 12 and is illustrated in Figure 3.3 [6]. In dual connectivity, the mobile still communicates through multiple cells on different radio frequencies. This time, however, the cells are controlled by two base stations instead of one.

The original base station is now known as a master evolved Node B (MeNB). The master typically controls macrocells on a low carrier frequency, so as to offer good coverage and reliable communications. The mobile communicates with the master through one primary cell and a number of optional secondary cells, as before. Together, those cells are known as the *master cell group* (MCG).

Dual connectivity introduces the secondary evolved Node B (SeNB). The SeNB typically controls microcells on a high carrier frequency, so as to offer a high data rate for mobiles that are in its coverage area, and a high network capacity. The *secondary cell group* (SCG) contains one *primary SCG cell* (PSCell), which operates in both the uplink and downlink and, in the case of a secondary eNB, only carries traffic for the user. Optionally, it can also contain one or more secondary cells, which operate in the same way as before. Through simultaneous communications with both base stations, the mobile gains the best of both worlds: reliable communications with the master, together with a high data rate while in the coverage area of the secondary.

In dual connectivity, the core network can exchange traffic with the SeNB over the S1 user plane (S1-U). However, there are no signalling communications between the secondary and the core: instead, the master controls the secondary using signalling messages over the X2 control plane (X2-C). The two reference points have a latency of several milliseconds, which is much longer than the timescales associated with radio communications, so the two base stations schedule their radio transmissions independently. In LTE dual connectivity, the

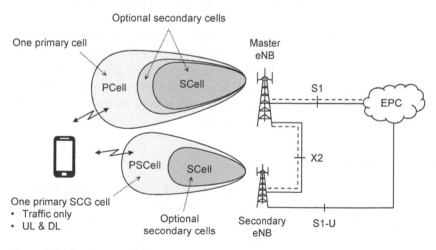

Figure 3.3 Dual connectivity.

base stations are generally installed at different sites: if they were at the same site, then the same benefits could be achieved more efficiently by means of carrier aggregation.

As before, the terms *master* and *secondary* are simply labels in respect of a particular mobile: a base station can simultaneously act as a master towards one mobile and as a secondary towards another. On the other hand, a base station can be configured to act only as a master, only as a secondary or as both. Similarly, a cell can be configured to act only as a primary, a primary SCG or a secondary cell, or as any combination of the three.

Unfortunately the 3GPP specifications overuse the term *secondary*: a secondary cell relates to carrier aggregation, but a secondary eNB relates to dual connectivity. Adding still more terminology, a *special cell* (SpCell) is either a PCell or a PSCell, while a *serving cell* is a PCell, PSCell or SCell.

3.2 The Next-generation Node B

3.2.1 High Level Architecture

A 5G base station is known as a next-generation Node B (gNB). It communicates with the mobile over an air interface known as the New Radio (NR), and carries out the same tasks as an evolved Node B but with a different implementation [7]. Collectively, the two types of base station are known as nodes.

In the early stages of 5G, a gNB can be installed in the E-UTRAN. In that network, the gNB communicates with the evolved packet core over the S1 backhaul, and with other nodes over X2. The signalling protocols over both reference points are enhanced to support the new features of the gNB.

In the longer term, both types of node are installed in a new network known as the next-generation radio access network (NG-RAN). That network is characterized by two new reference points which carry new signalling protocols. The NG reference point connects a node to the 5G core network: the NG control plane is identical to the N2 reference point from the 5G core, and the user plane is identical to N3. The Xn reference point connects nearby nodes within the NG-RAN.

If a node supports all four of the reference points identified above, then it lies in both of the radio access networks. This is valuable during the rollout of the NG-RAN and the 5G core, as it helps to remove any need for a sudden changeover from one network to the other.

Officially, a cell controlled by an eNB is known as an *EUTRA cell*, while a cell controlled by a gNB is known as an *NR cell*. More colloquially, we can refer to the first as an *eNB cell* or a *4G cell*, and to the second as a *gNB cell* or a *5G cell*.

3.2.2 Internal Architecture

In LTE, the eNB was originally specified as a single network element, for installation at the local cell site. In contrast, the gNB has a number of internal components, which communicate across open interfaces that are defined by the specifications. 3GPP investigated several possible functional splits as part of its studies into 5G [8]: the one selected is shown in Figure 3.4 [9].

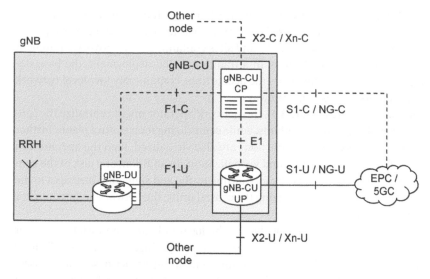

Figure 3.4 Internal architecture of the gNB. *Source:* Adapted from 3GPP TS 38.401.

The *gNB central unit* (gNB-CU) is intended for installation away from the cell site, for example at a nearby hub. Each central unit controls one or more *gNB distributed units* (gNB-DUs), which are intended for local installation. In turn, each distributed unit controls one or more cells using low-level functions such as digital signal processing, scheduling and re-transmission. The central and distributed units communicate over a new reference point, which is known as F1. (Release 16 introduces a similar split for the eNB [10].)

If a mobile is in carrier aggregation, then it communicates with one distributed unit and one central unit alone. In dual connectivity, a mobile has two distributed units, which are controlled either by the same central unit or by two different ones.

The central unit has two components of its own. The *gNB central unit control plane* (gNB-CU-CP) exchanges signalling messages with the mobile, the core network and other nodes, and processes those messages using high-level functions such as encryption and decryption. It also controls one or more *gNB central unit user planes* (gNB-CU-UPs), which have a similar role in respect of user plane traffic. The control and user planes communicate over another new reference point, known as E1.

Optionally, the tasks of radio transmission and reception can be implemented separately using an antenna panel that is known as a *remote radio head* (RRH). In a simple macrocell deployment, for example, each cell has one RRH, which is installed at the top of the mast and communicates with a distributed unit at the base. Communications take place over a high-speed, low-latency digital interface, a common choice being a version of the *Common Public Radio Interface* (CPRI) [11] that has been enhanced for use by 5G and is known as eCPRI [12].

3.2.3 Deployment Options

Network operators can use the internal architecture of the gNB in several different ways, to help them gain the benefits of network function virtualization and software-defined networking [13, 14]. Consider, for example, a network that contains four types of site: local cell sites (around 10 000 in a medium-sized country such as the United Kingdom), together with local access hubs, regional access hubs and sites containing core network equipment (with perhaps 1000, 100, and 10 of each). In a traditional LTE deployment, the base station is installed at the local cell site, while the access hubs contain only low-level network elements such as routers and security gateways.

In one example architecture (Figure 3.5), the network operator might centralize the central units' user planes at the local access hubs, while centralizing their control planes further at the regional hubs. If those network functions are also virtualized, then the operator can gain the benefits of RAN centralization and virtualization. In addition, one task of the central unit is to support dual connectivity by forwarding traffic and signalling messages to the appropriate distributed units. By centralizing the central units, the network architecture for dual connectivity is simplified.

Optionally, the operator can also cache data at the local hub, and can install local user plane functions there in support of multi-access edge computing. That can significantly reduce the user plane latency, although there is still a remaining latency of several milliseconds over the F1 interface to the distributed unit. Going further, the network operator might also centralize the distributed units at the local access hub, while retaining only remote radio heads at the cell sites. By adopting that approach, the operator can reduce the user plane latency still further, to support the most demanding low-latency applications.

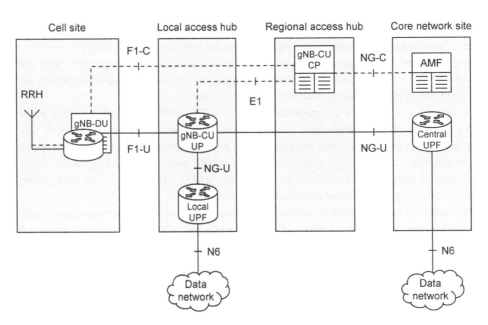

Figure 3.5 Example architecture for the deployment of the gNB. AMF: Access and mobility management function; UPF: user plane function.

3.3   Architectural Options

3.3.1   Multi-radio Dual Connectivity

In *multi-radio dual connectivity* (MR-DC), the mobile can communicate with two base stations using different *radio access technologies* (RATs). Generically, the two base stations are known as the *master node* (MN) and the *secondary node* (SN), with each one implemented as either an eNB or a gNB. MR-DC is an important concept for 5G, because it allows a mobile to communicate with both types of node at the same time.

Using this technique, the 3GPP technical reports defined a set of architectural options for 5G [15], some of which acquired names in the later specifications [16]. With these options, network operators can implement a phased roll-out, in which they gradually introduce the capabilities of 5G into their existing networks. Various migration paths are possible: the example chosen here is a typical one, based on migration path 2 from the 3GPP report.

3.3.2   Options 1 and 3 – EPC, E-UTRAN and MeNB

Each migration path begins with the legacy architecture from LTE, in which the mobile communicates with the EPC, the E-UTRAN, a master eNB and optionally a secondary eNB. In the 3GPP technical reports, that architecture is known as option 1. In the first step along the example path (Figure 3.6), the secondary eNB is replaced by a *secondary gNB* (SgNB). The resulting architecture is known in the 3GPP reports as option 3, and in the specifications as *EUTRA-NR dual connectivity* (EN-DC).

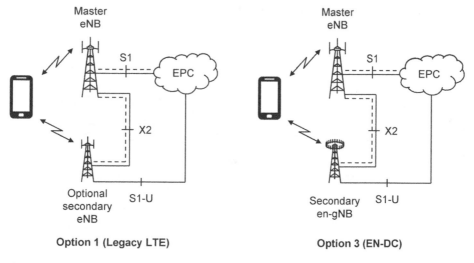

Figure 3.6   Architectural options 1 and 3.

The architecture is a non-standalone implementation of 5G, because the gNB can only act as a secondary node, not as a master. The gNB is known specifically as an *en-gNB*, to denote that it is backwards compatible with the S1 and X2 reference points from LTE. By using this architecture, a network operator can offer high 5G data rates in traffic hotspots without the need for wide-area 5G coverage or for the 5G core. Retention of the master eNB ensures that users' communications with the network are still reliable.

3.3.3 Options 5 and 7 – 5GC, NG-RAN and MeNB

The second migration step (Figure 3.7) introduces the NG-RAN and the 5G core network (5GC). In option 5, the mobile communicates with a master eNB, and optionally with a secondary eNB. In option 7, also known as *NG-RAN–EUTRA–NR dual connectivity* (NGEN-DC), the secondary eNB is replaced by a secondary gNB.

This architecture is still a non-standalone implementation of 5G. The eNB is known specifically as an *ng-eNB*, to denote that it supports the new reference points denoted NG and Xn. By using this architecture, a network operator gains the benefits of the 5G core network, such as network function virtualization, software-defined networking and network slicing, but still has no need for wide-area 5G coverage.

3.3.4 Options 2 and 4 – 5GC, NG-RAN and MgNB

In the final migration step (Figure 3.8), the gNB gains the ability to act as a master. In option 2, the mobile communicates with a *master gNB* (MgNB), and optionally with a secondary gNB through the use of *NR–NR dual connectivity* (NR-DC). In option 4, also known as *NR–EUTRA dual connectivity* (NE-DC), a secondary eNB is retained.

This architecture is now a standalone implementation of 5G. By using option 2, network operators free themselves from any reliance on LTE and can install 5G base stations in areas without any LTE coverage at all.

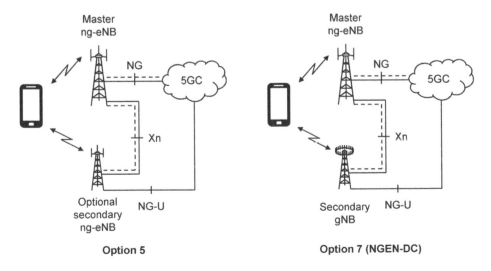

Figure 3.7 Architectural options 5 and 7.

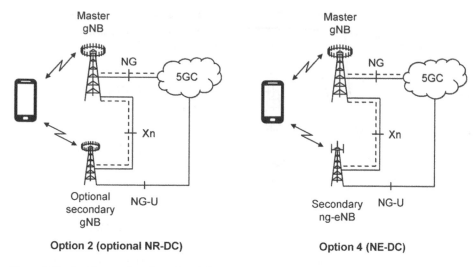

Figure 3.8 Architectural options 2 and 4.

3.3.5 Data Transport

A *data radio bearer* (DRB) is a data pipe in the NG-RAN which transports packets over a defined route, using a defined configuration of the underlying air interface protocols [17]. A DRB usually carries packets that belong to a single quality of service (QoS) flow, but it can also carry packets that belong to multiple QoS flows in the same protocol data unit (PDU) session.

From the viewpoint of the core network, there are two types of data radio bearer: an *MN-terminated bearer*, in which the core network exchanges packets with the master node; and an *SN-terminated bearer*, in which it exchanges packets directly with the secondary. From the viewpoint of the mobile, there are three types of DRB: an *MCG bearer*, in which the mobile exchanges packets with the master node; an *SCG bearer*, in which it exchanges packets with the secondary; and a *split bearer*, in which individual packets can be routed through either node or duplicated through both to improve the reliability.

The two classifications apply independently, so a DRB can be routed in one of six ways, illustrated in Figure 3.9, with the choice made by negotiation between the master and secondary nodes. A mobile is in dual connectivity if one or more of its DRBs is routed through the secondary node, either as an SN-terminated bearer, or as an SCG or split bearer. If a mobile has more than one bearer, then those bearers can be routed in different ways. However, a common choice in non-standalone implementations is the use of SN-terminated split bearers, in which most of the packets travel through the secondary gNB alone, but the master cell group is still available if required [18].

The distinction between DRBs and QoS flows is new to the 5G core network and the NG-RAN. It was introduced to uncouple the treatments of quality of service in the core and radio access networks, and to support the use of non-3GPP access networks (for example, WiFi) in which the support for QoS can be very different. There is less of a distinction when using the EPC and the E-UTRAN, but the available routes through the RAN are the same.

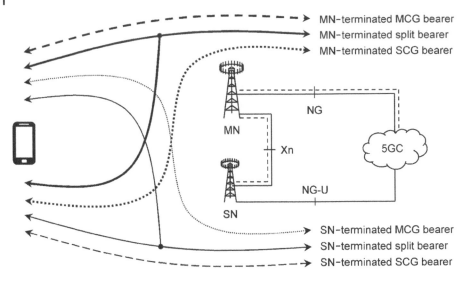

Figure 3.9 Data radio bearers.

In the 3GPP technical reports into 5G, each architectural option was associated with a few different sub-options such as 3, 3a and 3x. Each of those sub-options referred to one of the available routes from Figure 3.9, but the terminology is now obsolete.

3.4 Network Areas and Identities

3.4.1 Tracking Areas

The NG-RAN is organized into *tracking areas* (TAs) [19, 20]. At its heart, a tracking area is a small geographical region containing perhaps 100 nodes, although the exact number might vary considerably. Each tracking area serves one or more types of network slice, so an operator can restrict a network slice to certain tracking areas only. The NG-RAN's tracking areas do not overlap: a cell lies in just one tracking area, but a node can lie in multiple tracking areas by virtue of controlling multiple cells.

Each tracking area is identified by a 24-bit *tracking area code* (TAC), which is unique within the public land mobile network (PLMN). Adding the public land mobile network identity (PLMN-ID) gives the *tracking area identity* (TAI), which is globally unique.

Tracking areas are the basis of the registration areas that we introduced in Chapter 2, in that each registration area is implemented as a mobile-specific *tracking area list* that contains a maximum of 16 tracking areas. The use of tracking area lists avoids problems that would otherwise result if a mobile were restricted to just a single tracking area, as the network would then receive frequent location updates from mobiles that are moving back and forth across a tracking area boundary.

Similar issues apply to the E-UTRAN, except that network slicing is not supported, and the tracking area codes are 16 bits long. In that context, it is worth noting that the type of TAC depends on the radio access network that a cell is in, and not on its radio access technology. If a gNB or eNB lies in both radio access networks, then its cells have one tracking area code of each type.

3.4.2 RAN Areas

In the NG-RAN, each tracking area can be divided into one or more *radio access network areas* (RAN areas) [21]. As before, a cell lies in just one RAN area, but a node can lie in more than one. A RAN area is identified by an 8-bit *RAN area code*, which is unique within the tracking area.

RAN areas are the basis of mobile-specific *RAN-based notification areas* (RNAs). These are similar to tracking area lists, but are defined in a more flexible way: each one spans either a list of RAN areas, or a list of individual cells, within a single tracking area. RNAs are used by mobiles in the state of RRC_INACTIVE, which is introduced in Section 3.5.

3.4.3 Cell Identities

Every NR cell has a 36-bit identity that is unique within the PLMN, which is known as the *New Radio Cell Identity* (NCI) [22]. By combining that with the PLMN-ID, we arrive at the *New Radio Cell Global Identity* (NCGI), which is globally unique.

The NCI is rather long, so a cell is more often identified using its *physical cell identity*, which ranges from 0 to 1007 [23]. Nearby cells on the same radio frequency must have different physical cell identities to ensure that the mobile can distinguish them. Physical cell identities are organized into 336 *cell identity groups*, each of which contains three identities. Optionally, network operators can configure a three-sector gNB so that its sectors have different physical cell identities within the same group, but that is not essential.

Similar issues apply in the case of EUTRA cells. The *EUTRA cell identity* (ECI) is 28 bits long, and the corresponding global identity is the *EUTRA cell global identification* (ECGI). The physical cell identity ranges from 0 to 503, and there are 168 cell identity groups [24]. Unlike tracking areas, the type of cell identity refers to the radio access technology that the cell is using, and not to the RAN.

3.5 RRC State Diagram

3.5.1 5G State Diagram

The *radio resource control* (RRC) state diagram describes whether the mobile is active or on standby, from the viewpoint of the radio access network and the mobile's access stratum [25]. Figure 3.10 shows the 5G version of the state diagram, which is applicable to a mobile that is served by the NG-RAN and a master gNB under architectural options 2 and 4. There are three states: RRC_IDLE maps onto the core network state of CM-IDLE, while RRC_INACTIVE and RRC_CONNECTED both map onto the state of CM-CONNECTED.

The state of RRC_CONNECTED is intended for mobiles that are communicating with the network. The mobile is served by a master gNB, which knows the mobile's master cell group. In dual connectivity, the mobile is also served by a secondary node, which knows the secondary cell group. Mobility is controlled by the radio access network: the mobile measures the signals that arrive from its serving cells and their neighbours, and reports the

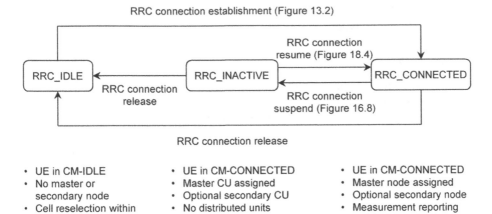

RRC connection establishment (Figure 13.2)

RRC connection resume (Figure 18.4)

RRC_IDLE ← RRC_INACTIVE ← → RRC_CONNECTED

RRC connection release

RRC connection suspend (Figure 16.8)

RRC connection release

- UE in CM-IDLE
- No master or secondary node
- Cell reselection within registration area

- UE in CM-CONNECTED
- Master CU assigned
- Optional secondary CU
- No distributed units
- Cell reselection within RAN-based notification area

- UE in CM-CONNECTED
- Master node assigned
- Optional secondary node
- Measurement reporting and handovers

Figure 3.10 5G radio resource control (RRC) state diagram, for a mobile served by a gNB. *Source:* Adapted from 3GPP TS 38.331.

results to its master and secondary nodes. In turn, the NG-RAN can change the mobile's master and secondary nodes, and can change the two cell groups.

RRC_IDLE is intended for mobiles that are on standby. In a procedure known as *camping*, the mobile continues to monitor a single serving cell. Mobility is handled by the mobile itself: the mobile measures the incoming signals as before, but this time uses them in the internal procedure of *cell reselection* so as to trigger changes to its serving cell. If that takes the mobile outside its registration area, then it informs the core network in a *mobility registration update*. The NG-RAN has no knowledge of the mobile at all.

RRC_INACTIVE is an optimization for mobiles that are using low data rate services such as machine-type communications. In this state, the central units maintain their knowledge of the mobile, but the distributed units tear their records down. Using the cell reselection procedure, the mobile can move throughout a RAN-based notification area without the need to inform the network.

The mobile's state affects the configuration of its PDU sessions. In RRC_IDLE, a mobile's PDU sessions are all *inactive*, and the core network cannot deliver any downlink data packets to the NG-RAN. In RRC_CONNECTED, each individual PDU session is either inactive or *active*, such that traffic can freely travel between the mobile and the core. In RRC_INACTIVE, each individual PDU session is either inactive or *suspended*. In that state, the core network can only deliver downlink data as far as the central unit, but the PDU session is easily re-activated when that occurs.

3.5.2 Interworking with 4G

In the NG-RAN, a mobile can also be served by a master ng-eNB under architectural options 5 and 7. In that situation, the mobile uses a 4G version of the same RRC state diagram, which has been enhanced as part of 3GPP Release 15 so that it contains the same three states. Figure 3.11 shows the relationships between the two state diagrams. In the states

Serving gNB (Options 2 and 4)

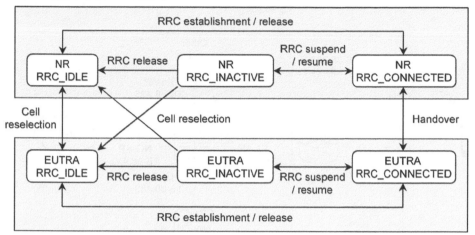

Figure 3.11 Relationship between the 4G and 5G RRC state diagrams, for a mobile within the NG-RAN. *Source:* Adapted from 3GPP TS 38.331.

of RRC_CONNECTED and RRC_IDLE, the mobile can move between a master gNB and a master ng-eNB using the procedures for handover and cell reselection. In the state of RRC_INACTIVE, however, any reselection between a master gNB and a master ng-eNB triggers a transition to RRC_IDLE.

In the E-UTRAN, the mobile is served by a master eNB under architectural options 1 and 3. In that situation, the mobile uses the original version of the 4G state diagram, which only contains two states: RRC_CONNECTED and RRC_IDLE. We will address inter-operation between the two radio access networks as part of Chapter 19.

3.6 Signalling Protocols

3.6.1 Signalling Protocol Architecture

Figure 3.12 shows the access stratum signalling protocols that are used within the NG-RAN. Each node communicates with the 5G core network using the *NG application protocol* (NG-AP) [26], and with other nodes using the *Xn application protocol* (Xn-AP) [27]. (In the E-UTRAN, those roles are filled by the *S1 application protocol* (S1-AP) [28] and the *X2 application protocol* (X2-AP) [29] respectively.)

In dual connectivity, a secondary node is configured using signalling communications with the master: it does not exchange any dual-connectivity signalling messages with the core. Nevertheless, a secondary node can still exchange other types of signalling message with the core network, for example if it is acting as a master node in respect of another mobile.

Inside the gNB, the central unit controls each of its distributed units using the *F1 application protocol* (F1-AP) [30]. Inside the central unit, the control plane controls each of its user planes using the *E1 application protocol* (E1-AP) [31].

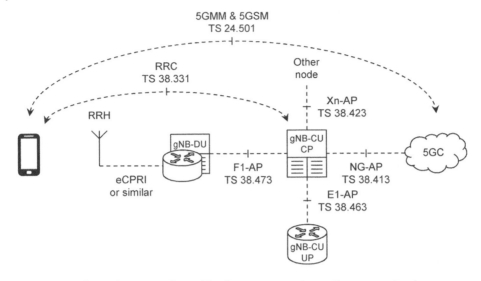

Figure 3.12 Signalling protocols used by the next generation radio access network.

These signalling messages are all transported over an IP transport network by means of an IP-based protocol stack. The layer 4 protocol is the *Stream Control Transmission Protocol* (SCTP) [32], which is similar to the transmission control protocol (TCP), but includes extra features that make it more suitable for the delivery of signalling messages. Any protocols can be used in layers 1 and 2, for example Ethernet.

The NG-RAN controls the mobile's radio communications using the radio resource control (RRC) protocol. A gNB uses the 5G version of the protocol [33]: if the gNB is split, then the messages are composed by the central unit's control plane, and are transported between the central and distributed units by embedding them into F1-AP signalling. An eNB uses the 4G protocol [34]. Two specific roles of the protocol are to host the RRC state diagram from earlier, and to transmit *system information* messages, which the base station broadcasts over the whole of the cell so as to tell its mobiles about how the cell is configured.

In dual connectivity, most of the RRC signalling is between the mobile and the master node. In particular, the mobile has one RRC state diagram, namely the 5G state diagram in the case of a master gNB, and the 4G state diagram in the case of a master ng-eNB. There is also a limited amount of RRC signalling between the mobile and the secondary node, whose sole purpose is to configure the mobile's communications with the secondary.

The 5G core network controls the mobile's non-access stratum using the 5G mobility management (5GMM) and 5G session management (5GSM) protocols from Chapter 2. Those messages are transported by embedding them into NG-AP signalling messages between the core network and the master node, and into RRC signalling messages between the master node and the mobile.

3.6.2 Signalling Radio Bearers

RRC signalling messages are transported by means of *signalling radio bearers* (SRBs) [35, 36]. These are similar to DRBs, in that each one delivers RRC signalling messages

Table 3.1 Signalling radio bearers.

Bearer	Node	Routing	Configured by	Used by
SRB 0	MN	MCG	System information	RRC signalling before setup of SRB 1
SRB 1	MN	MCG initially can split later	Signalling on SRB 0	Subsequent RRC signalling with MN plus optional NAS signalling
SRB 2	MN	MCG or split	Signalling on SRB 1	Stand-alone NAS signalling
SRB 3	SgNB	SCG	Signalling on SRB 1	RRC signalling with SgNB

between the mobile and one of its nodes, over a defined route and using a defined configuration of the air interface protocols. The specifications speak of four numbered SRBs, which are listed in Table 3.1. They also state which bearer should be used to transport each individual message.

Signalling radio bearer 0 (SRB 0) has a simple configuration, which is defined by the cell's system information and is understood by every mobile in the cell. It is used by mobiles in the states of RRC_IDLE and RRC_INACTIVE for procedures such as RRC connection establishment. SRB 0 connects the mobile with the master node, and is always implemented as an MCG bearer.

During RRC connection establishment, the master node tells the mobile how to set up SRB 1. That bearer conveys all RRC signalling messages between the mobile and the master node, except for the ones handled by SRB 0 and SRB 2. Some of those messages can also transport NAS messages between the mobile and the core. SRB 1 is initially implemented as an MCG bearer, but can be reconfigured as a split bearer if the mobile enters dual connectivity.

SRB 2 is set up later as part of a general-purpose procedure known as *RRC connection reconfiguration*. It is used for RRC messages known as uplink and downlink information transfers, which convey NAS messages between the mobile and the core. It can be implemented as either an MCG bearer or a split bearer from the beginning.

SRB 3 is also set up by means of RRC connection reconfiguration. It is implemented as an SCG bearer within the primary SCG cell, and delivers signalling messages between the mobile and a secondary gNB that do not involve the master node. It is only supported in the case of a secondary gNB, not a secondary eNB, and even then it is optional. If SRB 3 is not used, then the relevant signalling messages are forwarded by way of the master node, by embedding them into X2-AP or Xn-AP signalling, and into the RRC signalling messages that are delivered using SRB 1.

References

1 3GPP TS 38.300 (2019) NR; NR and NG-RAN overall description; Stage 2 (Release 15), December 2019.
2 3GPP TS 38.401 (2019) NG-RAN; Architecture description (Release 15), December 2019.

3 3GPP TS 37.340 (2019) Evolved Universal Terrestrial Radio Access (E-UTRA) and NR; Multi-connectivity; Stage 2 (Release 15), December 2019.

4 3GPP TS 36.300 (2019) Evolved Universal Terrestrial Radio Access (E-UTRA) and Evolved Universal Terrestrial Radio Access Network (E-UTRAN); Overall description; Stage 2 (Release 15), December 2019, Section 4.

5 3GPP TS 36.300 (2019) Evolved Universal Terrestrial Radio Access (E-UTRA) and Evolved Universal Terrestrial Radio Access Network (E-UTRAN); Overall description; Stage 2 (Release 15), December 2019, Section 5.5.

6 3GPP TS 36.300 (2019) Evolved Universal Terrestrial Radio Access (E-UTRA) and Evolved Universal Terrestrial Radio Access Network (E-UTRAN); Overall description; Stage 2 (Release 15), December 2019, Section 4.9.

7 3GPP TS 38.300 (2019) NR; NR and NG-RAN overall description; Stage 2 (Release 15), December 2019, Section 4.

8 3GPP TR 38.801 (2017) Study on New Radio access technology: Radio access architecture and interfaces (Release 14), March 2017, Section 11.

9 3GPP TS 38.401 (2019) NG-RAN; Architecture description (Release 15), December 2019, Section 6.1.

10 3GPP TR 36.756 (2017) Study on architecture evolution for Evolved Universal Terrestrial Radio Access Network (E-UTRAN) (Release 15), September 2017.

11 CPRI (2015) Common Public Radio Interface (CPRI): Interface specification, Version 7.0, September 2015.

12 CPRI (2019) Common Public Radio Interface: eCPRI interface specification, Version 2.0, May 2019.

13 3GPP TR 38.912 (2018) Study on New Radio (NR) access technology (Release 15), June 2018, Section 5.

14 Sutton, A. (2018) 5G network architecture, design and optimisation. IET 5G State of Play Conference, January 2018. https://www.slideshare.net/3G4GLtd/5g-network-architecture-design-and-optimisation (accessed 18 January 2020).

15 3GPP TR 38.801 (2017) Study on New Radio access technology: Radio access architecture and interfaces (Release 14), March 2017, Sections 7.2, 14.

16 3GPP TS 37.340 (2019) Evolved Universal Terrestrial Radio Access (E-UTRA) and NR; Multi-connectivity; Stage 2 (Release 15), December 2019, Section 4.1.

17 3GPP TS 37.340 (2019) Evolved Universal Terrestrial Radio Access (E-UTRA) and NR; Multi-connectivity; Stage 2 (Release 15), December 2019, Sections 4.2.2, 4.3.2.

18 GSM Association (2019) 5G implementation guidelines.

19 3GPP TS 23.501 (2019) System architecture for the 5G system (5GS); Stage 2 (Release 15), December 2019, Section 5.3.2.3.

20 3GPP TS 23.003 (2019) Numbering, addressing and identification (Release 15), September 2019, Sections 19.4.2.3, 28.6.

21 3GPP TS 38.300 (2019) NR; NR and NG-RAN overall description; Stage 2 (Release 15), December 2019, Section 9.2.2.3.

22 3GPP TS 23.003 (2019) Numbering, addressing and identification (Release 15), September 2019, Sections 19.6, 19.6A.

23 3GPP TS 38.211 (2019) NR; Physical channels and modulation (Release 15), December 2019, Section 7.4.2.1.

24 3GPP TS 36.211 (2019) Evolved Universal Terrestrial Radio Access (E-UTRA); Physical channels and modulation (Release 15), January 2020, Section 6.11.

25 3GPP TS 38.331 (2019) NR; Radio resource control (RRC) protocol specification (Release 15), December 2019, Section 4.2.1.

26 3GPP TS 38.413 (2019) NG-RAN; NG application protocol (NGAP) (Release 15), December 2019.

27 3GPP TS 38.423 (2019) NG-RAN; Xn application protocol (XnAP) (Release 15), December 2019.

28 3GPP TS 36.413 (2019) Evolved Universal Terrestrial Radio Access Network (E-UTRAN); S1 application protocol (S1AP) (Release 15), December 2019.

29 3GPP TS 36.423 (2019) Evolved Universal Terrestrial Radio Access Network (E-UTRAN); X2 application protocol (X2AP) (Release 15), December 2019.

30 3GPP TS 38.473 (2019) NG-RAN; F1 application protocol (F1AP) (Release 15), December 2019.

31 3GPP TS 38.463 (2019) NG-RAN; E1 application protocol (E1AP) (Release 15), December 2019.

32 IETF RFC 4960 (2007) Stream Control Transmission Protocol, September 2007.

33 3GPP TS 38.331 (2019) NR; Radio resource control (RRC) protocol specification (Release 15), December 2019.

34 3GPP TS 36.331 (2019) Evolved Universal Terrestrial Radio Access (E-UTRA); Radio resource control (RRC); Protocol specification (Release 15), December 2019.

35 3GPP TS 38.331 (2019) NR; Radio resource control (RRC) protocol specification (Release 15), December 2019, Section 4.2.2.

36 3GPP TS 37.340 (2019) Evolved Universal Terrestrial Radio Access (E-UTRA) and NR; Multi-connectivity; Stage 2 (Release 15), December 2019, Sections 4.2.1, 7.5, 7.6.

4

Spectrum, Antennas and Propagation

This is the first of three chapters that address the underlying technologies that are used by the 5G air interface. One of those technologies is a move to higher radio frequencies in pursuit of a wider bandwidth and a higher data rate. That has several impacts on the coverage of the air interface, which we will review as part of this chapter. We will also introduce some of the phenomena that appear in a multipath propagation environment, which will be useful for later chapters. There are several valuable references for further reading, for example Reference [1], while others are listed in the Bibliography.

4.1 Radio Spectrum

4.1.1 Radio Waves

Electromagnetic waves consist of oscillating electric and magnetic fields, whose frequency f and wavelength λ are related as follows:

$$c = f\lambda \tag{4.1}$$

where c is the speed of light, approximately $300\,000\,\mathrm{km\,s^{-1}}$. Waves with different frequencies are created and absorbed in different ways, so they are often grouped into a half-dozen categories that are illustrated in Figure 4.1. Nevertheless, there are no precise boundaries between those categories: instead, they blur into each other.

Radio waves are most easily created by supplying an alternating electric current to some type of transmit antenna, and absorbed by means of a similar receive antenna. They have frequencies up to about 300 GHz, and various classifications. Figure 4.1 shows some of the bands in two such classifications by the International Telecommunication Union (ITU) and the Institute of Electrical and Electronics Engineers (IEEE).

Electromagnetic waves are a type of transverse wave, that being a wave in which the oscillations are at a right angle to the wave's direction of travel. If the direction of oscillation is clearly defined, then the wave is said to be *polarized*. A suitably designed transmitter can emit radio waves with two independent polarizations, for example vertical and horizontal,

An Introduction to 5G: The New Radio, 5G Network and Beyond, First Edition. Christopher Cox.
© 2021 John Wiley & Sons Ltd. Published 2021 by John Wiley & Sons Ltd.

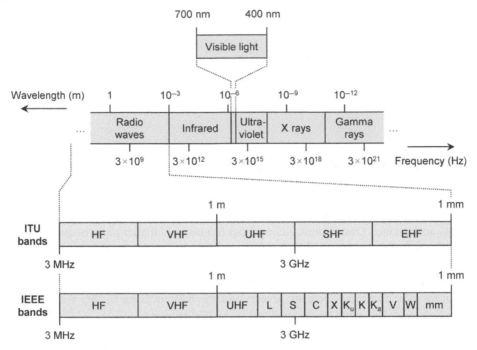

Figure 4.1 The electromagnetic spectrum.

or oriented at angles of ±45° to the vertical. Any other radio wave can be expressed as a combination of those two polarizations.

4.1.2 Use of Radio Spectrum

Traditionally, mobile cellular networks have operated in the ITU's *ultra high frequency* (UHF) band, which extends from 300 to 3000 MHz. That is the result of a trade-off between two competing issues. Firstly, the *capacity* of a wireless communication system depends upon its bandwidth, in accordance with the Shannon limit. More bandwidth is available at high frequencies, so the capacity is greatest if the radio frequency is high. Secondly, the *coverage* of a wireless communication system is the greatest distance at which the base station and mobile can communicate successfully. Coverage tends to be greatest at low carrier frequencies, an issue that is discussed further in this chapter.

Recently, however, two problems have arisen. Firstly, the UHF band is shared with many other applications, for example television broadcasting, civil aviation and defence, and has become increasingly congested and fragmented. Secondly, the data rates in mobile cellular networks have greatly increased and now require more spectrum than the UHF band can provide.

To help deal with these problems, 5G can operate not only in UHF but also in the *super high frequency* (SHF) band, which extends from 3 to 30 GHz, and the lower reaches of the *extremely high frequency* (EHF) band, from 30 to 300 GHz [2, 3]. More precisely, the 5G specifications support two *frequency ranges*, namely frequency range 1 (FR1), which extends from 410 to 7125 MHz, and frequency range 2 (FR2), from 24.25 to 52.6 GHz [4]. There is no fundamental reason for the gap between the two ranges: it is simply that network operators have not yet expressed an interest in frequencies within the gap.

Radio waves in the EHF band have wavelengths of 1–10 mm and are often known as *millimetre waves*. Extending that definition in a manner consistent with other 5G literature, we will use the term *millimetre wave* to describe frequencies throughout FR2.

4.1.3 Spectrum Allocations for 5G

The ITU is a specialized agency of the United Nations, whose tasks include co-ordinating the worldwide use of radio spectrum. To help that process, the ITU divides the world into three regions. ITU region 1 covers Europe, Africa and North and West Asia (including the Middle East and the former Soviet Union); region 2 covers the Americas; and region 3 covers the rest of Asia (including India and China) and Australasia.

The ITU takes many of its decisions at its *World Radiocommunication Conferences* (WRCs), which are held every three to four years. These conferences have assigned several frequency bands to mobile telecommunications, either worldwide or in individual regions or countries, and either for sole use or shared with other services. Two decisions at the 2015 conference (WRC-15) were particularly relevant to 5G, namely the assignment of the 3400–3600 MHz band to cellular networks in ITU regions 1 and 2, and the opening up of the 694–790 MHz cellular band so as to span the entire world [5]. The ITU built on those decisions at WRC-19 by identifying the bands 24.25–27.5, 37–43.5, 45.5–47, 47.2–48.2 and 66–71 GHz for use by 5G [6–8].

National regulators then assign radio spectrum to individual services, in a manner that is constrained by the ITU's resolutions and is often guided by regional organizations. In Europe, for example, the European Commission has identified three frequency bands for use by 5G: 694–790 MHz, which is suitable for wide-area communications; 3400–3800 MHz to offer a higher capacity; and 24.25–27.5 GHz for applications that can exploit the use of millimetre waves [9–13]. In response, European national regulators have been gradually freeing up those bands and making them available to network operators, generally by means of auctions.

Other countries use similar procedures, but many of their frequencies are different, and the situation is constantly changing. To illustrate the use of radio spectrum, Table 4.1 lists the frequency bands that regulators have been assigning for use by 5G in four countries: China, Japan, the UK and the USA [14–16]. (Several of these allocations are technology-neutral, so they might be used by other cellular technologies such as 4G, while others are shared with other services or are only licensed for local use. There are more details in the References.)

4.2 Antennas and Propagation

4.2.1 Antenna Gain

Conceptually, the simplest type of antenna is an *isotropic* antenna, which transmits and receives equally in all directions. Practical antennas are not isotropic: instead, their signals

Table 4.1 Example spectrum allocations for 5G.

Frequency range	China	Japan	UK	USA
Low FR1 (MHz)	—	—	694–791	**614–698**
Mid FR1 (MHz)	2515–2675	—	—	**2476–2690**
High FR1 (MHz)	3300–3400	**3600–4100**	**3410–3600**	3550–3700
	3400–3600	**4500–4600**	3600–3800	*3700–4200*
	4400–4500	4800–4900	*3800–4200*	
	4800–5000			
FR2 (GHz)	**24.75–27.5**	**27.0–28.2**	24.25–26.5	**24.25–24.45**
	37.0–43.5	*28.2–29.1*	*26.5–27.5*	**24.75–25.25**
		29.1–29.5		*25.25–27.5*
				27.5–28.35
				37.6–40.0
				47.2–48.2

Bold: Licensed, including for tests and trials. Plain text: Licensing planned. *Italics: Under consideration or consultation.*

are focussed within a beam. The *antenna gain* describes how strongly the signal is focussed and is measured in decibels relative to an isotropic antenna (dBi). It can be difficult to compute the exact gain of an antenna, but the following approximation is often useful:

$$G \approx \frac{41000}{\theta^{\circ}_{3dB,h}\,\theta^{\circ}_{3dB,v}} \tag{4.2}$$

where $\theta^{\circ}_{3dB,h}$ and $\theta^{\circ}_{3dB,v}$ are the antenna's horizontal and vertical beamwidths respectively at half the maximum power, measured in degrees; and G is the antenna gain.

Because of the diffraction effects that we will discuss in Section 4.3.1, the beamwidth of an antenna depends upon its physical size. To illustrate that relationship, Figure 4.2 shows

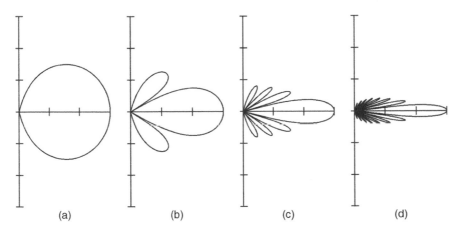

(a) (b) (c) (d)

Figure 4.2 Radiation pattern of a parabolic reflector antenna, with a diameter of (a) one, (b) two, (c) four and (d) eight wavelengths.

how the radiation pattern of a parabolic reflector antenna changes with its diameter [17]. (The scale is in units of decibels, with a range of 30 dB and tick marks at intervals of 10 dB, and the radiation pattern is normalized so that the main beam spans the full range of the chart.) We can see that a small antenna creates a wide-angle beam with a low gain while, for a large antenna, the beam is narrow and the gain is high.

We can quantify this relationship by introducing an antenna's *effective area*, A_{eff}, which is related to its gain as follows [18]:

$$A_{\text{eff}} = \frac{\lambda^2 G}{4\pi} \tag{4.3}$$

This quantity is easiest to understand in the context of a receive antenna, which intercepts the incoming radio waves that lie within its effective area.

4.2.2 Radio Propagation in Free Space

We can now investigate the transmission and reception of radio waves in free space [19]. At a distance r from a transmitter, the outgoing wave occupies a sphere with a surface area of $4\pi r^2$. At the centre of the beam, the power of the radio signal per unit area is therefore:

$$S = \frac{G_T L_T P_T}{4\pi r^2} \tag{4.4}$$

where P_T is the power at the transmit antenna connector, G_T is the transmit antenna gain and L_T expresses any losses in the transmitter due to analogue components such as cables and connectors. Similarly, the power at the receive antenna connector is the following:

$$P_R = L_R A_{\text{eff},R} S \tag{4.5}$$

where $A_{\text{eff},R}$ is the effective area of the receive antenna, and L_R expresses the losses within the receiver. Combining the above equations gives the following relationship:

$$\frac{P_T}{P_R} = \frac{1}{G_T G_R L_T L_R} \left(\frac{4\pi f r}{c} \right)^2 \tag{4.6}$$

The ratio P_T/P_R is the *coupling loss* between the transmit and receive antenna connectors. Equation (4.6) is a form of the *Friis transmission equation*, which defines the coupling loss in free space. Going further, it is often useful to combine the transmitted and received signal powers with the corresponding cable losses and antenna gains. In doing so, we introduce the *equivalent isotropic radiated power*, P_{TI}, and the *isotropic receive level*, P_{RI}. These are defined as follows:

$$P_{TI} = G_T L_T P_T \tag{4.7}$$

$$P_{RI} = G_R L_R P_R \tag{4.8}$$

Combining Equations (4.6), (4.7), and (4.8) gives the following result:

$$\frac{P_{TI}}{P_{RI}} = \left(\frac{4\pi f r}{c} \right)^2 \tag{4.9}$$

The ratio P_{TI}/P_{RI} is known as the *propagation loss* or *path loss*. It is a similar quantity to the coupling loss, but excludes components such as cables and antennas so as to focus exclusively on the propagation environment. Equation (4.9) defines the propagation loss in free space.

Propagation loss would not be a problem by itself, but the received signal is also distorted by thermal noise and by interference from other transmitters. We can describe the quality of the received signal by means of the signal-to-interference plus noise ratio (SINR), which equals the power of the received signal divided by the power due to noise and interference. If the propagation loss rises, or the levels of noise and interference rise, then the received SINR falls, and the quality of the incoming signal degrades.

4.2.3 Antenna Arrays for 5G

Equation (4.9) shows that the coverage of a radio communication system falls as the frequency increases. It is important to note, however, that this result has nothing to do with radio propagation. Instead, it only arises because we made an assumption: that the gain and beamwidth of each antenna are independent of frequency. It is easiest to understand that assumption by thinking about the receive antenna. If we increase the frequency, but keep the receive antenna's gain and beamwidth unchanged, then Equation (4.3) shows that its effective collecting area falls. That reduces the received signal power, which in turn reduces the coverage of the system. The effect is illustrated in Figure 4.3a,b.

We can deal with the problem by means of antenna arrays, as illustrated in Figure 4.3c [20]. As the frequency rises, let us add more antennas to the base station, so as to keep the total effective collecting area the same. In the case of free space propagation, the received signal power will then stay the same as the frequency rises, so the propagation loss will stay the same as well.

Collectively, the antennas create a beam that has a larger gain than before, and a narrower beamwidth. The beam no longer spans the whole of the cell, so we have to steer it so that

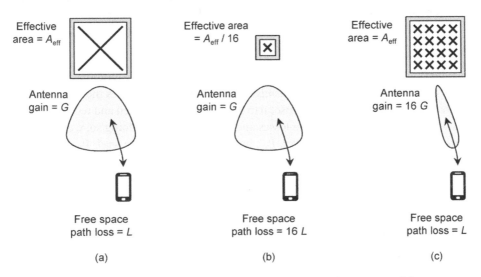

Figure 4.3 Benefit of using a base station antenna array in a high-frequency mobile telecommunication system. (a) Single antenna at a frequency f_c. (b) Single antenna at a frequency $4f_c$. (c) Array of 16 antennas at a frequency $4f_c$.

it points towards the mobile of interest. Fortunately, we can do that task electronically by combining the incoming signals in the correct way within the base station's uplink receiver, and by adjusting the outgoing signals within the base station's downlink transmitter. These techniques are a crucial way to improve the coverage of 5G at high radio frequencies, and we will discuss them as part of Chapter 6.

At the highest frequencies, however, this measure does not fully compensate for the increase in propagation loss. One reason is simply the extra cost and complexity of a large antenna array. In the 3400–3800 MHz band, for example, current macrocell designs call for base stations with around 64 antennas [21]. In the European 24.25–27.5 GHz band, the radio frequencies are around seven times greater and would require arrays containing over 3000 antennas to compensate for the propagation loss in full. That figure is too large to be feasible using current technology. We can improve the situation by using antenna arrays at the mobile as well, but the number of antennas is limited by its small size and by regulatory limits on its equivalent isotropic radiated power.

Another reason is the appearance of other propagation issues that degrade the coverage at high radio frequencies still further, which we will discuss in more detail in Section 4.3 [22, 23]. Bringing all of these issues together, we find that the cell sizes for millimetre waves are usually small, typically 100 m or so [24].

4.3 Radio Propagation Issues for Millimetre Waves

4.3.1 Diffraction and Reflection

Radio propagation in a mobile cellular network is more complex than it is in free space. Two issues are particularly important. Firstly, every point on the wavefront itself acts as a source of radio waves. That allows the waves to spread into the shadow behind an obstruction, in a phenomenon known as *diffraction*. Secondly, radio waves can be scattered or reflected by obstructions such as buildings or the ground. Combining these issues, we generally find that the received signal power falls away with distance more quickly in a cellular network than it does in free space. The relationship is roughly as follows:

$$PL \propto f^2 r^m \tag{4.10}$$

where PL is the propagation loss, and m is the *path loss exponent* whose value depends on the geometry of the cell but is typically around 3.5–4.

The amount of diffraction depends on the ratio of two quantities, namely the wavelength of the radio waves and the sizes of the obstructions. In the UHF band, the wavelength is between 10 cm and 1 m. At those wavelengths, there is usually enough diffraction to allow *non-line-of-sight* (NLOS) communication with a mobile behind a building, as illustrated in Figure 4.4a. At millimetre waves, the wavelength is shorter, and diffraction is weaker.

There is some compensation: millimetre waves tend to be scattered and reflected more strongly than waves in the UHF band. In a cellular network, we can sometimes carry out NLOS communication at millimetre waves by relying on scattering and reflection, as illustrated in Figure 4.4b. There are, however, limitations: for example, the reflections can be highly directional, which causes large variations in the received signal power as the mobile

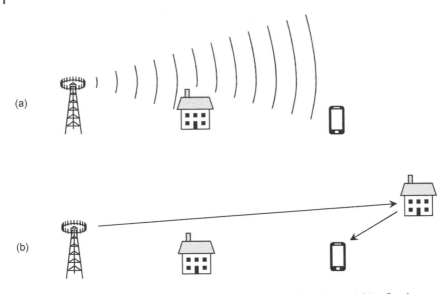

Figure 4.4 Non-line-of-sight communications using (a) diffraction and (b) reflection.

moves and can make it difficult to maintain the communication link. Because of these limitations, the compensation is only partial, and millimetre wave communications are often limited to the *line-of-sight* (LOS) case.

4.3.2 Penetration Losses

Penetration loss is the reduction in power as the signal travels through obstructions such as walls and windows. Penetration losses for millimetre waves can be high, for the same underlying reasons as the strong reflections that we discussed in Section 4.3.1. Some materials do not present a problem, but the situation is particularly severe for two cases: heavy building materials such as brick, stone and concrete; and energy-efficient metal-coated windows. These materials can have penetration losses of tens of decibels, which in some cases can increase rapidly as the frequency rises.

To illustrate the problem, Figure 4.5 shows measurements of penetration loss by Aalborg University, in a variety of building materials at frequencies from 1 to 18 GHz [25]. 3GPP has reported similar results at frequencies of 28, 39 and 73 GHz [26], and several other datasets have been collated by the ITU [27].

These penetration losses can make it difficult or even impossible to carry out outdoor-to-indoor communications using millimetre waves. There are three main alternatives: stick to lower frequencies, install a repeater or relay that receives the base station's downlink transmissions outside the building and re-transmits them inside, and install an indoor base station that controls a picocell or femtocell.

4.3.3 Foliage Losses

Another problem is foliage losses, in other words absorption by vegetation. To illustrate this, Figure 4.6 shows estimates of the foliage loss for four different depths of vegetation,

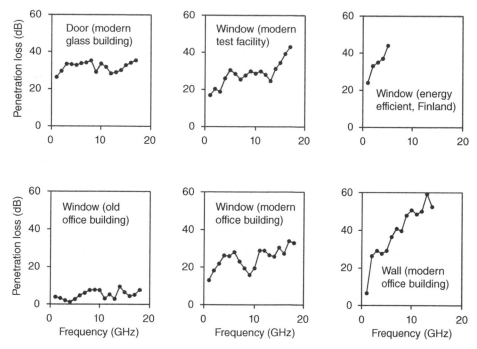

Figure 4.5 Example measurements of penetration loss at frequencies from 1 to 18 GHz. *Source:* Modified from Aalborg University/IEEE 2014.

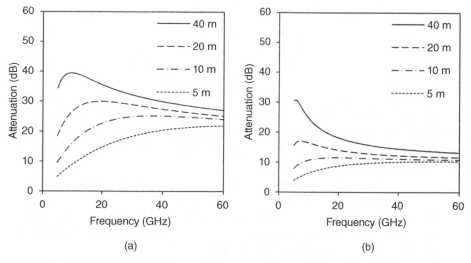

Figure 4.6 Estimates of foliage losses due to different depths of vegetation: An example illumination area of 2 square metres. (a) In leaf. (b) Out of leaf. *Source:* Modified from ITU–R P.833-4 (2003) Attenuation in vegetation.

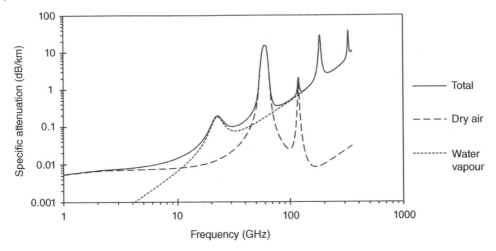

Figure 4.7 Estimates of atmospheric losses due to absorption by the air. *Source:* Modified from ITU-R P.676-10.

computed using an ITU model [28]. The model is only an approximate one, but suggests that foliage losses for millimetre waves might reach a few tens of decibels. Those losses are greater when the trees are in leaf, leading to differences in temperate climates between summer and winter.

4.3.4 Atmospheric Losses

Atmospheric absorption is not usually a problem in a cellular network, but it can occasionally be noticeable. There are two issues. Firstly, radio waves are absorbed by molecules of oxygen and water vapour at frequencies that change their rotational motion. To illustrate that issue, Figure 4.7 shows the losses that are computed using another ITU model [29, 30]. The results for water vapour assume an absolute humidity of $7.5\,\mathrm{g\,m^{-3}}$ which, at one atmosphere and 15 centigrade, implies a relative humidity of 58%.

At frequencies around 60 GHz, atmospheric absorption is severe, so those frequencies are unsuitable for wide-area 5G communications. However, they are ideal for shorter-range systems such as WiFi, in which atmospheric absorption helps to isolate a cell from its neighbours and reduce the levels of interference, so they are assigned to those technologies in the form of unlicensed spectrum.

Secondly, millimetre waves are scattered by raindrops, which have a similar size to the wavelength of the radio waves. That issue is illustrated in Figure 4.8, which is computed for three different rainfall rates using another ITU model [30, 31]. For comparison, the UK Meteorological Office classifies rain showers as slight, moderate, heavy and violent for rainfall rates of 0–2, 2–10, 10–50 and over $50\,\mathrm{mm\,h^{-1}}$ respectively.

Rainfall is generally not a problem in temperate climates if the cell is small enough. However, the rates of rainfall in tropical climates can be much greater and can sometimes reduce a cell's coverage area during a storm.

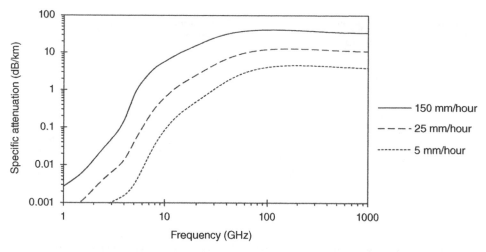

Figure 4.8 Estimates of atmospheric losses due to rainfall. *Source:* Modified from ITU-R P.838-3.

4.4 Multipath, Fading and Coherence

4.4.1 Introduction

As a result of reflections, rays can take several different routes from the transmitter to the receiver. This phenomenon is known as *multipath*. The number of significant rays tends to be larger at low radio frequencies than at high ones, for two reasons. Firstly, the coverage at low frequencies is greater, so long-distance reflections make a larger contribution to the received signal power. Secondly, the beams are wider, so rays can reach the receiver from a wider range of angles. In the case of millimetre waves, the propagation environment tends to be dominated by just a small number of rays, or even just one.

When the incoming rays reach the receiver, they can add together in different ways that are shown in Figure 4.9. If the peaks of the incoming rays coincide, then they reinforce each other, a situation known as *constructive interference*. If, however, the peaks of one ray coincide with the troughs of another, then the result is *destructive interference*, in which the rays cancel. Destructive interference can make the received signal power drop to a very low level, a situation known as *fading*.

If the mobile moves or the radio frequency changes, then the interference pattern can change between constructive and destructive, and the received signal can alternately fade and reappear. Figure 4.10 shows some examples. We will now examine these issues in more detail, using simple models to illustrate the principles that arise [32]. The results will be useful when we look at orthogonal frequency division multiple access (OFDMA) in Chapter 5 and at multiple antennas in Chapter 6.

4.4.2 Angular Spread and Coherence Distance

First, let us consider the situation in Figure 4.11, in which two mobiles are separated by a distance D. As a result of reflections, there are two incoming rays at each mobile, at angles

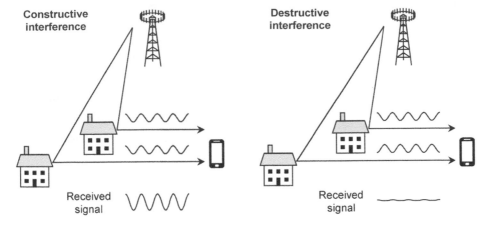

Figure 4.9 Generation of constructive interference, destructive interference and fading in a multipath environment.

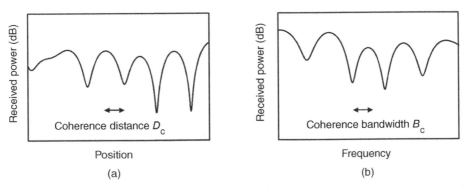

Figure 4.10 Examples of fading as a function of (a) position and (b) frequency.

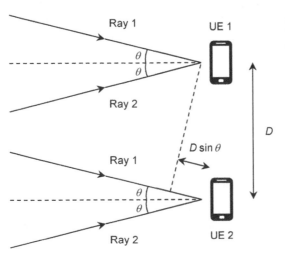

Figure 4.11 Example used to estimate the relationship between angular spread and coherence distance.

of $+\theta$ and $-\theta$ to the dashed horizontal line. (We assume that the base station is far away, so the equivalent rays at the two mobiles are parallel.) If we move from mobile 1 to mobile 2, then ray 1 becomes longer by a distance of $D \sin \theta$, while ray 2 shortens by the same amount. The interference pattern changes from constructive to destructive if:

$$2D \sin \theta = \frac{\lambda}{2} \tag{4.11}$$

so that:

$$D = \frac{\lambda}{4 \sin \theta} \tag{4.12}$$

We can interpret Equation (4.12) as a rough estimate of the *coherence distance*, in other words the greatest distance over which we can predict the amplitude and phase of the received signal. More accurate analyses use a precise definition of the coherence distance, as well as several incoming rays. Equation (4.13) is often used to estimate the distance over which the correlation between the incoming rays falls to 50% of its original value [33, 34]:

$$D_c \approx \frac{\lambda}{5\theta_{\text{rms}}} \tag{4.13}$$

where D_c is the coherence distance, and θ_{rms} expresses the angular spread of the incoming rays by means of its standard deviation, measured in radians. In a multipath environment, we can often assume that rays reach the mobile equally from all angles, a situation known as *Rayleigh fading*. In that situation, the coherence distance is the following:

$$D_c \approx \frac{9\lambda}{16\pi} \tag{4.14}$$

At a frequency of 3600 MHz, for example, the wavelength is about 83 mm. In the case of Rayleigh fading, the resulting coherence distance is about 15 mm. At the base station, the angular spread is usually less than it is at the mobile, so the coherence distance is greater. If, for example, the rays at the base station have an angular spread of 5°, then the coherence distance there is about 190 mm. In the case of LOS communication, the angular spreads can be very much smaller, and the coherence distances are even larger.

4.4.3 Doppler Spread and Coherence Time

If the receiver is moving with a speed v, then the coherence distance translates directly into a *coherence time*, in other words the greatest time interval over which we can predict the received signal's amplitude and phase. Equation (4.13) becomes the following:

$$T_c \approx \frac{1}{5f_D\theta_{\text{rms}}} \tag{4.15}$$

where T_c is the coherence time, and f_D is the maximum Doppler shift that can result at a speed v:

$$f_D = \frac{v}{c}f_C \tag{4.16}$$

where f_C is the carrier frequency, and c is the speed of light. In the case of Rayleigh fading, the mobile receives rays from all angles, which are spread over every Doppler shift between $-f_D$ and $+f_D$. The coherence time is then the following:

$$T_c \approx \frac{9}{16\pi f_D} \tag{4.17}$$

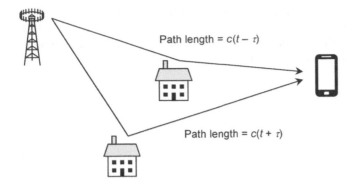

Figure 4.12 Example used to estimate the relationship between delay spread and coherence bandwidth.

As an example, a car might travel on a motorway at a speed of 30 ms^{-1} (108 km h^{-1}). At a carrier frequency of 700 MHz, the resulting Doppler spread is ±70 Hz, and the coherence time in the case of Rayleigh fading is about 2.6 ms. At higher carrier frequencies of 3.6 and 28 GHz, the respective Doppler spreads are ±360 and ±2800 Hz, and the corresponding coherence times are about 0.5 and 0.06 ms.

4.4.4 Delay Spread and Coherence Bandwidth

Finally, let us consider the situation in Figure 4.12. There are two incoming rays at the receiver, with propagation delays that we denote as $(t - \tau)$ and $(t + \tau)$, and corresponding path lengths of $c(t - \tau)$ and $c(t + \tau)$. The rays interfere constructively if the difference between the two path lengths is a whole number of wavelengths, such that:

$$2c\tau = n\lambda \tag{4.18}$$

where λ is the wavelength, and n is any integer. If the wavelength falls to λ', but the geometry remains the same, then the rays interfere destructively if:

$$2c\tau = \left(n + \frac{1}{2}\right)\lambda' \tag{4.19}$$

Combining Equations (4.18) and (4.19) yields the following result:

$$\delta f = \frac{1}{4\tau} \tag{4.20}$$

where δf is the change in carrier frequency over which the interference pattern changes from constructive to destructive, such that:

$$\delta f = \frac{c}{\lambda'} - \frac{c}{\lambda} \tag{4.21}$$

We can interpret Equation (4.20) as a rough estimate of the *coherence bandwidth*, in other words the greatest change in frequency over which we can predict the received signal's amplitude and phase. As before, a more accurate analysis, using a large number of rays, gives the following result:

$$B_c \approx \frac{1}{5\tau_{\text{rms}}} \tag{4.22}$$

where τ_{rms} is the standard deviation of the *delay spread*, in other words the difference between the rays' arrival times; and B_{c} is the coherence bandwidth. For example, the rays in a typical macrocell might vary in length by about 300 m, which gives a delay spread of around 1 μs and a coherence bandwidth of around 200 kHz. In a microcell, the delay spread is smaller, and the coherence bandwidth is larger. If, for example, the rays vary in length by only 15 m, then the delay spread falls to 50 ns, and the coherence bandwidth rises to around 4 MHz.

4.4.5 Channel Reciprocity

Channel reciprocity is a state in which we can predict the amplitude, phase and/or SINR of the received signal on the downlink from measurements that have been made on the uplink, and vice versa. For channel reciprocity to apply in full, we have to meet several conditions. Firstly, the radio frequencies of the uplink and downlink must be closer together than the coherence bandwidth. In practical terms, that usually means that the two frequencies must be the same. Secondly, the base station's transmit and receive antennas must be closer together than the coherence distance, and similarly for the antennas at the mobile. That usually means that the transmit and receive antennas must be the same, a conclusion that is particularly notable in a multiple-antenna system. Thirdly, the levels of interference on the uplink and downlink must be low, with the mobile close to the base station. Finally, the transceivers' uplink and downlink analogue processing chains must be properly balanced to ensure that they do not introduce any unexpected attenuations or phase shifts. That can be achieved by means of careful engineering design.

If channel reciprocity does apply, then it allows us to make some useful simplifications in the management of the air interface. We will highlight some of these in the chapters that follow.

References

1 Saunders, S. and Aragón-Zavala, A. (2007). *Antennas and Propagation for Wireless Communication Systems*, 2e. Chichester: Wiley.
2 Rappaport, T.S., Sun, S., Mayzus, R. et al. (2013). Millimeter wave mobile communications for 5G cellular: it will work! *IEEE Access* 1: 335–349.
3 ITU-R M.2376-0 (2015) Technical feasibility of IMT in bands above 6 GHz.
4 3GPP TS 38.104 (2019) NR; Base station (BS) radio transmission and reception (Release 15), December 2019, Section 5.1.
5 Ofcom (2016) UK Report on the outcome of the World Radiocommunication Conference 2015 (WRC-15). www.ofcom.org.uk/__data/assets/pdf_file/0018/50067/uk_report_of_wrc-15.pdf (accessed 18 January 2020).
6 ITU (2017) World Radiocommunication Conference 2019 (WRC-19) Agenda and relevant resolutions.
7 ITU (2019) WRC-19 identifies additional frequency bands for 5G. https://news.itu.int/wrc-19-agrees-to-identify-new-frequency-bands-for-5g (accessed 18 January 2020).
8 ITU (2019) World Radiocommunication Conference 2019 (WRC-19) provisional final acts, Resolutions COM4/7, COM4/8, COM4/9 and COM4/10.

9 Ofcom (2017) Update on 5G spectrum in the UK. www.ofcom.org.uk/__data/assets/pdf_file/0021/97023/5G-update-08022017.pdf (accessed 18 January 2020).

10 Ofcom (2018) Award of the 700 MHz and 3.6–3.8 GHz spectrum bands. www.ofcom.org.uk/__data/assets/pdf_file/0019/130726/Award-of-the-700-MHz-and-3.6-3.8-GHz-spectrum-bands.pdf (accessed 18 January 2020).

11 European Union (2017) Decision (EU) 2017/899 of the European Parliament and of the Council of 17 May 2017 on the use of the 470–790 MHz frequency band in the Union, Document 32017D0899. https://eur-lex.europa.eu/eli/dec/2017/899/oj (accessed 18 January 2020).

12 European Union (2019) Commission Implementing Decision (EU) 2019/235 of 24 January 2019 on amending Decision 2008/411/EC as regards an update of relevant technical conditions applicable to the 3 400–3 800 MHz frequency band, Document 32019D0235. https://eur-lex.europa.eu/eli/dec_impl/2019/235/oj (accessed 18 January 2020).

13 European Union (2019) Commission Implementing Decision (EU) 2019/784 of 14 May 2019 on harmonisation of the 24,25–27,5 GHz frequency band for terrestrial systems capable of providing wireless broadband electronic communications services in the Union, Document 32019D0784. https://eur-lex.europa.eu/eli/dec_impl/2019/784/oj (accessed 18 January 2020).

14 GSA (2019) Spectrum for Terrestrial 5G Networks: Licensing Developments Worldwide, August 2019. https://gsacom.com/paper/5g-global-spectrum-report-august-2019 (accessed 18 January 2020).

15 5G Americas (2019) 5G spectrum vision.

16 Lee, J., Tejedor, E., Ranta-aho, K. et al. (2018). Spectrum for 5G: global status, challenges, and enabling technologies. *IEEE Communications Magazine* 56 (3): 12–18.

17 Saunders, S. and Aragón-Zavala, A. (2007). *Antennas and Propagation for Wireless Communication Systems*, 2e. Chichester: Wiley, Section 4.5.8.

18 Balanis, C.A. (2009). *Antenna Theory: Analysis and Design*, 3e. Chichester: Wiley.

19 Saunders, S. and Aragón-Zavala, A. (2007). *Antennas and Propagation for Wireless Communication Systems*, 2e. Chichester: Wiley, Section 4.3.

20 Roh, W., Seol, J.-Y., Park, J. et al. (2014). Millimeter-wave beamforming as an enabling technology for 5G cellular communications: theoretical feasibility and prototype results. *IEEE Communications Magazine* 52 (2): 106–113.

21 GSM Association (2019) 5G implementation guidelines.

22 Rappaport, T.S., Heath, R.W., Daniels, R.C. et al. (2014). *Millimeter Wave Wireless Communications*. Upper Saddle River, NJ: Prentice Hall, Chapter 3.

23 Hemadeh, I.A., Satyanarayana, K., El-Hajjar, M. et al. (2018). Millimeter-wave communications: physical channel models, design considerations, antenna constructions, and link-budget. *IEEE Communications Surveys and Tutorials* 20 (2): 870–913.

24 ITU-R M.2376-0 (2015) Technical feasibility of IMT in bands above 6 GHz.

25 Rodriguez Larrad, I., Nguyen, H.C., Jørgensen, N.T.K. et al. (2014) Radio propagation into modern buildings: Attenuation measurements in the range from 800 MHz to 18 GHz. IEEE 80th Vehicular Technology Conference (VTC2014-Fall).

26 3GPP R1-161642 (2016) Building penetration loss measurement for mmWave with a range of materials.

27 ITU-R P.2346-2 (2017) Compilation of measurement data relating to building entry loss.

28 ITU-R P.833-4 (2003) Attenuation in vegetation.

29 ITU-R P.676-10 (2013) Attenuation by atmospheric gases, Annex 2.

30 ITU-R M.2376-0 (2015) Technical feasibility of IMT in bands above 6 GHz, Section 4.1.2.

31 ITU-R P.838-3 (2005) Specific attenuation model for rain for use in prediction methods.

32 Marzetta, T.L., Larsson, E.G., Yang, H. et al. (2016). *Fundamentals of Massive MIMO*. Cambridge: Cambridge University Press, Section 2.1.

33 Ghosh, A., Zhang, J., Andrews, J.G. et al. (2010). *Fundamentals of LTE*. Upper Saddle River, NJ: Prentice Hall, Section 3.4.

34 Du, K.-L. and Swamy, M.N.S. (2010). *Wireless Communication Systems: From RF Subsystems to 4G Enabling Technologies*. Cambridge: Cambridge University Press, Section 2.4.

5

Digital Signal Processing

This chapter addresses the principles of digital signal processing on the 5G air interface. There are three main topics. The first covers the underlying techniques for transmission and reception that 5G has inherited from 2G and 3G systems, for example modulation, demodulation and channel estimation. The second covers orthogonal frequency division multiple access (OFDMA), a technique inherited from LTE through which the base station can communicate with several different mobiles at the same time. The third covers the management of errors in the received data stream by means of forward error correction and re-transmission. We will also discuss a number of issues which appear at high radio frequencies and are especially relevant to 5G.

This chapter contains more mathematics than most of the others in this book, but it has been kept lightweight with the aim of making the material accessible to those without a mathematical background. The Bibliography has several references for further reading, for both the material in this chapter and the underlying maths.

5.1 Modulation and Demodulation

5.1.1 Carrier Signal

In a wireless communication system, the radio wave from Chapter 4 is known as a *carrier signal*. Mathematically, we can express the carrier signal as follows:

$$I(t) = a\cos(2\pi f_C t + \phi) \tag{5.1}$$

Here, a is the amplitude of the radio wave, ϕ is its initial phase angle and f_C is the carrier frequency. The angles are measured in radians, with π radians equal to 180°. It is usually easiest to manipulate this signal by treating it as the real part, or *in-phase component*, of the following complex number:

$$z(t) = I(t) + jQ(t) \tag{5.2}$$

Here, $z(t)$ is the complex representation of the signal at time t, and j is the square root of -1. $Q(t)$ is the imaginary part of that complex number, which is also known as the *quadrature component*:

$$Q(t) = a\sin(2\pi f_C t + \phi) \tag{5.3}$$

An Introduction to 5G: The New Radio, 5G Network and Beyond, First Edition. Christopher Cox.
© 2021 John Wiley & Sons Ltd. Published 2021 by John Wiley & Sons Ltd.

(a) QPSK waveform

(b) QPSK constellation

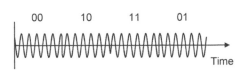

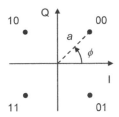

Figure 5.1 Quadrature phase shift keying (QPSK). (a) Example QPSK waveform. (b) QPSK constellation diagram.

By using that complex representation, we can exploit the following identity, which makes the calculations easier than they would otherwise be:

$$a \cos \phi + ja \sin \phi = a \exp(j\phi)$$
$$= ae^{j\phi} \tag{5.4}$$

Here, e is approximately 2.718, and the two right-hand sides are simply two different notations for the same quantity. The quadrature component is not actually transmitted, on the grounds that the real world does not understand imaginary numbers: it is just used for internal bookkeeping.

One use of the complex representation is to distinguish positive and negative frequencies. We cannot do that using real numbers alone, so negative frequencies do not appear in the real world, but we can using the real and imaginary parts of a complex number. More precisely, we can create a negative-frequency replica of our complex signal by changing the sign of its imaginary part, as follows:

$$z^*(t) = I(t) - jQ(t) \tag{5.5}$$

Here, $z^*(t)$ is the *complex conjugate* of $z(t)$, and is at a frequency of $-f_C$.

5.1.2 Modulation

A *modulator* encodes a sequence of bits onto the carrier signal by adjusting the parameters that describe it. In 5G, we choose to adjust the signal's amplitude a and/or its initial phase ϕ. Figure 5.1 shows an example modulation scheme known as *quadrature phase shift keying* (QPSK). A QPSK modulator takes the incoming bits two at a time and transmits them using a radio wave that can have four different states, which are known as *symbols*. Each symbol is described using two numbers. These can either be the amplitude and initial phase of the resulting radio wave, or the initial values of its in-phase and quadrature components:

$$I_0 = a \cos \phi$$
$$Q_0 = a \sin \phi \tag{5.6}$$

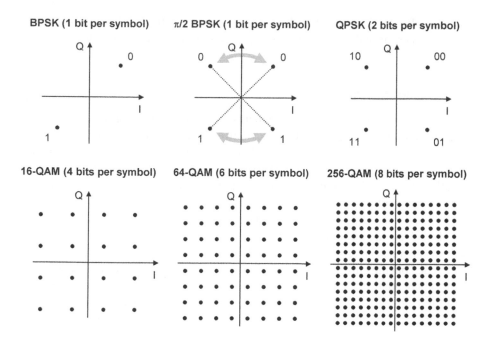

Figure 5.2 Modulation schemes used by 5G.

In QPSK, the symbols have the same amplitude and have initial phases of 45°, 135°, 225° and 315° (Figure 5.1a), which correspond to bit combinations of 00, 10, 11 and 01 respectively. We can also represent the four QPSK symbols using the *constellation diagram* shown in Figure 5.1b. In this diagram, the distance of each symbol from the origin represents the amplitude of the transmitted wave, while the angle (measured anti-clockwise from the x-axis) represents its initial phase.

As shown in Figure 5.2, 5G uses a number of different modulation schemes. *Binary phase shift keying* (BPSK) sends one bit at a time, using two symbols with initial phases of 45° and 225°. $\pi/2$-BPSK is a variant of BPSK, in which symbol phases of 45° and 225° alternate with phases of 135° and 315°. *16-quadrature amplitude modulation* (16-QAM) sends four bits at a time, using 16 symbols that have different amplitudes and phases. Similarly, 64-QAM and 256-QAM send six and eight bits at a time, using 64 and 256 different symbols respectively.

Modulation also affects the power spectrum of the transmitted signal. The original carrier was at a single frequency, so its power was confined to that frequency alone. However, the modulated signal has a different waveform from the carrier, so its power is spread over a range of frequencies that is known as the bandwidth. Roughly speaking, the bandwidth B and symbol duration T are related as follows:

$$B \approx \frac{1}{T} \tag{5.7}$$

If, for example, the duration of each symbol is 1 μs, then the symbol rate is 1 Msps (megasamples per second), and we can expect the transmitted signal to occupy a bandwidth of around 1 MHz.

5.1.3 The Modulation Process

Figure 5.3 shows the most important components of a modulator, using the example of a QPSK signal. In outline, the transmitter accepts two bits at a time from the higher-layer protocols, calculates the resulting symbols and modulates the carrier signal by mixing the symbols and carrier together. A real modulator also has other components, such as filters that smooth the abrupt phase transitions in the output signal, but we will ignore these so as to focus on the most important parts of the process.

In Figure 5.3, we have chosen to do the modulation process in two stages, by first mixing the symbols with an *intermediate frequency* (IF) carrier at a frequency f_{IF}, and then mixing the result with a much larger *radio frequency* (RF) carrier f_{RF}. The low-frequency calculations are carried out digitally, so there is a *digital-to-analogue* (D/A) converter between the two stages. There are two justifications for this choice: practical modulators often use this technique, and the concept will be useful when we discuss OFDMA later in this chapter. As a result of this process, the radio wave is eventually transmitted with the following carrier frequency:

$$f_C = f_{IF} + f_{RF} \tag{5.8}$$

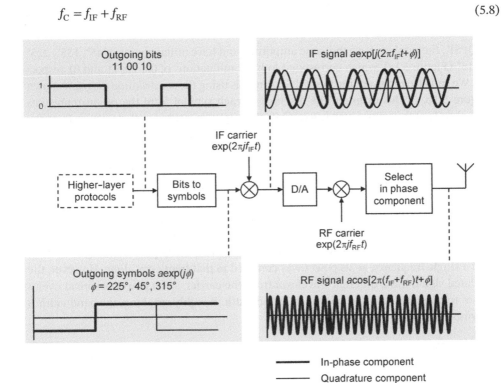

Figure 5.3 Block diagram of the modulator in a wireless communication system.

To mix two signals, we simply multiply their complex representations together, subject to the condition that $j^2 = -1$. Following that rule, the transmitted signal is as follows:

$$I(t) = \text{Re}[a\exp(j\phi)\exp(2\pi jf_{\text{IF}}t)\exp(2\pi jf_{\text{RF}}t)] \qquad (5.9)$$

where the three terms on the right-hand side represent the transmitted symbols and the IF and RF carriers, and Re[] is the real part of a complex number. We can then reach the following result:

$$I(t) = \text{Re}[a\exp(2\pi jf_C t + j\phi)]$$
$$= a\cos(2\pi f_C t + \phi) \qquad (5.10)$$

which is the signal that we require.

5.1.4 The Demodulation Process

The receiver accepts the incoming radio wave and recovers the bits by means of a *demodulator*. Figure 5.4 shows the most important parts of the demodulation process. In outline, the demodulator mixes the incoming signal with negative frequency replicas of the radio and intermediate frequencies, and digitizes it by means of an *analogue-to-digital*

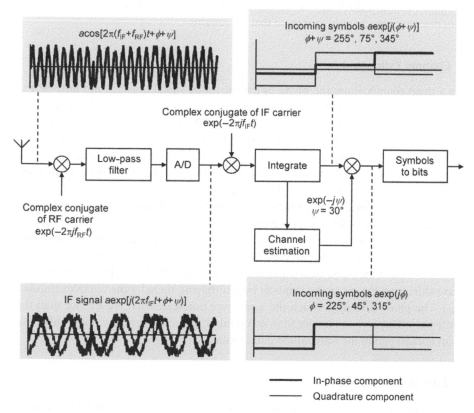

Figure 5.4 Block diagram of the demodulator in a wireless communication system.

(A/D) *converter*. By doing so, the demodulator mixes the incoming signal down to zero frequency, so as to recover the transmitted symbols and hence the transmitted bits.

There are a few complications. Firstly, the incoming signal may be distorted by thermal noise and by interference from other transmitters. To overcome that problem, the integration stage adds up the in-phase and quadrature components of the incoming symbols, I_0 and Q_0, for the duration of each individual symbol. During this time, the contributions from the signal itself accumulate while the contributions from noise and interference average out, so the received signal-to-interference plus noise ratio (SINR) improves.

The second complication arises because the incoming signal is a real quantity, which we can write as follows:

$$I(t) = a\cos(2\pi f_C t + \phi + \psi)$$
$$= \frac{a\exp[j(2\pi f_C t + \phi + \psi)] + a\exp[-j(2\pi f_C t + \phi + \psi)]}{2} \tag{5.11}$$

where ψ is a phase shift that we will discuss later in this chapter. This equation states that the incoming signal actually has two complex components, one with a positive frequency of f_C, and the other with a negative frequency of $-f_C$. After mixing the signal with the replica of the radio frequency, the first component moves to the desired intermediate frequency f_{IF}. However, the second component moves to a frequency of $-(f_{IF} + 2f_{RF})$ and causes rapid, unwanted fluctuations in the incoming signal. The low-pass filter removes those fluctuations, leaving only the signal that we require.

5.1.5 Channel Estimation

There is a third complication: the phase of the incoming signal depends not only on the phase of the transmitted signal, but also on the receiver's exact position. If the receiver moves through half a wavelength of the carrier signal (a distance of 5 cm at a carrier frequency of 3000 MHz, for example), then the phase of the received signal changes by 180°. When using QPSK, this phase change turns bit pairs of 00 into 11 and vice versa, and completely destroys the received information. In Equation (5.11), we expressed this issue by including an unknown phase shift ψ in the received signal.

To deal with the problem, the transmitter inserts occasional *demodulation reference symbols* into the data stream, which have a transmission time, amplitude and phase that are defined in the relevant specifications. In the receiver, a *channel estimation* function measures the incoming reference symbols, compares them with the ones that the specifications defined and estimates the phase shift ψ that the air interface introduced. It can then remove this phase shift from the incoming symbols by multiplying them by the complex number $\exp(-j\psi)$. The phase shift stays much the same over timescales that are shorter than the coherence time. That is much longer than one symbol, so the reference symbols only have to take up a small part of the transmitted data stream.

5.1.6 Adaptive Modulation

Our final complication concerns the effect of noise and interference. If they are large enough, then the receiver can mistake one incoming symbol for a different one, which results in bit errors. The error rate depends on two factors: the SINR at the receiver, and the choice of modulation scheme.

In a fast modulation scheme such as 256-QAM, the signal can be transmitted in many different ways, using states in the constellation diagram that are packed closely together. As a result, 256-QAM is vulnerable to errors and can only be used if the SINR is high. In contrast, QPSK only has a few states, so it is less vulnerable to errors and can be successfully used at a lower SINR. 5G exploits this by switching dynamically between different modulation schemes: it uses 256-QAM or 64-QAM at high SINR to give a high data rate, but falls back to 16-QAM or QPSK at lower SINR to reduce the number of errors.

5.2 Radio Transmission in a Mobile Cellular Network

5.2.1 Multiplexing and Multiple Access

The techniques described so far work well for one-to-one communications. However, a cellular network contains a large number of base stations, each of which has to communicate with a large number of different mobiles.

A device can handle multiple streams of traffic using a technique known as *multiplexing*, in which it shares the resources of the air interface amongst those different streams. Generalizing that idea, a base station can communicate with several different mobiles by means of *multiple access*. GSM used two implementations, known as *frequency division multiple access* (FDMA) and *time division multiple access* (TDMA), in which different mobiles communicate on different frequencies and at different times. 3G made use of *code division multiple access* (CDMA), which distinguishes the mobiles' signals by labelling them with codes. LTE and 5G both use orthogonal frequency division multiple access (OFDMA), which we will discuss in detail in Section 5.3.

5.2.2 FDD and TDD Modes

We still need a way to distinguish the mobiles' transmissions from those of the base stations. To achieve this, a mobile communication system can operate in two transmission modes (Figure 5.5). When using *frequency division duplex* (FDD), the base station and mobile transmit and receive at the same time, but use different carrier frequencies. Using *time division duplex* (TDD), they transmit and receive on the same carrier frequency but at different

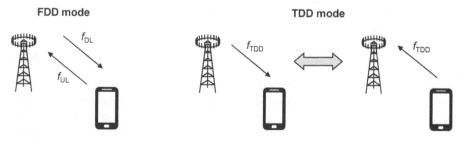

Figure 5.5 Operation of FDD and TDD modes.

times. Individual frequency bands are assigned either to FDD in the form of *paired spectrum*, in which there are separate allocations for the uplink and downlink, or to TDD in the form of *unpaired spectrum*.

FDD and TDD modes have different advantages and disadvantages. TDD mode can be badly affected by interference if, for example, one base station is transmitting while a nearby base station is receiving. We can minimize the problem by synchronizing the transmissions so that nearby base stations transmit at the same time, but that is a complication which has traditionally been avoided by the use of FDD. In the lower part of frequency range 1, most frequency bands have already been assigned to FDD mode, and 5G continues to use them.

At higher radio frequencies, there are fewer existing allocations, and two other issues become important. Firstly, the cells become progressively smaller and more isolated, so the interference problems fall away. That reduces the need for timing synchronization in TDD, for example by synchronizing nearby base stations only within small, isolated TDD hotspots. That in turn lets a TDD hotspot respond dynamically to changes in traffic levels, for example by increasing the amount of time that it allocates to the downlink when handling applications such as video downloads.

Secondly, high-frequency 5G systems make extensive use of multiple antennas. It is easier to control these systems in cases of channel reciprocity, for example by configuring the antennas on the downlink using measurements that have been previously made on the uplink. The discussion in Chapter 4 implies that channel reciprocity is only available in TDD mode. Because of these advantages, TDD mode is preferred in the upper part of frequency range 1, and is the only mode used in frequency range 2.

5.3 Orthogonal Frequency Division Multiple Access

5.3.1 Subcarriers

Earlier, we saw how a traditional communication system transmits data by modulating a carrier signal. Like LTE, 5G uses a modified version of this technique, known as *orthogonal frequency division multiplexing* (OFDM). As shown in Figure 5.6, an OFDM transmitter takes a block of symbols from the outgoing information stream, and transmits each symbol on a different radio frequency that is known as a *subcarrier*. The bandwidth of each individual subcarrier is small, so it can only support a low symbol rate. Collectively, however, the subcarriers occupy the same bandwidth as a traditional single-carrier system. If other issues remain unchanged, their collective symbol rate is the same.

In OFDM, the subcarrier spacing Δf is related to the symbol duration on each individual subcarrier, T, as follows:

$$\Delta f = \frac{1}{T} \tag{5.12}$$

For the moment this is just an arbitrary choice: the reasons will become clear later on in this chapter, but it is consistent with the idea that the subcarrier's bandwidth and symbol rate should be roughly the same. In LTE, the subcarrier spacing was fixed at 15 kHz, so the symbol duration was 66.7 μs. 5G can still use a 15 kHz subcarrier spacing, but it can use other values as well.

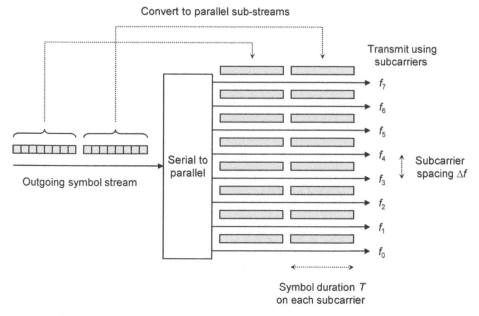

Figure 5.6 Division of the frequency band into subcarriers using OFDM.

On the basis of these figures, we would expect the symbol rate on each subcarrier to be identical with the subcarrier spacing, for example 15 ksps at a subcarrier spacing of 15 kHz. In practice the symbol rate is slightly less, because consecutive symbols are separated by a small gap known as the *cyclic prefix*. We will see the reasons for introducing the cyclic prefix in due course.

5.3.2 The OFDM Transmitter

Figure 5.7 shows the most important components of an OFDM transmitter, for an example subcarrier spacing of 15 kHz. The transmitter accepts a stream of bits from higher-layer protocols and converts them to symbols using the chosen modulation scheme, for example QPSK. The serial-to-parallel converter then takes a block of N symbols, eight in this example, and directs them onto N parallel sub-streams.

The transmitter mixes each sub-stream with a subcarrier, whose frequency is an integer multiple of the subcarrier spacing. By analogy with the modulator from earlier in this chapter, we can interpret each subcarrier as an intermediate frequency, whose modulated signal will eventually be transmitted at a slightly different frequency from the carrier. For consistency with the 5G specifications, it is convenient to place these subcarriers at frequencies of -60, -45, $\cdots$, $+45$ kHz so that the carrier frequency is roughly in the centre of the transmission band. The symbol duration is the reciprocal of the subcarrier spacing, so the 15 kHz subcarrier goes through one cycle during each 66.7 μs symbol, while the subcarriers at 30 and 45 kHz go through two and three cycles respectively. (If the subcarrier spacing is different from 15 kHz, then these figures all scale accordingly.)

We now have eight sine waves, whose amplitudes and phases represent the eight transmitted symbols. By adding these sine waves together and dividing by a scale factor of N, we

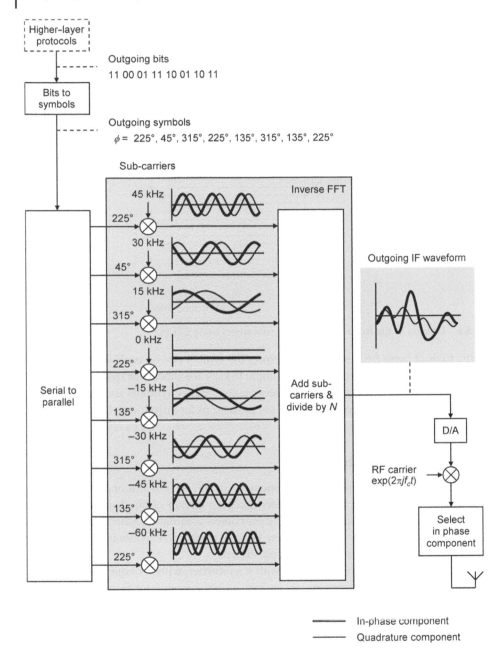

Figure 5.7 Processing steps in an OFDM transmitter.

can generate a single time-domain waveform, which is an intermediate-frequency representation of the signal that we need to send. The only remaining task is to mix the waveform up to radio frequency for transmission. We can therefore interpret the OFDM transmitter as a bank of intermediate-frequency modulators, each of which is tuned to the frequency offset of the corresponding subcarrier.

5.3.3 The OFDM Receiver

Figure 5.8 shows the most important components of an OFDM receiver. We can understand how it works by analogy with the demodulator from earlier in this chapter. The receiver accepts the incoming signal, mixes it with a complex conjugate replica of the original carrier

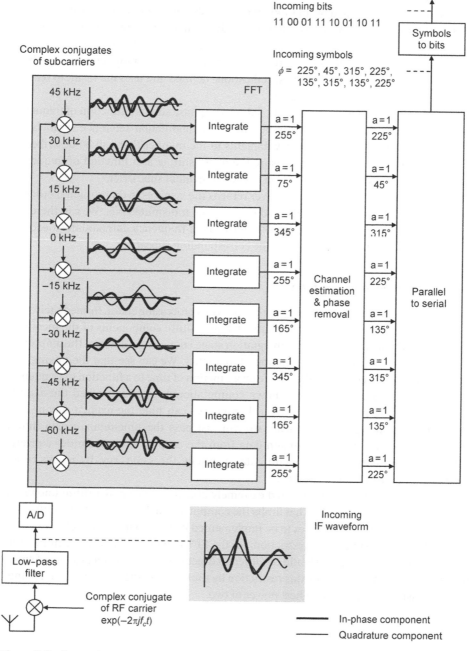

Figure 5.8 Processing steps in an OFDM receiver.

frequency and filters the result. It then directs the signal to eight separate paths and mixes each one with a complex conjugate replica of one of the original subcarriers. The information that arrived on that subcarrier is now at a frequency of zero. By integrating the result for the duration of one symbol, the receiver can extract the in-phase and quadrature components of the symbol that arrived on that subcarrier, while rejecting any noise and interference.

Each subcarrier is modified by an unknown phase shift ψ before it reaches the receiver. To deal with this, the OFDMA transmitter injects demodulation reference symbols into the data stream with an amplitude and phase that are defined in the relevant specifications. In the process of channel estimation, the receiver measures the incoming reference symbols, compares them with the ones that the specifications defined and uses the result to remove the phase shift from the incoming signal.

In the presence of frequency-dependent fading, the phase shift is a function of frequency as well as time. To account for this issue, the 5G demodulation reference symbols are scattered across the time and frequency domains in the manner that will be described in Chapter 11. The receiver can then measure the phase shift ψ as a function of frequency, and can apply a different phase correction to each individual subcarrier.

After the phase removal stage, the receiver returns the symbols to their original order by means of a parallel-to-serial converter, and recovers the transmitted bits. We can therefore interpret the OFDM receiver as a bank of intermediate-frequency demodulators, each of which is tuned to the frequency offset of the corresponding subcarrier.

5.3.4 The Fast Fourier Transform

Now let us look at two important points in the receiver's processing chain. Immediately after the low-pass filter, the data are the in-phase and quadrature components of the incoming signal, as a function of time. After the integration stage later on, the data are the amplitude and phase of each subcarrier, as a function of frequency. We can see that the mixing and integration steps have converted the data from a function of time to a function of frequency.

This conversion is actually a well-known computational technique called the *discrete Fourier transform* (DFT). By using this technique, we can hide the explicit mixing and integration steps in Figure 5.8: instead, we can just pass the time-domain data into a DFT and pick up the frequency-domain data from the output. The transmitter converts frequency-domain data to time-domain data in exactly the same way, by means of an *inverse discrete Fourier transform* (IDFT).

In turn, the DFT can be implemented extremely quickly using an algorithm known as the *fast Fourier transform* (FFT). That limits the computational load on the transmitter and receiver, and allows the two devices to be implemented in a computationally efficient way. However, there is one important restriction: for the FFT to work efficiently, the number of data points in each calculation should be either an exact power of two or a product of small prime numbers alone. We handle this restriction by rounding up the number of data points in the FFT, usually to the next highest power of two.

5.3.5 Block Diagram of the OFDMA Downlink

Figure 5.9 is a block diagram of an OFDMA transmitter and receiver. We assume that the system is operating on the downlink, so that the transmitter is in the base station and the receiver is in the mobile.

The base station is sending streams of bits to three different mobiles, so the diagram is an implementation of OFDMA. The base station modulates each bit stream independently, possibly using a different modulation scheme for each mobile. It then passes each symbol stream through a serial-to-parallel converter to divide it into sub-streams. The number of sub-streams per mobile depends on the mobile's required data rate: for example, a voice application might only require a few sub-streams, while a video application might require many more.

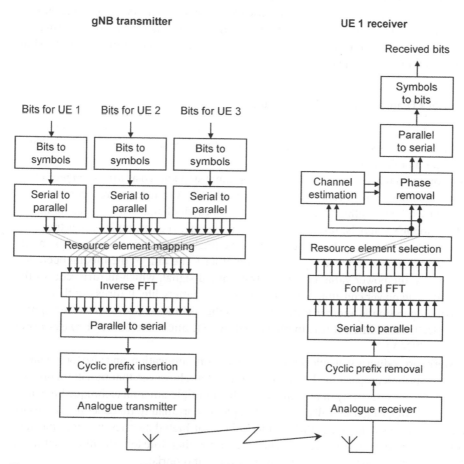

Figure 5.9 Block diagram of the OFDMA downlink.

The *resource element mapper* takes the individual sub-streams and chooses the subcarriers on which to transmit them. A mobile's subcarriers may lie in one contiguous block (as in the case of mobiles 1 and 3), or they may be divided (as for mobile 2). The resulting information is the amplitude and phase of each subcarrier as a function of frequency. By passing it through an inverse FFT, we can compute the in-phase and quadrature components of the corresponding time-domain waveform, and can place them in the correct order by means of a parallel-to-serial converter (a stage that was hidden within Figure 5.7). After adding the cyclic prefix, the resulting signal can be mixed up to radio frequency and converted to analogue form for transmission.

The mobile reverses the process. It starts by sampling the incoming signal, converting it down to baseband and filtering it. It then passes the data through a forward FFT, to recover the amplitude and phase of each subcarrier. We now assume that the base station has already told the mobile which subcarriers to use, through scheduling techniques that we will cover in Chapter 11. Using this knowledge, the mobile selects the required subcarriers and recovers the transmitted information, while discarding the remainder. After the steps of channel estimation and phase removal, the mobile can recover the transmitted bits and can pass them to the higher-layer protocols.

5.3.6 Block Diagram of the OFDMA Uplink

The uplink (Figure 5.10) differs from the downlink in two ways. The first is that the mobile transmitter only uses some of the subcarriers: the others are set to zero, and are available for the other mobiles in the cell.

The second difference is the inclusion of an optional step known as *transform precoding*, which is implemented by means of another forward FFT. A transform precoder mixes the transmitted symbols together, to ensure that each subcarrier handles a linear combination of all the mobile's symbols, rather than just a single one. (For example, when transmitting two symbols x_1 and x_2 on two subcarriers, we might send their sum $(x_1 + x_2)$ on one subcarrier, and their difference $(x_1 - x_2)$ on the other.) The aim of the mixing operation is to minimize the power variations in the transmitted signal, which in turn reduces the mobile's average power consumption and increases its battery life. It turns out that a suitable mixing operation is another forward FFT, so the technique is known as *DFT spread OFDMA* (DFT-s-OFDMA). It was implemented in LTE under the name of *single-carrier FDMA* (SC-FDMA).

Transform precoding is valuable for machine-type communication, and also for scenarios in which the uplink propagation loss is large. It turns out, however, that transform precoding does not work effectively in conjunction with a multiple-antenna technique known as single-user MIMO, which we will discuss in Chapter 6. Fortunately, that does not cause any problems, because single-user MIMO is a technique for high data rates, whereas the data rates for machine-type devices are low. The base station therefore tells the mobile either to use the transform precoder or to omit it, whichever is appropriate.

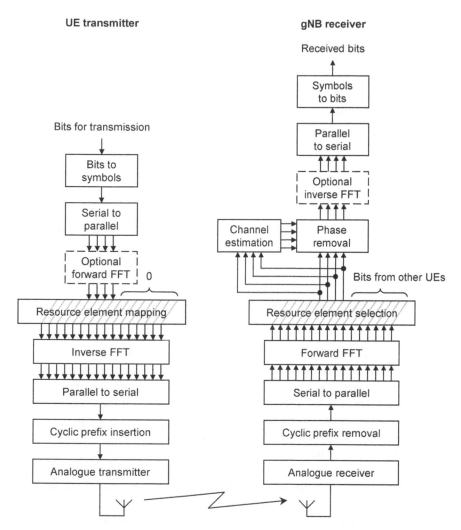

Figure 5.10 Block diagram of the OFDMA uplink.

5.4 Other Features of OFDMA

5.4.1 Frequency-specific Scheduling

Earlier, we saw that a base station can communicate with several mobiles at the same time by allocating them different groups of subcarriers. The base station's exact choice of subcarriers is influenced by the existence of frequency-dependent fading, in the manner shown in Figure 5.11.

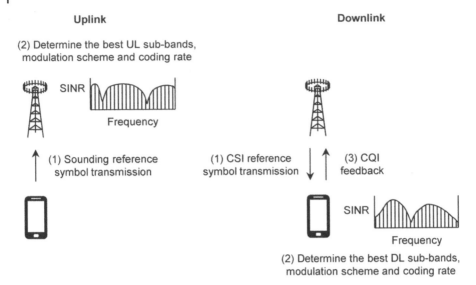

Figure 5.11 Frequency-specific scheduling in 5G.

On the uplink, the mobile transmits additional reference symbols that are known as *sounding reference symbols*. These are similar to the demodulation reference symbols from earlier, but differ in two ways: they are only transmitted occasionally, and they are transmitted over a much wider bandwidth. The base station measures the incoming sounding reference symbols and uses the information in three ways: to schedule the mobile using the block of uplink subcarriers on which the received SINR is the strongest; to determine the fastest uplink modulation scheme that the mobile can handle; and to determine a related quantity known as the coding rate that we will introduce later in this chapter. (The unwanted subcarriers can be used by other mobiles, which are in different locations and therefore have different fading patterns.)

There is a similar procedure on the downlink. The base station transmits reference symbols that are known as *channel state information* (CSI) *reference symbols*, over a much wider bandwidth than the demodulation reference symbols, but only occasionally. The mobile measures the incoming reference symbols, deduces the blocks of downlink subcarriers on which the received SINR is the strongest, and determines the fastest downlink modulation scheme and coding rate that it can handle. The information is actually needed by the base station, so the mobile feeds it back using a quantity known as the *channel quality indicator* (CQI), for use during downlink scheduling.

In cases of channel reciprocity, the best choices of uplink and downlink subcarriers are the same, so shortcuts are possible. For example, the base station might determine the best choice of uplink subcarriers by measuring the sounding reference symbols, and then apply that same choice to the downlink. (It might still use the CSI reference symbols to determine the best choice of downlink modulation scheme and coding rate, on the grounds that the base station's transmit power is greater than that of the mobile.)

5.4.2 Subcarrier Orthogonality

Figure 5.12 adds some detail to our treatment of the OFDMA receiver, using an example in which information is arriving on the 15 kHz subcarrier alone. When the incoming signal is mixed with a complex conjugate replica of the 15 kHz subcarrier, the result is at a frequency of zero. The integration process then extracts the amplitude and phase of that subcarrier, as required.

In the branch of the receiver that lies one level below, the incoming signal is mixed with a complex conjugate replica of the 0 kHz subcarrier, so the result remains at a frequency of 15 kHz. During the integration period, the resulting signal goes through exactly one cycle, so its in-phase and quadrature components both sum to zero. We therefore conclude that no information has arrived on the 0 kHz subcarrier at all. The same argument applies to all the other subcarriers in the receiver, so we can draw the following conclusion: if the transmitter sends a signal on one subcarrier, then the receiver picks up the signal on that one subcarrier alone, and does not pick up interference on any of the others. Subcarriers with this property are said to be *orthogonal*.

That argument works fine if the mobile is stationary. If the mobile is moving, and the propagation environment is line-of-sight, then the incoming signal is Doppler shifted to a higher or lower frequency. In Chapter 4, for example, we estimated that a moving vehicle might have a Doppler shift of about 360 Hz, so the 15 kHz subcarrier actually arrives at a frequency of 15.360 kHz. When the signal is mixed with a complex conjugate replica of the 0 kHz subcarrier, the result remains at 15.360 kHz. During the integration period, that signal goes through slightly more than one cycle, so its in-phase and quadrature components sum to a result that is no longer zero. We therefore receive interference on the 0 kHz subcarrier – and, by extension, all of the others – so we have lost orthogonality.

The situation is worse in a multipath environment, because a mobile can be moving towards some rays which are shifted to higher frequencies, but away from others whose frequencies are lower. As a result, the incoming signals are not simply shifted: instead, they are blurred across a range of frequencies. Even so, the amount of interference will still be acceptable if the subcarrier spacing is much greater than the Doppler shift. We therefore need to choose the subcarrier spacing Δf as follows:

$$\Delta f >> f_D \tag{5.13}$$

where f_D is the Doppler shift from Equation (4.16). If the mobile is fast-moving or the carrier frequency is high, then the subcarrier spacing has to be wide.

5.4.3 Inter-symbol Interference and the Cyclic Prefix

In a multipath environment, different rays can have different propagation delays, so the symbols that travel on those rays can reach the receiver at different times. The symbols then overlap at the receiver, causing a problem known as *inter-symbol interference* (ISI) that is illustrated in Figure 5.13a.

We can prevent the creation of ISI by inserting a *guard period* before each symbol, in which nothing is transmitted. If the guard period is longer than the delay spread, then the receiver can be confident of reading information from just one symbol at a time, without

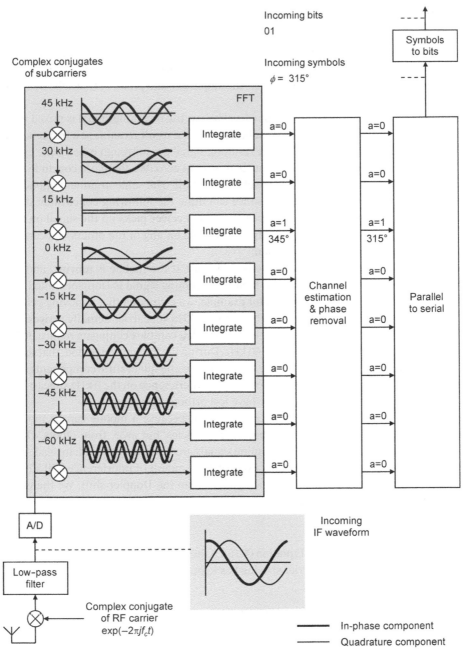

Figure 5.12 Processing steps in an OFDM receiver, in which the information arrives on a 15 kHz subcarrier alone.

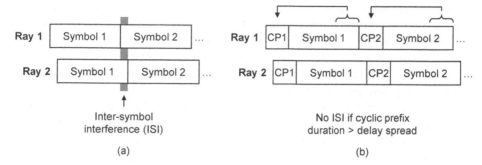

Figure 5.13 Inter-symbol interference (ISI). (a) Creation of ISI in a multipath environment. (b) Avoidance of ISI by means of the cyclic prefix.

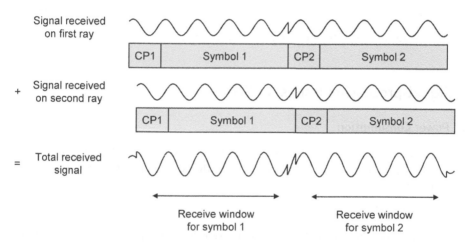

Figure 5.14 Operation of the cyclic prefix on a single subcarrier.

any overlap with the symbols that precede or follow. In practice, 5G uses a more complex technique that is known as *cyclic prefix* (CP) *insertion*. As shown in Figure 5.13b, the transmitter starts by inserting a guard period before each symbol, but then copies data from the end of the symbol following so as to fill up the guard period.

We can see how cyclic prefix insertion works by looking at one subcarrier (Figure 5.14). The transmitted signal is a sine wave whose amplitude and phase change from one symbol to the next. Each symbol contains an exact number of cycles of the sine wave, so the amplitude and phase at the start of each symbol equal the amplitude and phase at the end. Because of this, the transmitted signal changes smoothly as we move from each cyclic prefix to the symbol following.

In a multipath environment, the receiver picks up multiple copies of the transmitted signal with multiple arrival times. These add together at the receive antenna, giving a sine wave with the same frequency but a different amplitude and phase. The received signal still changes smoothly at the transition from a cyclic prefix to the symbol that follows. There are a few glitches, but these are only at the start of the cyclic prefix and the end of the symbol, where the preceding and following symbols interfere.

The receiver processes the received signal within a window whose length equals the symbol duration, and discards the remainder. If the window is correctly placed, then the received signal is exactly what was transmitted, without any glitches, and subject only to an amplitude change and a phase shift. But the receiver can compensate for these by using the channel estimation techniques described earlier in this chapter, so it can handle the cyclic prefix without any extra processing at all.

Admittedly the system uses multiple subcarriers, not just one. However we have already seen that the subcarriers do not interfere with each other and can be treated independently, so the existence of multiple subcarriers does not affect this argument at all.

The existence of ISI places another constraint on the parameters used by 5G. The symbols should be much longer than the cyclic prefixes, which in turn should be longer than the delay spread τ. We therefore need to choose the symbol duration T as follows:

$$T >> \tau \tag{5.14}$$

If the delay spread is large, then the symbol duration has to be long.

5.5 Signal-processing Issues for 5G

5.5.1 Power Consumption

Power consumption can be a problem in a millimetre wave receiver. To illustrate that issue, References [1, 2] quote the following rule of thumb for the power consumed by the receiver's A/D converters:

$$\text{Power} \propto \text{Speed} \times \text{Accuracy}^2 \tag{5.15}$$

In this relationship, the speed of the A/D converter depends upon its digitization rate, and therefore on the bandwidth of the radio signal. The accuracy depends on the number of bits to which the incoming signal is resolved, while the constant of proportionality depends on the underlying manufacturing technology. There is a similar relationship for the D/A converters in the transmitter, but those are simpler devices, so their power consumption is lower.

At the wide bandwidths used by millimetre waves, the resulting power consumption can be high enough to be a problem [3]. The problem is lessening as the technology improves [4], but in the meantime there are two solutions. The first is to limit the number of bits resolution in the receiver's A/D converters, at the expense of increasing the quantization noise.

The second solution is to limit the number of A/D and D/A converters that the system is using. Recalling that multiple antennas are an important technique for millimetre waves, we prefer to avoid the use of millimetre wave transceivers in which each antenna has an A/D or D/A converter of its own. Instead, we prefer receivers that combine the incoming signals from several antennas using purely analogue techniques and only then digitize them, and we prefer a similar design in the transmitter. In Chapter 6, that will drive the use of analogue or hybrid beamforming techniques for millimetre waves, and it will limit the use of purely digital techniques to low radio frequencies.

5.5.2 Timing Jitter and Phase Noise

Earlier, we saw that the transmitter and receiver both contain local oscillators, which generate the carrier signals that are required for modulation and demodulation. The signals from those local oscillators are not perfect sine waves: instead, they are subject to random timing jitter whose size is roughly independent of the carrier frequency. At high frequencies, that causes two problems.

Firstly, the timing jitter results in *phase noise*, in other words random errors in the phase of the received symbols. That in turn leads to bit errors at the output from the receiver's demodulator. If the carrier frequency increases but the timing jitter remains roughly the same, then the phase noise becomes progressively more severe.

When using OFDMA, the phase errors on nearby subcarriers are almost identical, so the problem is often known as *common phase error*. We can compensate for it by introducing *phase-tracking reference symbols*, which have a similar role to the demodulation reference symbols from earlier in this chapter, but differ in two ways. Firstly, they have a high density as a function of time, to help the receiver track the rapid phase variations in the incoming signal. Secondly, they have a low density as a function of frequency, on the grounds that nearby subcarriers behave in much the same way.

The second problem arises because frequency is the rate of change of phase. If there are phase errors in the transmitter and the receiver, then there are frequency errors as well, so the receiver is no longer tuned to the exact frequency that the transmitter is using. In an OFDMA system, those frequency errors destroy the orthogonality between the subcarriers, in the same way as the Doppler shifts from Section 5.4.2. That in turn leads to inter-frequency interference.

This second problem appears in the analogue parts of the receiver, so the phase-tracking reference symbols do not help at all. However, we can lessen its impact by the use of a wide subcarrier spacing, which reduces the amount of inter-frequency interference that arises for a given frequency error.

5.5.3 Choice of Symbol Duration and Subcarrier Spacing

We have now seen three constraints on the symbol duration and subcarrier spacing. Firstly, we prefer a wide subcarrier spacing and a short symbol duration at high carrier frequencies, to reduce the inter-frequency interference that arises due to local oscillator timing jitter. Secondly, we also prefer a wide subcarrier spacing at high carrier frequencies, to reduce the impact of Doppler shifts from fast-moving mobiles.

Thirdly, we prefer a narrow subcarrier spacing and a long symbol duration if the delay spread is large, to prevent inter-symbol interference from late incoming rays. This is mainly an issue at low radio frequencies, in which the cells are large and the base station is using a wide-angle beam. It is less important at high frequencies, because the cells are smaller and because the number of reflected rays is reduced by the use of beamforming.

In LTE, we could meet all these constraints using a fixed subcarrier spacing of 15 kHz, and a fixed symbol duration of 66.6 µs. In 5G, the largest carrier frequencies are 10–20 times greater, which makes it harder to meet the first two constraints. That in turn makes the use of a fixed subcarrier spacing impossible.

Instead, 5G uses an adjustable subcarrier spacing. The usual values are 15, 30, 60 and 120 kHz, so the corresponding symbol durations range from 66.7 down to 8.33 μs. Narrow subcarrier spacings are preferred at low carrier frequencies, and wide subcarrier spacings are preferred if the carrier frequency is high. We also prefer wide subcarrier spacings when carrying out low-latency communications, because the short symbol duration allows many of the other timing intervals to be shortened as well. We will examine these issues further in Chapter 7.

5.6 Error Management

5.6.1 Forward Error Correction

In the earlier sections of this chapter, we saw that noise and interference lead to errors in a wireless communication receiver. These are not much of a problem during voice and video calls, but are damaging for other applications such as web browsing, emails, machine-type communications and especially ultra-reliable low-latency communication (URLLC). Fortunately, there are several ways to solve the problem.

The most important technique is *forward error correction*. In this technique, the transmitted information is represented using a *codeword* that typically contains two or three times as many bits. The extra bits supply additional, redundant data that help the receiver to recover the original information sequence. For example, a transmitter might represent the information sequence 101 using the codeword 110010111. After an error in the second bit, the receiver might recover the codeword 100010111. If the coding scheme has been well designed, then the receiver can conclude that this is not a valid codeword, and that the most likely transmitted codeword was 110010111. The receiver has therefore corrected the bit error and can recover the original information. The effect is very like written English, which contains redundant letters that allow the reader to understand the underlying information, even in the presence of spelling mistakes.

The *coding rate* is the number of information bits divided by the number of transmitted bits (1/3 in the example above). Changes in the coding rate have a similar effect to changes in the modulation scheme. If the coding rate is low, then the transmitted data contain many redundant bits. That allows the receiver to correct a large number of errors and to operate successfully at a low SINR, but at the expense of a low information rate. If the coding rate is close to 1, then the information rate is higher but the system is more vulnerable to errors. 5G exploits this with a similar trade-off to the one we saw earlier, by transmitting with a high coding rate if the received SINR is high and vice versa.

Many forward error correction algorithms can only handle a limited set of coding rates. Even so, a transmitter can adjust the coding rate further by processing the information in two stages. In the first stage, the information bits are passed through a coder as before. In the second stage, known as *rate matching*, some of the coded bits are selected for transmission, with the exact choice depending on a parameter known as the *redundancy version* (RV). The others are discarded in a process known as *puncturing*, and look just like bit errors. The receiver knows the redundancy version, so it can insert dummy bits at the points where information was discarded. It can then pass the result through the corresponding decoder for error correction.

5.6.2 Automatic Repeat Request

Automatic repeat request (ARQ) is another error management technique, in which the transmitter takes a block of information bits and uses them to compute some extra bits that are known as a *cyclic redundancy check* (CRC). It appends these to the information bits, and then transmits the two sets of data in the usual way.

In a process known as *error detection*, the receiver separates the two fields and uses the information bits to compute the expected CRC bits. If the observed and expected CRC bits are the same, then it concludes that the information has been received correctly and returns a positive acknowledgement (ACK) to the transmitter. If the CRC bits are different, it concludes that an error has occurred and returns a negative acknowledgement (NACK). On receiving a NACK, the transmitter sends the original block of information again, and the process continues until the information arrives correctly.

5.6.3 Hybrid ARQ

A wireless communication system often combines these two error management techniques in a technique known as *hybrid acknowledgement request* (hybrid ARQ or HARQ). The simplest implementations correct most of the bit errors by the use of forward error correction, and then use re-transmissions to handle the remaining errors that leak through.

Like LTE, 5G uses a more powerful implementation known as *hybrid ARQ with soft combining*, which is illustrated in Figure 5.15. The transmitter takes a block of information bits, adds a CRC and computes the corresponding codeword, as before. The rate-matching algorithm stores the codeword in a buffer and uses the redundancy version to select just some of the coded bits for transmission. The modulator then computes the corresponding symbols.

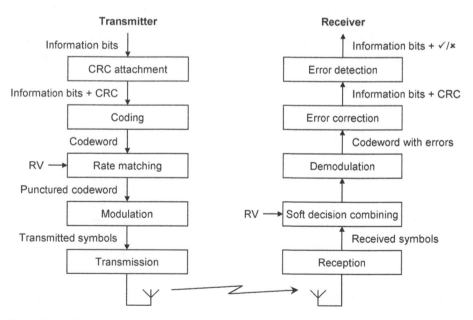

Figure 5.15 Block diagram of a transmitter and receiver using hybrid ARQ with soft combining.

(To simplify the modulation procedure, the rate-matching algorithm either selects all of the bits that map onto a particular symbol or selects none of them, so each symbol is transmitted either in its entirety or not at all.)

In the stage of soft decision combining, the receiver stores the incoming symbols in the correct places in a reception buffer, using its knowledge of the redundancy version to insert zeros for the symbols that have not yet arrived. Having done so, the receiver demodulates the buffered symbols, so as to compute the bits of the corresponding codeword. It then passes the codeword through the stages of error correction and error detection, and sends an acknowledgement back to the transmitter.

If the CRC fails, then the transmitter sends the data again, using a different redundancy version so as to supply at least some of the symbols that were originally missing. The receiver can then combine the soft decisions from the first transmission and the re-transmission simply by adding the corresponding symbols together. That increases the signal energy at the receiver, so it increases the likelihood of a CRC pass. As a result, this scheme performs better than the simplest implementation of hybrid ARQ, in which an unsuccessful transmission is discarded.

5.6.4 Hybrid ARQ Processes

Normally, hybrid ARQ uses a re-transmission technique called *stop-and-wait*, in which the receiver sends its acknowledgement right away, and the transmitter waits for that acknowledgement to arrive before sending new data or a re-transmission. That leads to a simple design with low time delays, but it means that the transmitter has to pause while waiting for the acknowledgement to arrive. To prevent the throughput from falling, the system shares its data amongst several *hybrid ARQ processes*, which are multiple copies of Figure 5.15. One process can then transmit while the others are waiting for acknowledgements, in the manner shown in Figure 5.16.

The use of multiple hybrid ARQ processes means that the receiver decodes data blocks in a different order from the one in which they were transmitted. In Figure 5.16, for example,

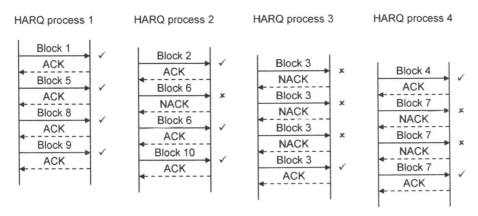

Figure 5.16 Operation of a stop-and-wait re-transmission scheme using multiple hybrid ARQ processes.

block 3 is transmitted four times and is only decoded a long time after block 4. To deal with that problem, the receiver includes a re-ordering function that accepts the decoded blocks and returns them to their initial order.

In 5G, hybrid ARQ is implemented in the air interface's physical layer, and introduces time delays of a few milliseconds into the received data stream. That is acceptable for all forms of mobile broadband application, including real-time voice and video. In the case of URLLC, however, there is only enough time to carry out a few hybrid ARQ transmissions, or even just one. To solve the problem, URLLC applications correct most or all of their errors by means of forward error correction, and use a slow modulation scheme and a low coding rate to ensure that the reliability of each individual transmission is high.

5.6.5 Higher-layer Retransmissions

Hybrid ARQ is not completely reliable, mainly because the receiver applies little or no error protection when returning its acknowledgements [5]. Sometimes, the receiver returns a NACK, but there is a bit error, and the transmitter misinterprets the acknowledgement as an ACK. It then moves on to the next transmission, and the original information is lost.

5G solves that problem by using a second procedure which re-transmits the original information bits if the underlying hybrid ARQ procedure fails. This second procedure uses a technique called *selective re-transmission* (Figure 5.17), in which the receiver waits for several blocks of data to arrive before acknowledging them all. (The acknowledgements are protected by means of hybrid ARQ, so their reliability is high.) The transmitter can continue sending data without waiting for an acknowledgement, but any re-transmitted data can take a long time to arrive.

In 5G, this second level of re-transmission is implemented in a higher-layer protocol known as radio link control. It introduces time delays of several tens of milliseconds, which are acceptable for non-real-time applications such as web pages, emails and machine-type communications. However, the time delays are too high for real-time applications such as conversational voice and video, which operate using hybrid ARQ alone.

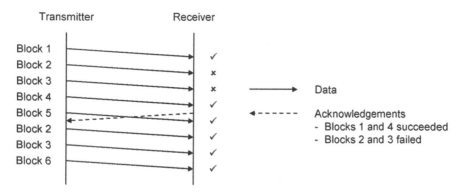

Figure 5.17 Operation of a selective re-transmission scheme.

References

1 Rappaport, T.S., Heath, R.W., Daniels, R.C. et al. (2014). *Millimeter Wave Wireless Communications*. Prentice Hall, Section 6.5.

2 Uyttenhove, K. and Steyaert, M.S.J. (2002). Speed–power–accuracy tradeoff in high-speed CMOS ADCs. *IEEE Transactions on Circuits and Systems II: Analog and Digital Signal Processing* 29 (4): 280–287.

3 ITU-R M.2376-0 (2015) Technical feasibility of IMT in bands above 6 GHz. ITU.

4 Bjornson, E., Van der Perre, L., Buzzi, S. et al. (2019). Massive MIMO in Sub-6 GHz and mmWave: physical, practical, and use-case differences. *IEEE Wireless Communications* 26 (2): 100–108.

5 Dahlman, E., Parkvall, S., and Sköld, J. (2018). *5G NR: The Next Generation Wireless Access Technology*. Academic Press Chapter 13.

6

Multiple-antenna Techniques

Multiple antennas are an important part of the 5G air interface, which uses them in two distinct ways. At high radio frequencies, for example millimetre waves, analogue beam selection compensates for the loss of coverage that would otherwise result from the radio propagation issues that we introduced in Chapter 4. At low radio frequencies, for example the 700 MHz band, digital beamforming is the basis for a technique known as multiple input multiple output (MIMO) antennas, which increases the capacity of the air interface in situations where the bandwidth is low. At intermediate radio frequencies, for example the 3500 MHz band, hybrid beamforming systems employ a mixture of the two. We will examine all of those techniques in this chapter, starting with the simplest and gradually progressing to the more complex ones.

Multiple antennas are a rather technical subject, and some previous knowledge of matrices would help the reader to understand the more mathematical parts of this chapter. There are some valuable introductions to matrix algebra in the Bibliography, as well as some more detailed treatments of multiple-antenna techniques in general. Once again, however, we will keep the maths as lightweight as we reasonably can, and we will introduce the important concepts and terminology as we go along.

6.1 Analogue Beam Selection

6.1.1 Spatial Filtering

To begin our discussion, Figure 6.1 shows a base station that is equipped with four transmit antennas. To help keep the diagram clear, we have made a number of simplifications. We assume that the antennas are arranged in a horizontal line, so that the figure is a view from above. We also assume that the antennas are uniformly spaced, in an architecture known as a *uniform linear array* (ULA), and that they transmit using a single polarization. Finally, we assume line-of-sight propagation to a mobile that is a long way from the base station, so that the outgoing rays are parallel.

When viewed from face-on to the antenna array, the rays from the different antennas reinforce each other and interfere constructively. When viewed from other directions, they do not. From certain directions, for example the one shown by the dotted lines, the rays cancel out completely, so the received signal power is zero. We have therefore created a

An Introduction to 5G: The New Radio, 5G Network and Beyond, First Edition. Christopher Cox.
© 2021 John Wiley & Sons Ltd. Published 2021 by John Wiley & Sons Ltd.

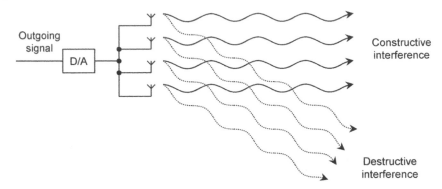

Figure 6.1 Constructive and destructive interference from a base station array.

synthetic radio beam whose energy is focussed in the direction of reinforcement and which has nulls in the directions of cancellation.

Let us now apply a phase ramp to the signals from the different transmit antennas, so that they are phase-shifted through phase angles of 0, ϕ, 2ϕ, 3ϕ and so on. If we do that, then the directions of reinforcement and cancellation change, for example as shown in Figure 6.2. We can therefore steer the beam to any direction we please, simply by changing the phase difference ϕ between the signals from adjacent antennas.

This process is also known as *spatial filtering*, a term that is used throughout the 3GPP specifications for 5G. We will refer to the resulting beams as *spatially filtered beams*, to help distinguish them from the other types of beam that appear later on.

Spatial filtering can also be implemented in the receiver. To do so, we simply apply a phase ramp to the signals that arrive at the different receive antennas, and then add the phase-shifted signals together. That process creates a synthetic reception beam which points in the direction at which those phase-shifted signals reinforce.

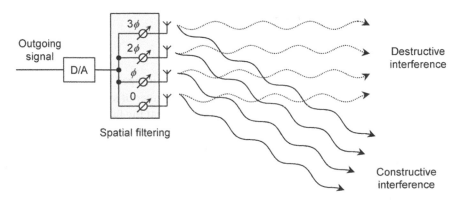

Figure 6.2 Spatial filtering by means of a phase ramp.

6.1.2 Beam Steering

Figure 6.3 shows some examples of the resulting beams. The results are computed using eight antennas that are spaced half a wavelength apart, a spacing that is often used in practical systems. In common with the other radiation patterns later in this chapter, the scale is in units of decibels, with a range of 30 dB and tick marks at intervals of 10 dB, and the radiation pattern is normalized so that the main beam spans the full range of the chart. Each beam contains a main lobe, nulls and sidelobes, and can be steered by adjusting the phase ramp.

To steer the main lobe towards an azimuth angle θ_0, we require the following phase difference between adjacent antennas:

$$\phi = \frac{2\pi d \sin \theta_0}{\lambda} \tag{6.1}$$

Here, λ is the wavelength of the radio waves, d is the distance between the adjacent antennas, an azimuth angle of zero is face-on to the antenna array and the phase difference ϕ is measured in radians. Figure 6.3a–6.3c used phase differences between adjacent antennas of $0°$, $-45°$ and $-90°$, giving steering directions of $0°$, $-14.5°$ and $-30°$ respectively.

There are two issues to note. Firstly, the exact direction of the main lobe in Equation (6.1) depends on the wavelength of the radio waves, and therefore on frequency. That is not a problem if the transmitted signal is a pure sine wave, or if the bandwidth is much less than the carrier frequency. In a wideband system, however, the main lobes of different radio frequencies will point in slightly different directions. We can solve the problem by replacing our phase shifts with time delays, which can sometimes be valuable if the bandwidth is sufficiently large. In this chapter, however, we will keep things simple by sticking with phase shifts.

Secondly, we assumed in computing Figure 6.3 that the individual base station antennas were omnidirectional. In practice, each individual antenna might have a wide enough

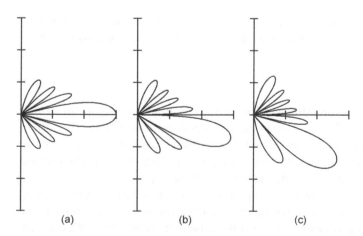

(a)   (b)   (c)

Figure 6.3 Radiation pattern from an array of eight antennas, with a spacing of half a wavelength and a phase difference between adjacent antennas of (a) $0°$, (b) $-45°$ and (c) $-90°$.

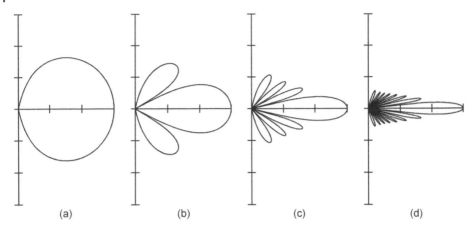

Figure 6.4 Radiation pattern from an array of (a) two, (b) four, (c) eight and (d) 16 antennas, with a spacing of half a wavelength.

beamwidth to span a sector of 120°. That would attenuate the transmitted signal at large azimuth angles, but would otherwise have little effect on the results that we discuss in this chapter. The use of directional antennas at the mobile will be more significant, and we will address that point towards the end of the chapter.

6.1.3 Beamwidth of the Antenna Array

The nulls in the radiation pattern appear at azimuth angles θ that can be calculated as follows:

$$\sin\theta - \sin\theta_0 = \frac{m\lambda}{Md} \quad m = \pm1, \pm2, \cdots \tag{6.2}$$

where M is the number of antennas, and θ_0 is the direction of the main lobe. For small angles, $\sin\theta \approx \theta$, so the angle between the main lobe and the first null is roughly the following:

$$\theta - \theta_0 \approx \frac{\lambda}{Md} \tag{6.3}$$

We can use this quantity as an estimate of the half-power beamwidth of the main lobe, θ_{3dB}. To illustrate that point, Figure 6.4 shows the beams that are created using arrays of two, four, eight and 16 antennas, with a spacing of half a wavelength as before. As the number of antennas increases, so the beamwidth falls. (It might be useful to compare this example with the one for a parabolic reflector antenna in Figurenbsp;6.2.)

Equation (6.3) shows that the transmitted signal power is focussed within a beam whose width is inversely proportional to M, the number of antennas in the array. In turn, that implies that the received signal power at the mobile is also proportional to M. We can identify M with a quantity known as the *array gain*, which is the increase in the received signal power due to spatial filtering.

By increasing the received signal power in this way, the use of multiple antennas increases the coverage of the system. But that is exactly what we require to overcome the problem of high-frequency propagation loss from Chapter 4. We therefore conclude that antenna arrays will be a key part of high-frequency radio communications for 5G.

6.1.4 Grating Lobes

More generally, the transmitted signals reinforce each other at azimuth angles θ that are calculated as follows:

$$\sin\theta - \sin\theta_0 = \frac{n\lambda}{d} \quad n = 0, \pm1, \pm2, \cdots \tag{6.4}$$

If the antenna spacing is less than one wavelength, then the only direction of complete reinforcement is the one where n equals zero, which is the direction of the main lobe from Equation (6.1). If the antennas are spaced more widely, then other directions of reinforcement appear, which are known as *grating lobes*.

To illustrate that point, Figure 6.5 shows the beams that are created using an array of four antennas. Figure 6.5a uses an antenna spacing of half a wavelength, as before, while the others show the grating lobes that appear when using wider antenna spacings. Grating lobes are not usually an issue in an analogue system, but they will become important when we discuss hybrid beamforming later on.

6.1.5 Analogue Signal-processing Issues

As implied by Figure 6.2, the transmitter can implement spatial filtering after digital-to-analogue (D/A) conversion, by means of analogue signal processing. Similarly, the receiver can implement spatial filtering before analogue-to-digital (A/D) conversion. That makes the process appealing in the case of millimetre waves, when we wish to limit the numbers of D/A and A/D converters in order to limit the system's power consumption.

However, there is a limitation: the base station can only transmit or receive using one spatially filtered beam at a time. To use different beams, the base station has to switch from one beam to another by means of time division multiplexing. Nevertheless, the base station

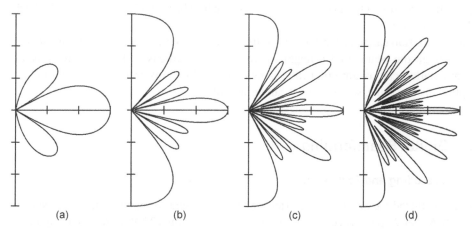

(a)   (b)   (c)   (d)

Figure 6.5 Radiation pattern from an array of four antennas, with a spacing of (a) one-half, (b) one, (c) two and (d) four wavelengths.

can still communicate simultaneously with different mobiles that are in the same beam, by scheduling them on different radio frequencies by means of orthogonal frequency division multiple access (OFDMA).

Two other issues arise. Firstly, the base station's antennas are often cross-polarized, with each polarization driven by an antenna connector of its own. That creates two independent polarized beams, which can be steered either in the same direction or in different directions by means of two independent phase ramps. Secondly, the base station can implement analogue beam steering in elevation as well as in azimuth, by means of a two-dimensional antenna array. In fact, base stations have long used vertical arrays containing around four to eight antennas, so as to deliver the narrow elevation beamwidth that is appropriate for a terrestrial cellular network. The difference in 5G is that the base station can change the beam direction dynamically, in a so-called *active antenna system* (AAS).

6.1.6 Beam Management

In principle, the base station could allow the phase difference between adjacent antennas to vary continuously. That would allow it to steer the main lobe in any direction, so as to point exactly towards a mobile of interest. However, that procedure requires a complex analogue control loop, so is not generally used.

Instead, the phase difference is generally restricted to a limited set of values, so as to create a set of fixed beams that point in different but fixed directions. Those directions are most often chosen so that the main lobe of one beam coincides with the nulls of the others. (We used just that choice when computing the beams in Figure 6.3.) If the base station has an array containing M antennas, then it can create M such beams.

The process of beam steering then reduces to one of *beam selection*, in which we simply choose the beam in which the received signal-to-interference plus noise ratio (SINR) is the greatest. On the downlink, the base station can only transmit using one spatially filtered beam at a time, but it can still sweep progressively through each of its transmit beams, sending reference symbols in each one that indicate the beam's identity. A mobile can then inspect the incoming reference symbols, identify the best beam and report its identity back to the base station.

The base station can use a similar procedure on the uplink by sweeping its receive spatial filter through the beams that it might use for reception. In cases of channel reciprocity, the base station can assume that the beams used for downlink transmission and uplink reception are the same, so the process can be simplified.

6.2 Digital Beamforming

6.2.1 Precoding and Postcoding

Now let us introduce one D/A converter for each individual transmit antenna, as shown in Figure 6.6, and one A/D converter for each receive antenna. That increases the system's complexity and power consumption, so the architecture is only suitable for low and intermediate radio frequencies, not for millimetre waves.

Figure 6.6 Precoding by means of antenna-specific amplitude scaling factors and phase shifts.

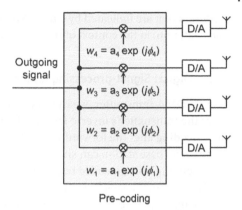

Outgoing signal

$w_4 = a_4 \exp{(j\phi_4)}$

$w_3 = a_3 \exp{(j\phi_3)}$

$w_2 = a_2 \exp{(j\phi_2)}$

$w_1 = a_1 \exp{(j\phi_1)}$

D/A

Pre-coding

However, the benefit is increased flexibility. Instead of using a simple analogue phase ramp, we can now process the symbols digitally by means of an antenna-specific amplitude scaling factor and phase shift. In the transmitter, the relevant step is known as *precoding*, and the scaling factors and phase shifts are often known as *precoding weights*. We will refer to the corresponding beams as *digitally precoded beams*, to help distinguish them from the spatially filtered beams from before. The equivalent step in the receiver is sometimes known as *postcoding*, or simply as signal detection or reception.

We can now generalize the earlier task of beam steering to the more complex procedure of *beamforming*. Using a linear array containing M antennas, we can not only steer the main beam towards the target of interest, but also steer $(M - 2)$ nulls. In doing so, we can avoid delivering interference to other receivers on the downlink (for example, mobiles in other nearby cells) and avoid receiving interference from other transmitters on the uplink.

Figure 6.7 shows an example. Figure 6.7a directs the main beam towards the target mobile, which is indicated by an arrow, while Figure 6.7b and 6.7c direct nulls towards

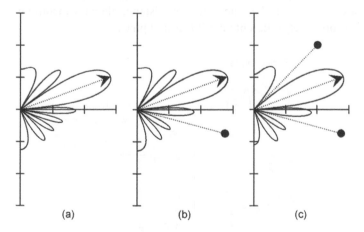

(a)   (b)   (c)

Figure 6.7 Example radiation patterns from an array of eight antennas using digital beamforming, with a target mobile at an azimuth of 20°. (a) No interferer. (b) One interferer, at −15°. (c) Two interferers, at −15° and +45°.

interferers that are indicated by circles. We will discuss the algorithms used to create these beams later on, in the context of spatial multiplexing and MIMO.

6.2.2 Digital Signal-processing Issues

In an OFDMA transmitter, we can implement digital precoding in the frequency domain, before the transmitter's inverse fast Fourier transform (IFFT). We can also implement digital postcoding after the forward FFT in the receiver. That brings two further benefits.

Firstly, the base station can simultaneously communicate with different mobiles that lie in different digitally precoded beams, by using a different set of precoding weights for each one. That is more flexible than the situation which arose when using analogue beam steering, in which the base station could only communicate using one spatially filtered beam at a time.

Secondly, let us recall that the direction of the main lobe when using analogue beam steering depended on the frequency of the radio waves. When using digital beamforming, we can compensate for that problem by applying slightly different phase shifts to different sets of subcarriers. That ensures that the main lobe and nulls will point in the same direction across the whole of the frequency band, even in a wideband system.

6.2.3 Diversity Processing

So far, we have limited ourselves to the case of line-of-sight propagation. Figure 6.8 shows the geometry for multipath propagation in the case of the uplink, from a mobile with a single transmit antenna, to a base station with two receive antennas.

Around the mobile, the outgoing rays depart in all directions. At the base station, the angular spread of the incoming rays is usually less, partly due to the use of sectors and partly because the base station is usually sited with few reflectors nearby. Even so, the coherence distance from Equation (4.13) is usually small, perhaps a few wavelengths long. If the base station's receive antennas are further apart than the coherence distance, then the incoming signals will reach them with completely different amplitudes and phases.

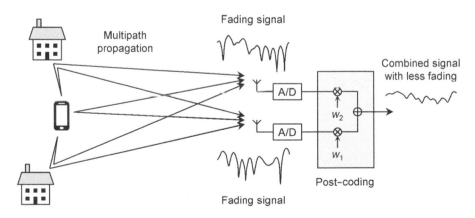

Figure 6.8 Uplink diversity reception.

If the mobile moves, then the signal arriving at each individual antenna will be subject to fading, in accordance with the coherence time from Equation (4.15). With the antennas far apart, however, the signals at different antennas are likely to fade at different times. If the base station combines those signals in a suitable postcoding receiver, then the fades in the combined signal will be shallower than those at the individual receive antennas. That in turn reduces the bit error ratio, and increases the coverage of the system.

The whole procedure is known as *diversity reception*, and the resulting benefit is known as *diversity gain*. This acts in addition to the array gain from Section 6.1.3. The downlink can use a similar procedure, known as *diversity transmission*, in which the base station adjusts the amplitudes and phases of its transmitted signals by means of precoding.

6.3 Spatial Multiplexing

6.3.1 Principles of Spatial Multiplexing

Spatial multiplexing is the delivery of multiple streams of traffic on the same subcarriers and at the same time, using multiple antennas at both the transmitter and receiver. The different streams of traffic are known as *layers* (which are completely different from the layers of a protocol stack). Spatial multiplexing is also known as the use of MIMO antennas, in which the transmit antennas are the inputs to the air interface, and the receive antennas are the outputs. However, the term MIMO can also be used more generally, to encompass the delivery of a single stream of traffic using diversity processing in both the transmitter and receiver.

Figure 6.9 shows the underlying architecture, in a system where the transmitter and receiver both have two antennas. Following on from the previous section, we will assume the use of digital signal processing throughout. For the time being, the same description will apply to the uplink and downlink. The transmitted signals travel over four separate radio paths, so we can write the received signals as follows:

$$y_1 = H_{11}x_1 + H_{12}x_2 + n_1$$
$$y_2 = H_{21}x_1 + H_{22}x_2 + n_2 \tag{6.5}$$

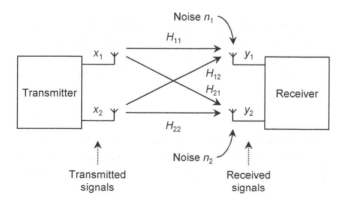

Figure 6.9 Principles of spatial multiplexing.

Here, x_1 and x_2 are the outgoing signals at the two transmit antennas; y_1 and y_2 are the incoming signals at the two receive antennas; and n_1 and n_2 represent the incoming noise and interference. These quantities are all complex numbers, whose real and imaginary parts are the in-phase and quadrature components of the corresponding signals. The four values of H_{ij} are also complex numbers, which denote the attenuations and phase shifts that are introduced by the air interface, as the signals travel to receive antenna i from transmit antenna j. (That order may look the wrong way round, but it is consistent with the mathematical notation for matrices.)

The receiver can estimate the values of H_{ij} by measuring the reference symbols that arrive from the transmitter. (To keep things simple, we will assume that those estimates are accurate.) Once it has done so, the receiver has enough information to recover the two transmitted signals x_1 and x_2. If the noise and interference are small enough, then the receiver's best estimate comes from ignoring n_1 and n_2, and solving the resulting pair of simultaneous equations as follows:

$$
\begin{aligned}
\tilde{x}_1 &= \frac{H_{22}y_1 - H_{12}y_2}{H_{11}H_{22} - H_{21}H_{12}} \\
\tilde{x}_2 &= \frac{H_{11}y_2 - H_{21}y_1}{H_{11}H_{22} - H_{21}H_{12}}
\end{aligned}
\tag{6.6}
$$

These equations define a postcoding receiver, in which $\tilde{x}_1$ and $\tilde{x}_2$ are the receiver's estimates of the transmitted signals. If the noise and interference are non-zero, then those estimates are corrupted as follows:

$$
\begin{aligned}
\tilde{x}_1 &= \frac{H_{22}y_1 - H_{12}y_2}{H_{11}H_{22} - H_{21}H_{12}} - \frac{H_{22}n_1 - H_{12}n_2}{H_{11}H_{22} - H_{21}H_{12}} \\
\tilde{x}_2 &= \frac{H_{11}y_2 - H_{21}y_1}{H_{11}H_{22} - H_{21}H_{12}} - \frac{H_{11}n_2 - H_{21}n_1}{H_{11}H_{22} - H_{21}H_{12}}
\end{aligned}
\tag{6.7}
$$

6.3.2 Matrix Representation

We can write Equation (6.5) in matrix notation, as follows:

$$
\begin{bmatrix} y_1 \\ y_2 \end{bmatrix} = \begin{bmatrix} H_{11} & H_{12} \\ H_{21} & H_{22} \end{bmatrix} \cdot \begin{bmatrix} x_1 \\ x_2 \end{bmatrix} + \begin{bmatrix} n_1 \\ n_2 \end{bmatrix}
\tag{6.8}
$$

Here, the values of x_i, y_i and n_i make three column vectors that describe the transmitted signals, received signals and noise at the corresponding antennas; similarly, the values of H_{ij} make a *channel matrix*, which describes the attenuations and phase shifts that the air interface has introduced. To support architectures where there are more antennas, we can generalize this equation as follows:

$$
y = H \cdot x + n
\tag{6.9}
$$

Here, x, y and n are column vectors that describe the transmitted signals, received signals and noise; while H is the channel matrix. If the numbers of transmit and receive antennas are the same, then we can generalize the receiver's best estimate of the transmitted signals from Equation (6.6) as follows:

$$
\tilde{x} = H^{-1} \cdot y
\tag{6.10}
$$

Here, $\tilde{x}$ is the receiver's estimate of the transmitted signal vector, and H^{-1} is the *inverse* of the matrix H. If the noise and interference are non-zero, then that estimate is corrupted in the same way as Equation (6.7):

$$\tilde{x} = H^{-1} \cdot y - H^{-1} \cdot n \tag{6.11}$$

6.3.3 MIMO and Coherence

Readers may already have spotted a problem: if $H_{11}H_{22} - H_{21}H_{12}$ in Equation (6.6) is zero, then we end up dividing by zero, which is nonsense. In the language of matrices, the problem arises if the channel matrix H is *singular*, with a *determinant* of zero, in which case its inverse does not exist. Furthermore, a similar problem arises if $H_{11}H_{22} - H_{21}H_{12}$ is small but non-zero. If that happens, then the noise in Equation (6.7) is amplified, and the result is so badly corrupted that it is completely unusable. In matrix language, the channel matrix H is *ill-conditioned*, with a determinant that is non-zero but small.

The problem arises because the receiver is expecting to measure two completely different pieces of information, namely the incoming signals y_1 and y_2. But now imagine what happens if the two receive antennas are in the same place. In that situation, the receiver no longer measures two different pieces of information: instead, it measures the same piece of information twice. As a result, it does not have enough information to recover the two transmitted signals. The channel matrix is singular in this example, because H_{11} equals H_{21} and H_{12} equals H_{22}, but the same argument applies to every other case with a singular matrix.

We can usually avoid this problem if the transmit and receive antennas are spaced more widely than their respective coherence distances, so that we cannot predict the amplitude and phase at one transmit or receive antenna, from the amplitude and phase at another. That is easy to achieve in a multipath environment, although problems sometimes appear due to random fluctuations in the elements of the channel matrix. However, it is harder to achieve in the case of line-of-sight communications, for which the coherence distance can be very large.

As an alternative, we can arrange the antennas so that they transmit and receive using different polarizations. If we do so, then the incoming signal phases are usually unrelated to each other, and the channel matrix is usually well-conditioned. However, the technique only works on its own for a maximum of two transmit and two receive antennas.

6.3.4 Uplink Multiple-user MIMO

To take things further, we need to look at some individual scenarios. Let us begin with uplink *multiple-user MIMO* (MU-MIMO), which is illustrated in Figure 6.10. There are K mobiles, each sending one layer from one transmit antenna, while the base station has M receive antennas. There is no need for any precoding, so the transmitted signals are simply the following:

$$x = s \tag{6.12}$$

where s is a vector containing each mobile's transmitted symbol. The uplink channel matrix, which we denote as H_{UL}, has M rows and K columns, and is known as an $M \times K$ matrix.

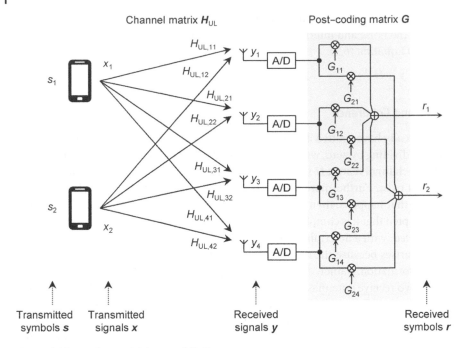

Figure 6.10 Uplink multiple-user MIMO.

The base station applies a $K \times M$ postcoding matrix, denoted $\boldsymbol{G}$, to the signals that arrive at its receive antennas. By doing so, the base station computes its best estimate of the received symbols as follows:

$$\boldsymbol{r} = \boldsymbol{G} \cdot \boldsymbol{y} \tag{6.13}$$

The channel matrix is usually well-conditioned, partly because the mobiles tend to be far apart, and partly because the base station is free to choose the mobiles that are involved. If it is, then the base station's best estimate of the transmitted symbols is a *zero-forcing* (ZF) receiver. That generalizes our earlier examples to support the case where the numbers of transmit and receive antennas can be different, as follows [1–4]:

$$\boldsymbol{G} = (\boldsymbol{H}_{\mathrm{UL}}^{\mathrm{H}} \cdot \boldsymbol{H}_{\mathrm{UL}})^{-1} \cdot \boldsymbol{H}_{\mathrm{UL}}^{\mathrm{H}} \tag{6.14}$$

where $\boldsymbol{H}_{\mathrm{UL}}^{\mathrm{H}}$ is the *Hermitian conjugate* of the uplink channel matrix. The Hermitian conjugate is the complex conjugate of the *transpose* matrix $\boldsymbol{H}_{\mathrm{UL}}^{\mathrm{T}}$, which is itself formed by interchanging the rows and columns of $\boldsymbol{H}_{\mathrm{UL}}$.

In a line-of-sight environment, the zero-forcing receiver is equivalent to the establishment of K parallel reception beams, with one beam per mobile. In each beam, we ensure a strong incoming signal by placing the target mobile somewhere in the main lobe: ideally at the centre, but not necessarily. More important is the complete removal of interference, which is achieved by placing all the other mobiles precisely within the nulls. We used just such a receiver in our examples of digital beamforming in Figure 6.7.

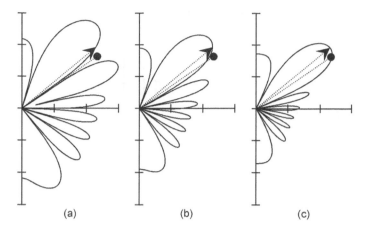

(a)   (b)   (c)

Figure 6.11 Example radiation patterns from an array of eight antennas, using (a) zero forcing, (b) minimum mean square error and (c) maximum ratio receivers.

If the noise is large, or the channel matrix is singular, then the base station's best chance of recovering the transmitted symbols is to use a *maximum ratio* (MR) receiver, which is defined as follows:

$$G = H_{\mathrm{UL}}^{\mathrm{H}} \tag{6.15}$$

In a line-of-sight environment, the maximum ratio receiver is still equivalent to the establishment of K parallel reception beams. This time, however, we maximize the received signal-to-noise ratio by directing each main lobe in the exact direction of the corresponding target mobile, while ignoring interference from the others altogether.

In intermediate cases, a *minimum mean square error* (MMSE) receiver provides a compromise between the two extremes:

$$G = (H_{\mathrm{UL}}^{\mathrm{H}} \cdot H_{\mathrm{UL}} + \beta I_{\mathrm{K}})^{-1} \cdot H_{\mathrm{UL}}^{\mathrm{H}} \tag{6.16}$$

where I_{K} is a $K \times K$ *identity* matrix, in other words a matrix that equals its own inverse; and β is a weighting factor. If β is small, then the equation approaches the zero-forcing receiver from earlier; if β is large, then it approaches the maximum likelihood receiver.

Figure 6.11 is an example. There are two mobiles, at azimuths of 40° and 35°: the figure shows the beams that are generated towards the first. In Figure 6.11a, the base station uses a zero-forcing receiver, which places the second mobile in a null but amplifies the noise. In Figure 6.11c, the base station uses a maximum ratio receiver, which directs the main lobe towards the first mobile but makes no attempt to remove interference from the second. The MMSE receiver in Figure 6.11b is a compromise between the two extremes.

6.3.5 Downlink Multiple-user MIMO

Downlink multiple-user MIMO is illustrated in Figure 6.12, and is superficially similar. The base station uses M antennas to transmit to each of K mobiles, with one layer per mobile.

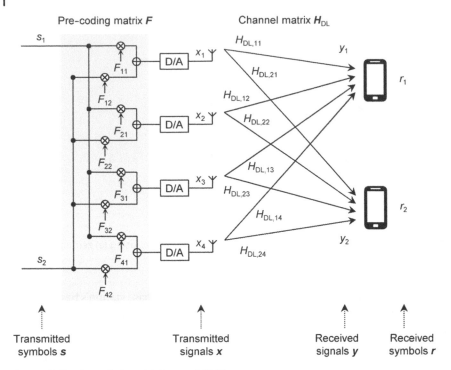

Figure 6.12 Downlink multiple-user MIMO.

The downlink channel matrix, H_{DL}, has K rows and M columns. In cases of channel reciprocity, the uplink and downlink channel matrices are related by a transpose operation:

$$H_{DL} = H_{UL}^{T} \tag{6.17}$$

This time, there is no opportunity for any postcoding: for example, the zero-forcing receiver from Equation (6.6) won't work at all, because none of the mobiles can access the signals that arrive at the others. Instead, the base station has to do all the work by itself, using a precoding matrix F that computes the transmitted signals as follows:

$$x = F \cdot s \tag{6.18}$$

There are various types of precoding matrix, which correspond to the three types of receiver that we introduced earlier in this chapter, and whose interpretations are exactly the same. If the channel matrix is well-conditioned and the noise levels are low, then the best choice is a zero-forcing precoder:

$$F = H_{DL}^{H} \cdot (H_{DL} \cdot H_{DL}^{H})^{-1} \tag{6.19}$$

If the noise is large or the channel matrix is singular, then the best choice is a maximum ratio precoder:

$$F = H_{DL}^{H} \tag{6.20}$$

In intermediate cases, an MMSE precoder is a compromise between the two extremes:

$$F = H_{DL}^{H} \cdot (H_{DL} \cdot H_{DL}^{H} + \beta I_{K})^{-1} \tag{6.21}$$

6.3.6 Management of Multiple-user MIMO

Now let us consider how the process is managed. In the uplink, the mobile transmits a set of sounding reference symbols. The base station measures the reference symbols that arrive at its receive antennas, compares them with the ones defined in the specifications and computes each mobile's contribution to the uplink channel matrix. The base station can then combine its measurements of all the mobiles to compute the full uplink channel matrix H_{UL}. That in turn allows it to compute the uplink postcoding matrix G, using the techniques of zero-forcing, maximum ratio reception or MMSE as appropriate. In cases of channel reciprocity, the base station can use the same measurements to compute the downlink precoding matrix F.

In the absence of channel reciprocity, the base station sends a distinct set of channel state information (CSI) reference symbols from each of its transmit antennas. The mobile measures the incoming reference symbols, computes its own part of the downlink channel matrix and feeds back a compressed description of that sub-matrix as the *precoding matrix indicator* (PMI). By combining the sub-matrices that it receives from all the mobiles, the base station can compute the full downlink channel matrix H_{DL} and the precoding matrix F. This technique is used by the 5G New Radio's type II PMI codebooks, which we will discuss in Chapter 10.

6.3.7 Single-user MIMO

Single-user MIMO (SU-MIMO) involves one mobile equipped with multiple antennas and is illustrated in Figure 6.13. The transmitter and receiver have M_T and M_R antennas respectively, which they use to exchange L layers of traffic. The maximum number of layers is constrained so that:

$$1 \leq L \leq \min\left(M_T, M_R\right) \tag{6.22}$$

with any additional antennas providing array and diversity gain. The architecture is known as an $M_T \times M_R$ MIMO system.

The receiver can only recover the maximum number of layers if the channel matrix is well-conditioned, so the antennas should be further apart than the corresponding coherence distances. This time, however, the mobile's antennas are on a single device, so they are inevitably close together. That implies either that there should be a maximum of two antennas using different polarizations, or that the propagation environment should involve multipath.

Even so, the channel matrix can occasionally become ill-conditioned, due to fluctuations in the amplitudes and phases of the incoming rays. If it does so, then the base station reacts by dynamically reducing the number of layers, to a value that the receiver can handle. The end-point is the exchange of a single layer by means of diversity transmission and reception.

That in turn makes precoding essential on the uplink as well as the downlink. There are two reasons: to map the outgoing layers onto the transmit antennas, and to prevent any risk of destructive interference at the receiver if the system falls back to diversity transmission. Postcoding is used as well, because the transmitter no longer has to direct its individual layers towards the individual receive antennas: instead, it can rely on the receiver to disentangle them.

6.3.8 Signal Processing for Single-user MIMO

The signal processing for single-user MIMO relies on a technique known as *singular value decomposition* (SVD). Using that technique, it is possible to write any channel matrix as follows:

$$H = U \cdot \Sigma \cdot V^{\mathrm{H}} \tag{6.23}$$

Here, H is the $M_R \times M_T$ channel matrix, U is an $M_R \times M_R$ matrix, V is an $M_T \times M_T$ matrix and V^{H} is the Hermitian conjugate of V. Both U and V are *unitary* matrices whose inverses equal their Hermitian conjugates. Σ is an $M_R \times M_T$ *diagonal* matrix whose elements σ_i are either positive or zero and are known as the *singular values* of H. In the example from Figure 6.13, with four transmit and two receive antennas, the diagonal matrix is:

$$\Sigma = \begin{bmatrix} \sigma_1 & 0 & 0 & 0 \\ 0 & \sigma_2 & 0 & 0 \end{bmatrix} \tag{6.24}$$

Now let us transmit the symbols using a precoding matrix F and receive them using a postcoding matrix G. If the precoding matrix is a good approximation to the second unitary matrix from earlier:

$$F \approx V \tag{6.25}$$

and the postcoding matrix is chosen as follows:

$$\begin{aligned} G &= (H \cdot F)^{\mathrm{H}} \\ &\approx (U \cdot \Sigma \cdot V^{\mathrm{H}} \cdot V)^{\mathrm{H}} \\ &\approx \Sigma^{\mathrm{T}} \cdot U^{\mathrm{H}} \end{aligned} \tag{6.26}$$

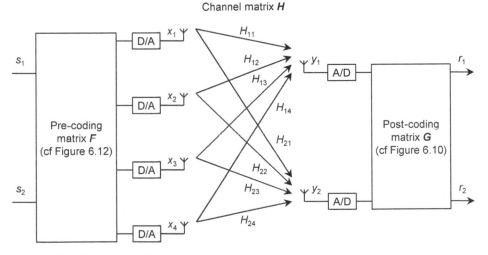

Figure 6.13 Single-user MIMO.

then the received symbol vector becomes the following:

$$r = G \cdot H \cdot F \cdot s + G \cdot n$$
$$\approx (\Sigma^{\mathrm{T}} \cdot U^{\mathrm{H}}) \cdot (U \cdot \Sigma \cdot V^{\mathrm{H}}) \cdot V \cdot s + (\Sigma^{\mathrm{T}} \cdot U^{\mathrm{H}}) \cdot n$$
$$\approx \Sigma^{\mathrm{T}} \cdot \Sigma \cdot s + \Sigma^{\mathrm{T}} \cdot U^{\mathrm{H}} \cdot n \qquad (6.27)$$

Ignoring the noise, we can now write the received symbols in our 4×2 MIMO example as follows:

$$\begin{bmatrix} r_1 \\ r_2 \end{bmatrix} = \begin{bmatrix} \sigma_1^2 & 0 \\ 0 & \sigma_2^2 \end{bmatrix} \cdot \begin{bmatrix} s_1 \\ s_2 \end{bmatrix} \qquad (6.28)$$

It is now easy for the receiver to recover a maximum of two transmitted symbols, as follows:

$$\tilde{s}_i = \frac{r_i}{\sigma_i^2} \qquad (6.29)$$

We can understand what is happening by thinking about a sparse multipath environment, illustrated in Figure 6.14. The matrix Σ defines a set of independent propagation paths, in which the individual signals are attenuated but do not interfere. In a sparse multipath environment, we can identify each of these paths with a distinct reflection. The matrix V^{H} maps the transmit antennas onto the outgoing propagation paths, and the matrix U maps the incoming paths onto the receive antennas.

If the precoding matrix F is well-chosen, then it cancels the effect of the matrix V^{H}, so as to map each outgoing layer onto a single propagation path. Similarly, the postcoding matrix G cancels the effect of the matrix U, so as to map each incoming path onto a single layer. The receiver can then recover the transmitted layers using Equation (6.29). (More advanced receivers can handle the case where F is not such a good approximation to V, by using a more sophisticated postcoding matrix.)

If the channel matrix H is singular, then some of the singular values σ_i are zero. If it is ill-conditioned, then some of the singular values are small, so the reconstructed symbols are badly corrupted by noise. The *rank* of H is the number of singular values that are large enough to be usable. In Figure 6.14, the rank of H is limited by three separate quantities,

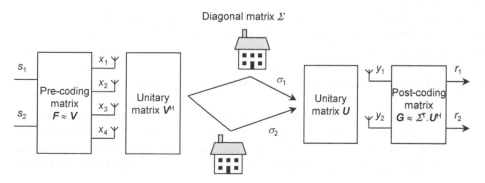

Figure 6.14 Single-user MIMO in a sparse multipath environment using singular value decomposition.

namely the number of transmit antennas, the number of receive antennas and the number of significant paths.

If the channel matrix is ill-conditioned, then the system responds by limiting the number of layers L to the rank of H. We retain only the rows and columns of Σ that contain the usable singular values, and the corresponding rows and columns of V^H and U. Once that is done, F becomes an $M_T \times L$ matrix that maps the L outgoing layers onto M_T transmit antennas, while G becomes an $L \times M_R$ matrix that maps the M_R receive antennas onto L incoming layers. Equations (6.25) and (6.26) still apply.

6.3.9 Management of Single-user MIMO

Now let us consider the management of single-user MIMO, beginning this time with the downlink. In the absence of channel reciprocity, the mobile measures the base station's CSI reference symbols and computes the downlink channel matrix H_{DL}. Using that matrix, the mobile estimates the number of layers that it can handle and feeds that quantity back to the base station as the *rank indication* (RI). The mobile also selects its preferred precoding matrix F from a codebook that is defined by the specifications, and feeds that back as the precoding matrix indicator (PMI). This technique is used by the 5G New Radio's type I PMI codebooks, which we will discuss in Chapter 10.

During a subsequent data transmission, the base station embeds a distinct set of demodulation reference symbols into each MIMO layer and precodes those symbols in the same way as the data. The mobile measures the signals that arrive at its receive antennas and compares them with the ones defined in the specifications. By doing so, the mobile measures the composite channel matrix $H_{DL} \cdot F$, which expresses the combined effects of precoding and the air interface. But that is exactly what we needed in Equation (6.26) to compute the postcoding matrix G. By applying that postcoding matrix, the mobile can reconstruct the transmitted symbols.

The uplink uses a similar procedure. In a technique known as *codebook-based precoding*, the base station measures the sounding reference symbols that arrive at its receive antennas, builds the uplink channel matrix H_{UL} and selects the best uplink precoding matrix from another codebook. It then tells the mobile which precoding matrix to use as part of an uplink scheduling command.

Shortcuts are possible in cases of channel reciprocity. Firstly, the base station can compute the best downlink precoding matrix from its measurements of the uplink channel matrix, without the need for any PMI feedback. Secondly, the mobile can compute the best uplink precoding matrix from its measurements of the downlink channel matrix, without the need for explicit downlink signalling. This latter technique is known as *non-codebook-based precoding*.

Finally, the base station can combine the use of single- and multiple-user MIMO, for example by exchanging two MIMO layers with each of two mobiles. In the absence of channel reciprocity, this is handled using the same type of precoding matrix indicator as for multiple-user MIMO, to ensure that the base station knows the full downlink channel matrix.

6.4 Massive MIMO

6.4.1 Architecture

In the technique of multiple-user MIMO that we discussed earlier in this chapter, there is a problem. As illustrated in Figure 6.11, multiple-user MIMO relies on placing interfering mobiles either in the nulls of the base station's transmission and reception beams, or very close to them. That works fine if the number of mobiles is small. However, if the number of mobiles rises, then it becomes increasingly difficult to place all the nulls in the correct directions: instead, the interference levels rise, and the technique quickly becomes unusable.

The solution lies in a modified version of multiple-user MIMO known as *massive MIMO* (mMIMO) [5–9]. The main feature of massive MIMO is that the number of base station antennas, M, is much greater than the number of mobiles that are involved, K, so that:

$$M >> K >> 1 \tag{6.30}$$

We also assume the use of digital signal processing in a rich scattering environment with lots of multipath, as shown in the illustration for the uplink in Figure 6.15. (The downlink is just the same but with the arrows reversed.) With these features in place, we can carry out the signal processing using maximum ratio techniques alone, without worrying much about interference. Further improvements are possible by means of MMSE, particularly if the propagation environment is not an ideal one, but we will stick with maximum ratio processing in the description here.

Massive MIMO frees us from the problems that arose with the basic procedure for multiple-user MIMO, and allows the base station to communicate with a much larger

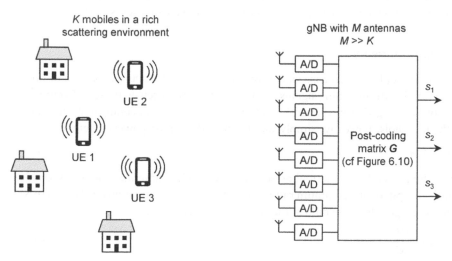

Figure 6.15 Massive MIMO.

number of mobiles than before. In doing so, it addresses two of the performance requirements for 5G that we introduced in Chapter 1, namely its energy efficiency and its spectral efficiency. Let us examine how.

6.4.2 Received Signal Power

To understand how massive MIMO works, let us focus on the uplink. The power received by a single base station antenna from a single mobile is the following:

$$p_R = \frac{P_T}{CL} \tag{6.31}$$

Here, P_T is the mobile's transmit power; and CL is the coupling loss from Equation (4.6), which includes the propagation loss and the gains and losses of the individual antenna elements, but does not yet include the base station's array gain.

To account for that last quantity, we combine the signals from the individual antennas in a maximum ratio receiver. When we do so, the signals add together coherently and reinforce each other, such that the amplitude of the combined signal equals the sum of the amplitudes of the individual contributions. Noting that the power of any signal depends on the square of its amplitude, we conclude that the power of the combined signal is the following:

$$P_R = \frac{M^2 P_T}{CL} \tag{6.32}$$

One the downlink, the transmitted signals add together coherently within a small patch around the target mobile, whose size is roughly equal to the coherence distance from Equation (4.13). The received signal power is different, but there is no impact on the more important quantities that we will discuss next.

6.4.3 Energy Efficiency

Returning to the uplink, each base station receive antenna also acts as a source of noise. When we combine the incoming signals in a maximum ratio receiver, those different noise contributions add together incoherently, on the average neither cancelling nor reinforcing. To find the total noise power, we simply add the individual noise powers together:

$$N = Mn \tag{6.33}$$

where n is the power due to thermal noise at each individual antenna, and N is the noise power at the output of the maximum ratio receiver. Combining Equations (6.32) and (6.33) shows that the signal-to-noise ratio at the base station receiver is:

$$SNR = \frac{MP_T}{nCL} \tag{6.34}$$

We can see that the signal-to-noise ratio depends on the number of base station receive antennas, M. If we increase the value of M, but keep the mobile's transmit power unchanged, then the signal-to-noise ratio increases without limit. Alternatively, we can increase the value of M while keeping the signal-to-noise ratio unchanged. If we do that, then the mobile can reduce its transmit power P_T, which reduces its overall power consumption and increases its battery life.

On the downlink, the calculations are different, but the end result is exactly the same: the signal-to-noise ratio is proportional to M. If we increase the number of transmit antennas at the base station, then we can greatly reduce the base station's transmit power while keeping the mobile's received signal-to-noise ratio unchanged. That greatly reduces the power consumption of the radio access network and improves its energy efficiency, measured in units of bits second^{-1} watt^{-1}. In doing so, we have addressed one of the key requirements for 5G.

6.4.4 Spectral Efficiency

Now let us return once again to the uplink. While decoding the transmission from our original mobile, the base station is also receiving interference from the $(K-1)$ other mobiles that are carrying out MIMO transmissions. The power received by a single base station antenna from a single mobile is the same as in Equation (6.31). This time, however, the interfering signals from different mobiles and different base station antennas add together incoherently, as in the case of thermal noise. At the output of the maximum ratio receiver for our original mobile, the power due to interference is the following:

$$I = \frac{M(K-1)P_T}{CL} \tag{6.35}$$

Combining Equations (6.32) and (6.35) suggests that the signal-to-interference ratio at the base station receiver is:

$$SIR \approx \frac{M}{K-1} \tag{6.36}$$

A more careful analysis takes account of other sources of interference, for example channel estimation errors, and leads to the following result:

$$SIR \approx \frac{M}{K} \tag{6.37}$$

If we increase the value of M, but keep the number of mobiles unchanged, then the signal-to-interference ratio increases. (A limit does arrive eventually, due to issues such as inter-cell interference that we have not accounted for here.) The same result applies on the downlink, although once again the calculations are slightly different. Alternatively, we can increase the value of K while keeping the signal-to-interference ratio unchanged. That increases the data rate on the air interface and improves its spectral efficiency, measured in units of bits second^{-1} Hz^{-1}. In doing so, we have addressed another of the key requirements for 5G.

These improvements mean that massive MIMO is an important aspect of 5G radio communications. Massive MIMO requires the use of digital signal processing, so it is most suitable for use at low radio frequencies. It can, however, be adapted for use at higher frequencies, in the manner described later in this chapter.

6.5 Hybrid Beamforming

6.5.1 Partly Connected Architecture

Digital beamforming is a powerful technique. However, it requires a large number of A/D and D/A converters, so its power consumption is too high for use at high radio frequencies, particularly millimetre waves. On the other hand, analogue beam selection has a lower consumption but is less flexible. *Hybrid beamforming* combines the two [10–12].

The most common implementation uses the partly connected architecture shown in Figure 6.16. In this architecture, the base station's antenna array is divided into M_1 panels, each driven through a digital connector by means of digital precoding. Each panel contains M_2 antennas that are driven by spatial filtering, so the total number of antennas is:

$$M = M_1 M_2 \tag{6.38}$$

To understand how the architecture works, let us think back to our earlier discussion of analogue beam steering. Each individual panel creates a spatially filtered beam, which is directed to an azimuth angle θ_0, and has nulls at the following angles:

$$\sin\theta - \sin\theta_0 = \frac{m\lambda}{M_2 d} \quad m = \pm 1, \pm 2, \cdots \tag{6.39}$$

Let us now assume that those beams are created using the same spatial filter, so they all point in the same direction. If they do, then the panels act in the same way as digitally precoded antennas that have a wider spacing of $M_2 d$. The panels therefore create a digitally precoded beam, whose half-power beamwidth is the one that we would expect from the

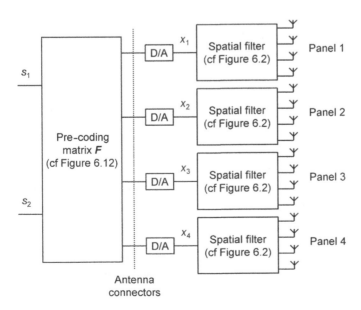

Figure 6.16 Partly connected architecture for hybrid beamforming.

complete antenna array:

$$\theta_{3dB} \approx \frac{\lambda}{M_1 M_2 d} \tag{6.40}$$

That beam repeats in grating lobes, which are centred at the following azimuth angles:

$$\sin\theta - \sin\theta_0 = \frac{n\lambda}{M_2 d} \quad n = 0, \pm 1, \pm 2, \cdots \tag{6.41}$$

However, the grating lobes with $n \neq 0$ are centred on the nulls of the original spatially filtered beam, so they are at least partly suppressed.

Using this technique, the base station can only communicate using one spatially filtered beam at a time. Within that beam, however, the base station can generate multiple digitally precoded beams, so as to communicate with one or more mobiles by means of single- and multiple-user MIMO. The technique therefore provides much of the flexibility of a purely digital system, with less complexity and with a lower power consumption.

Figure 6.17 shows the end result. In this example, the antenna array is divided into four panels, each containing four antennas. The antennas are spaced half a wavelength apart throughout, so the individual panels are two wavelengths apart. Solid lines depict the actual radiation patterns, created using digital precoding and spatial filtering, while dashed lines depict the underlying spatially filtered beams.

Figure 6.17a directs the main beam towards a target mobile that is face-on to the array. (It might be useful to compare this result with the earlier illustration of grating lobes in Figure 6.5c.) Figure 6.17b directs a null towards an interferer by means of digital precoding, while Figure 6.17c achieves a similar result using a different spatially filtered beam.

6.5.2 Fully Connected Architecture

An alternative is the fully connected architecture shown in Figure 6.18. When using this architecture, an antenna connector does not just drive a single panel: instead, it drives all the antennas in the array. In fact, we can identify each antenna connector with a spatially

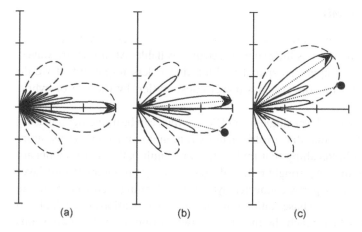

(a)   (b)   (c)

Figure 6.17 Example radiation patterns from a partly connected hybrid antenna array. (a) Target at 0° azimuth. (b) Target at 5° and interferer at −15°. (c) Target at 35° and interferer at 15°.

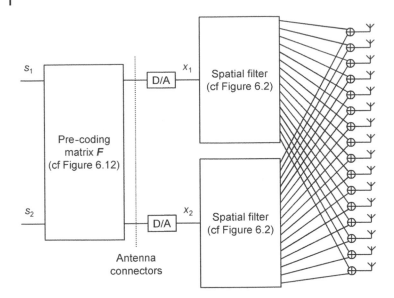

Figure 6.18 Fully connected architecture for hybrid beamforming.

filtered beam, with the choice of beam made by means of the spatial filter. The precoding matrix maps each outgoing layer onto all the beams that the spatial filters have chosen, by applying the amplitude and phase with which the layer should be transmitted within each beam. The architecture therefore allows the base station to transmit using more than one spatially filtered beam at the same time. However, the number of MIMO layers is still limited by the number of inputs to the precoder.

This architecture is more complex than the partly connected architecture described above, but also more flexible. It has fewer A/D and D/A converters than a fully digital system, so it still has a lower power consumption.

6.5.3 Millimetre Wave MIMO

Spatial multiplexing is an important technique at low radio frequencies, because the radio spectrum is heavily used and there is little spare spectrum available. At millimetre waves, these issues do not currently apply, so spatial multiplexing is less important. If we do implement it, then we can limit the system's power consumption by the use of hybrid beamforming [13–16].

The simplest technique is dual-layer single-user MIMO, which can be implemented using an array of cross-polarized antennas that are driven by two antenna connectors. To use more layers, we require multipath. Recalling that millimetre wave multipath environments are dominated by a small number of strongly reflected rays, we might implement four-layer single-user MIMO by means of a partly connected hybrid beamforming architecture, in the manner shown in Figure 6.19. The base station transmits two polarizations in each of two digitally precoded beams, which reach the mobile over two distinct rays. The two beams lie within the same spatial filter, so they can be transmitted at the same time. We can also direct the beams towards different mobiles in an implementation of multiple-user MIMO, gaining at least some of the benefits of mMIMO through the use of a large antenna array.

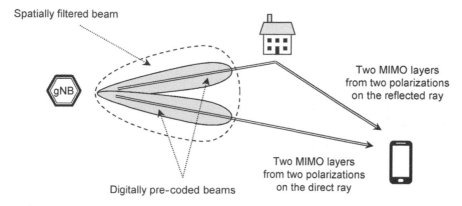

Spatially filtered beam

Two MIMO layers
from two polarizations
on the reflected ray

Two MIMO layers
from two polarizations
on the direct ray

Digitally pre-coded beams

Figure 6.19 Example of a sparse multipath environment supporting four-layer single-user MIMO.

6.6 Multiple Antennas at the Mobile

6.6.1 Architecture

Apart from our discussion of single-user MIMO, we have focussed so far on the use of multiple antennas at the base station. However, the mobile can also use multiple antennas for analogue beam steering and digital beamforming, particularly to improve the coverage of the system at high radio frequencies.

There are a few complications. Firstly, the lack of space limits the number of antennas that the mobile can support. Secondly, the mobile can be at any orientation, and may have to change from one beam to another if it rotates. Users can also obscure the mobiles' antennas with their hands.

Figure 6.20 shows the architecture that is typically used [17]. The mobile has a small number of antenna panels, each of which contains a small number of antennas. Those antennas are not omnidirectional: instead, each individual antenna has a wide-angle beam of its own. As shown in Figure 6.21, the mobile selects one panel at a time by means of a switch, the chosen panel being the one whose antennas point roughly towards the base station. The

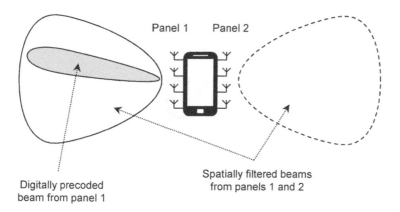

Panel 1   Panel 2

Digitally precoded
beam from panel 1

Spatially filtered beams
from panels 1 and 2

Figure 6.20 Architecture for multiple antenna transmission and reception at the mobile.

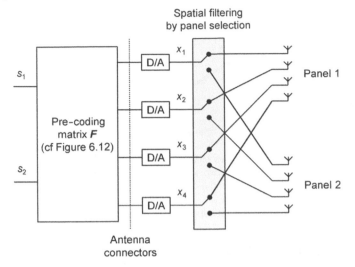

Figure 6.21 Digital precoding and analogue spatial filtering at the mobile.

switch acts as a simple spatial filter and selects the wide-angle antenna beam that corresponds to the chosen panel.

The mobile then drives the antennas within the selected panel using digital precoding, so as to create one or more digitally precoded beams that lie within the original spatially filtered beam. In a purely analogue system, the precoder might be replaced by a simple phase ramp, but the process is otherwise unchanged.

6.6.2 Beam Management

We now have a situation in which the base station and mobile are both communicating by means of spatially filtered beams, using one base station beam and one mobile beam at a time. That leads to the need for beam pair selection, as illustrated in Figure 6.22.

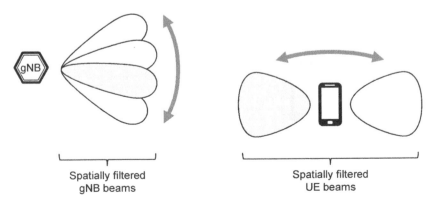

Figure 6.22 Beam pair selection.

The principles are as follows. On the downlink, the base station sweeps through its spatially filtered transmission beams, sending CSI reference symbols that identify its choice of beam. For each transmission beam, the mobile sweeps its receiver through the spatially filtered reception beams. The mobile identifies the best combination, reports the best choice of transmission beam back to the base station and records the corresponding reception beam. Later, the base station tells the mobile about its choice of transmission beam, using a field known as the *transmission configuration indication* (TCI) *state*. The mobile can then look up the corresponding beam for reception.

On the uplink, the mobile sweeps through its own spatially filtered transmission beams, sending sounding reference symbols in each one. For each transmission beam, the base station sweeps its receiver through the spatially filtered reception beams, identifies the best combination and records it. Later, the base station tells the mobile about its choice of transmission beam, using a field known as *spatial relationship information*. It can then receive using the corresponding reception beam. As before, shortcuts are possible in cases of channel reciprocity, because the uplink and downlink will be using the same pair of beams.

References

1 Analog Devices (2017) Massive MIMO and beamforming: The signal processing behind the 5G buzzwords. https://www.analog.com/en/analog-dialogue/articles/massive-mimo-and-beamforming-the-signal-processing-behind-the-5g-buzzwords.html (accessed 18 January 2020).

2 Gao, X. (2016) Massive MIMO in Real Propagation Environments. Doctoral Thesis, Lund University.

3 Joham, M., Nossek, J.A., and Utschick, W. (2005). Linear transmit processing in MIMO communications systems. *IEEE Transactions on Signal Processing* 53 (8): 2700–2712.

4 Ngo, H.Q. (2015) Massive MIMO: Fundamentals and System Design. Doctoral Thesis, Linköping University.

5 Marzetta, T.L. (2010). Noncooperative cellular wireless with unlimited numbers of base station antennas. *IEEE Transactions on Wireless Communications* 9 (11): 3590–3600.

6 Rusek, F., Persson, D., Lau, B.K. et al. (2013). Scaling up MIMO: opportunities and challenges with very large arrays. *IEEE Signal Processing Magazine* 30 (1): 40–60.

7 Larsson, E.G., Edfors, O., Tufvesson, F. et al. (2014). Massive MIMO for next generation wireless systems. *IEEE Communications Magazine* 52 (2): 186–195.

8 Lu, L., Li, G.Y., Swindlehurst, A.L. et al. (2014). An overview of massive MIMO: benefits and challenges. *IEEE Journal of Selected Topics in Signal Processing* 8 (5): 742–758.

9 Björnson, E., Larsson, E.G., and Marzetta, T.L. (2016). Massive MIMO: ten myths and one critical question. *IEEE Communications Magazine* 54 (2): 114–123.

10 Molisch, A.F., Ratnam, V.V., Han, S. et al. (2017). Hybrid beamforming for massive MIMO – a survey. *IEEE Communications Magazine* 55 (9): 134–141.

11 Ericsson (2018) Advanced antenna systems for 5G networks. https://www.ericsson.com/en/reports-and-papers/white-papers/advanced-antenna-systems-for-5g-networks (accessed 18 January 2020).

12 Ahmed, I., Khammari, H., Shahid, A. et al. (2018). A survey on hybrid beamforming techniques in 5G: architecture and system model perspectives. *IEEE Communications Surveys and Tutorials* 20 (4): 3060–3097.

13 Sun, S., Rappaport, T.S., Heath, R.W. et al. (2014). MIMO for millimeter-wave wireless communications: beamforming, spatial multiplexing, or both? *IEEE Communications Magazine* 52 (12): 110–121.

14 Alkhateeb, A., Mo, J., González-Prelcic, N. et al. (2014). MIMO precoding and combining solutions for millimeter-wave systems. *IEEE Communications Magazine* 52 (12): 122–131.

15 Heath, R.W., González-Prelcic, N., Rangan, S. et al. (2016). An overview of signal processing techniques for millimeter wave MIMO systems. *IEEE Journal of Selected Topics in Signal Processing* 10 (3): 436–453.

16 Bjornson, E., Van der Perre, L., Buzzi, S. et al. (2019). Massive MIMO in Sub-6 GHz and mmWave: physical, practical, and use-case differences. *IEEE Wireless Communications* 26 (2): 100–108.

17 Hong, W., Baek, K.-H., Lee, Y. et al. (2014). Study and prototyping of practically large-scale mmWave antenna systems for 5G cellular devices. *IEEE Communications Magazine* 52 (9): 63–69.

7

Architecture of the 5G New Radio

This is the first of six chapters that cover the air interface of 5G. In this chapter, we will begin by introducing the air interface's protocol stack, and identify the various different streams of information that are transmitted and received. We will then introduce the architecture of the physical layer, describe how it organizes information as a function of frequency and of time, and discuss its use of multiple input multiple output (MIMO) antennas. We will also lay out a framework for data transmission and reception on the 5G air interface, which will be applicable to several of the information streams that are covered in later chapters.

There are several relevant specifications, both for this chapter and for the ones that follow. The most important is the stage 2 specification for the 5G air interface, TS 38.300 [1], while References [2–5] define the details of the air interface's physical layer. Other introductions to the air interface include References [6, 7]. There are also several books containing more detailed accounts of its principles and implementation, which are listed in the Bibliography, and which will be valuable for readers who wish to take their understanding further.

7.1 Air Interface Protocol Stack

7.1.1 5G Protocol Stack

Figure 7.1 shows the simplest version of the protocol stack for the 5G air interface, from the viewpoint of the mobile [8]. The diagram assumes the use of architectural option 2, in which the mobile is controlled by the 5G core network, the next-generation radio access network (NG-RAN) and a master next-generation Node B (gNB) alone. Most of the protocols have the same names and objectives as their 4G equivalents, but their low-level implementations are different.

In the user plane, the mobile exchanges packets with an external data network such as the Internet. Those packets are created by applications in the mobile and the data network, and are transported over that network using protocols such as TCP and IP. In the control plane, the mobile exchanges non-access stratum signalling messages with the core network using the 5G mobility management (5GMM) and 5G session management (5GSM) protocols [9], and exchanges access stratum messages with the radio access network using the protocol for radio resource control (RRC) [10].

The next four protocols form layer 2 of an OSI protocol stack. The *service data adaptation protocol* (SDAP) [11] is new to 5G, and maps the quality of service (QoS) flows used by the 5G

An Introduction to 5G: The New Radio, 5G Network and Beyond, First Edition. Christopher Cox.
© 2021 John Wiley & Sons Ltd. Published 2021 by John Wiley & Sons Ltd.

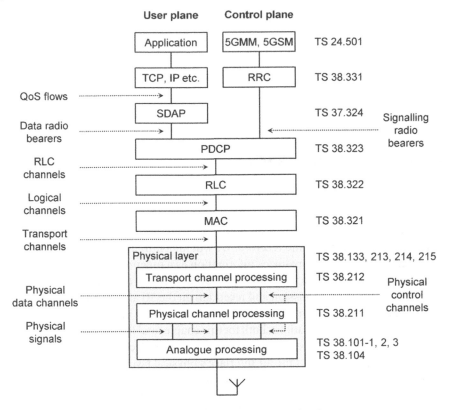

Figure 7.1 Air interface protocol stack for architectural option 2.

core network onto the data radio bearers used by the NG-RAN. The *packet data convergence protocol* (PDCP) [12] secures the air interface's traffic and signalling messages, and routes packets through the correct base station in cases of dual connectivity. The *radio link control* (RLC) protocol [13] segments any large data packets before transmission, and handles the high-level re-transmissions that we introduced in Chapter 5. The *medium access control* (MAC) protocol [14] schedules transmissions between the mobile and the base station, and controls the low-level operation of the physical layer. The MAC protocols in the mobile and network can also communicate with each other by the exchange of MAC *control elements* (CEs).

Most of the complexity is in the *physical layer*, which has three main parts. The *transport channel processor* [3] handles the remaining error management tasks from Chapter 5, including low-level re-transmissions by means of hybrid automatic repeat request (HARQ). The *physical channel processor* [2] handles the remaining digital signal processing tasks from Chapters 5 and 6, notably orthogonal frequency division multiple access (OFDMA) and multiple antenna processing. Finally, the *analogue processor* [15–19] converts the information to analogue form and transmits it. The physical layer also handles several procedures for measurement and control [4, 5, 20, 21].

Inside the core network, the 5GMM and 5GSM protocols are respectively implemented in the access and mobility management function (AMF) and the session management function (SMF). In the radio access network, the SDAP is in the gNB central unit's user plane, and the PDCP is split between its user and control planes. The RRC protocol lies mainly in the central unit's control plane, although the distributed unit can send and receive a small number of unciphered RRC signalling messages as well [22]. The RLC, MAC and physical layer protocols all lie in the distributed unit.

The information flows between the different protocols are known as channels and signals. Data and signalling messages travel on *RLC channels* between the PDCP and the RLC, *logical channels* between the RLC and MAC, *transport channels* between the MAC and physical layer, and *physical data channels* within the physical layer itself. Within the physical layer, the transport channel processors in the mobile and network communicate by exchanging uplink and downlink *control information*, which is delivered by means of *physical control channels*. Similarly, the two physical channel processors communicate by exchanging *physical signals*. At a higher level, the diagram also shows the locations in the protocol stack of the QoS flows and radio bearers.

7.1.2 Dual Connectivity

The protocol stack is more complex if the air interface is using carrier aggregation or dual connectivity, as shown in Figure 7.2 [23]. In cases of dual connectivity, the PDCP routes traffic and signalling messages through either the master cell group (MCG), the secondary cell group (SCG), or both. In cases of carrier aggregation, the MAC protocol routes traffic and signalling messages through a node's individual cells.

The PDCP is implemented in the master node (MN) if there are any MN-terminated data radio bearers, and in the secondary node (SN) if there are any SN-terminated bearers. Higher up in the protocol stack, the same applies to the SDAP, but only if the mobile is controlled by the 5G core network: there is no SDAP if the mobile is controlled by the evolved packet core. The RRC protocol is implemented in both the master and the secondary nodes, but the mobile acquires its RRC state from the master node alone. Lower down, the RLC, MAC and physical layer protocols are implemented in the master node if there are any MCG or split bearers, and in the secondary node if there are any SCG or split bearers.

In cases of multi-radio dual connectivity, an eNB uses the 4G versions of the RRC, RLC, MAC and physical layer protocols, while a gNB uses the 5G versions. There is only one version of the SDAP, which is shared by both air interface technologies. The PDCP is more complicated, as the air interface normally uses the 5G version of the protocol, but with a couple of exceptions. Firstly, a master eNB initially configures signalling radio bearer (SRB) 1 using the 4G version of the PDCP, but can subsequently configure SRB 1 and SRB 2 using either the 4G or 5G version. Secondly, a master eNB can configure MN-terminated MCG bearers using either the 4G or 5G version of the PDCP.

7.1.3 Channels and Signals

5G uses several different types of channel and signal, which are distinguished by the information that they carry and by the ways in which the information is processed [24–26]. They

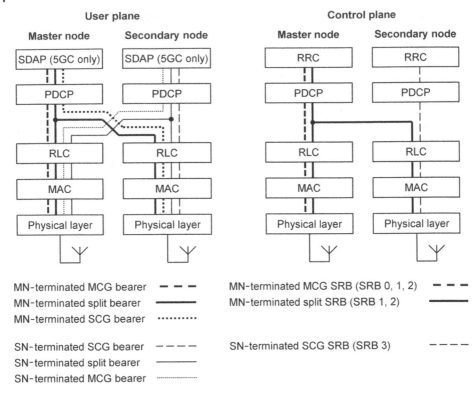

Figure 7.2 Air interface protocol stack in cases of dual connectivity. *Source:* Adapted from 3GPP TS 37.340.

are listed in Tables 7.1–7.5. Their names and purposes are almost identical to the equivalent channels and signals from LTE but, as in the case of the protocols, their implementations are different.

7.1.4 Information Flows

Figures 7.3 and 7.4 show the information carried by the channels and signals in 5G, and the relationships between them, for the uplink and downlink respectively. The arrows are

Table 7.1 Logical channels.

Channel	Name	Information carried	Direction
BCCH	Broadcast control channel	System information	DL
PCCH	Paging control channel	Paging messages	DL
CCCH	Common control channel	Signalling messages on SRB 0	UL, DL
DCCH	Dedicated control channel	Signalling messages on SRB 1–3	UL, DL
DTCH	Dedicated traffic channel	User plane traffic	UL, DL

Table 7.2 Transport channels.

Channel	Name	Information carried	Direction
BCH	Broadcast channel	Master information block	DL
PCH	Paging channel	Paging messages	DL
DL-SCH	Downlink shared channel	Downlink traffic and signalling	DL
RACH	Random access channel	Random access requests	UL
UL-SCH	Uplink shared channel	Uplink traffic and signalling	UL

Table 7.3 Control information.

Field	Name	Information carried	Direction
DCI	Downlink control information	Downlink scheduling commands	DL
		Uplink scheduling grants	
		Slot format indicators	
		Pre-emption indications	
		Uplink power control commands	
		Uplink sounding requests	
UCI	Uplink control information	Scheduling requests	UL
		Hybrid ARQ acknowledgements	
		Channel state information	

Table 7.4 Physical channels.

Channel	Name	Information carried	Direction
PBCH	Physical broadcast channel	BCH	DL
PDCCH	Physical downlink control channel	DCI	DL
PDSCH	Physical downlink shared channel	DL-SCH and PCH	DL
PRACH	Physical random access channel	RACH	UL
PUCCH	Physical uplink control channel	UCI	UL
PUSCH	Physical uplink shared channel	UL-SCH and UCI	UL

drawn from the viewpoint of the base station, so that uplink and downlink channels have arrows pointing upwards and downwards.

The 5G air interface delivers most of a mobile's traffic and signalling on three bi-directional logical channels. The *dedicated traffic channel* (DTCH) carries user plane traffic for a single mobile, while the *dedicated control channel* (DCCH) carries most of the control plane signalling. The *common control channel* (CCCH) carries signalling messages on SRB 0, for a mobile that is moving from RRC_IDLE or RRC_INACTIVE

Table 7.5 Physical signals.

Signal	Name	Use	Direction
PSS	Primary synchronization signal	Acquisition	DL
SSS	Secondary synchronization signal	Acquisition	DL
CSI-RS	Channel state information reference signal	Link adaptation	DL
SRS	Sounding reference signal	Link adaptation	UL
DM-RS	Demodulation reference signal	Demodulation	UL, DL
PT-RS	Phase-tracking reference signal	Demodulation	UL, DL

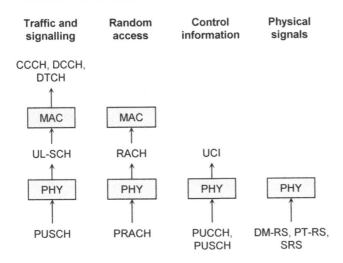

Figure 7.3 Uplink information flows. PHY: Physical layer.

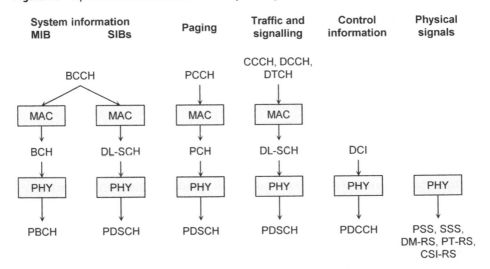

Figure 7.4 Downlink information flows. PHY: Physical layer.

into RRC_CONNECTED. Those channels are transported by the *downlink shared channel* (DL-SCH) and *physical downlink shared channel* (PDSCH) in the downlink, and by the *uplink shared channel* (UL-SCH) and *physical uplink shared channel* (PUSCH) in the uplink.

A base station can send RRC signalling messages known as known as *system information* to all the mobiles in the cell, which describe how the cell is configured. At the highest level, that information is divided into two unequal parts. A base station can transmit a *master information block* (MIB), which contains a few important parameters that the mobile needs to acquire a cell, and is delivered on the *broadcast control channel* (BCCH), *broadcast channel* (BCH) and *physical broadcast channel* (PBCH). It can also transmit several *system information blocks* (SIBs), which are numbered from SIB 1 to SIB 9. They are delivered first on the BCCH in the same way as the master information block, and then on the DL-SCH and PDSCH in the same way as the rest of the downlink signalling.

A base station can also compose RRC *paging* messages, if it needs to contact mobiles in the states of RRC_IDLE or RRC_INACTIVE. It delivers those messages on the *paging control channel* (PCCH) and *paging channel* (PCH), which maps onto the PDSCH.

The base station explicitly schedules most of a mobile's uplink data transmissions. However, that causes a problem, because the base station does not know the exact location of a mobile in the state of RRC_INACTIVE, and does not know that a mobile in RRC_IDLE is there at all. To solve the problem, the mobile can contact the base station without any prior scheduling by using the *random access channel* (RACH) and the *physical random access channel* (PRACH).

Downlink control information (DCI) mainly comprises scheduling commands and scheduling grants, by which the base station alerts the mobile to a forthcoming downlink transmission on the PDSCH, and gives it resources for an uplink transmission on the PUSCH. There are three main types of *uplink control information* (UCI): *hybrid ARQ acknowledgements* (HARQ-ACKs) of the base station's transmissions on the PDSCH, *scheduling requests* (SRs) if the mobile wishes to transmit on the PUSCH but does not have the resources to do so, and feedback about the nature of the downlink radio channel that is known as *channel state information* (CSI). Downlink control information is delivered on the *physical downlink control channel* (PDCCH), while uplink control information is delivered on the PUSCH if the mobile is transmitting data at the same time, and on the *physical uplink control channel* (PUCCH) otherwise.

As for the physical signals, the *primary synchronization signal* (PSS) and *secondary synchronization signal* (SSS) help the mobile to acquire a new cell by transmitting information such as the physical cell identity. The *demodulation reference signal* (DM-RS) is an amplitude and phase reference that helps the receiver to demodulate the incoming signal, while the *phase tracking reference signal* (PT-RS) supports that process by helping the receiver to compensate for timing jitter at high carrier frequencies. On the uplink, the *sounding reference signal* (SRS) helps the base station to estimate the best choice of uplink subcarriers, as well as the best modulation scheme, coding rate and antenna configuration. On the downlink, that role is filled by the *channel state information reference signal* (CSI-RS). (There is no equivalent to the downlink cell-specific reference signals from LTE: instead, their role is handled by the CSI-RS and DM-RS.)

Usually, a primary cell implements all of the information streams in Figures 7.3 and 7.4. A primary SCG cell omits the system information blocks, paging messages and CCCH, but implements the remainder. A secondary cell delivers downlink traffic on the PDSCH, but can omit any of the other information streams: for example, it can omit the PDCCH if it is scheduled from another cell, and the PUCCH if its uplink control information is delivered through another cell. It can also be configured as a downlink-only secondary cell, in which case it omits all of the uplink channels.

7.2 Frequency Bands and Combinations

7.2.1 Frequency Bands

As we discussed in Chapter 4, 5G can operate in two frequency ranges: frequency range 1 (FR1), which runs from 410 to 7125 MHz, and frequency range 2 (FR2), which runs from 24.25 to 52.6 GHz. Within these ranges, the 3GPP specifications support a large number of frequency bands [27], which are listed in Tables 7.6 and 7.7. These two tables also list the 3GPP releases in which the frequency bands were first introduced, and the releases in which they were first used by 5G.

Many of the bands in FR1 are inherited from earlier 3GPP systems, but are distinguished by the prefix 'n'. Some of the bands are new to 5G, while some of the old LTE bands are omitted. The bands in FR2 begin with n257, and are all new.

Each band can be used in one of four ways. Frequency division duplex (FDD) mode uses paired spectrum, in which the uplink and downlink can be transmitted continuously, but on different carrier frequencies. Time division duplex (TDD) mode uses unpaired spectrum, in which the uplink and downlink alternate on the same carrier. As we noted in Chapter 4, FDD mode is preferred in the lower parts of FR1, while TDD mode is preferred in the upper parts of FR1 and used exclusively in FR2.

A *supplementary downlink* (SDL) is a downlink-only secondary cell. It is used during carrier aggregation in conjunction with a normal FDD or TDD cell, to increase the downlink throughput for mobiles that are using services with a high downlink data rate, such as video streaming.

A *supplementary uplink* (SUL) is an uplink-only carrier frequency. It is also used in conjunction with a normal FDD or TDD cell, but for a different purpose: the supplementary uplink is on a lower carrier frequency than the normal uplink, so it has a lower propagation loss and improves the uplink coverage for mobiles that are a long way from the base station. Although the two names are similar, there are a couple of distinctions: the supplementary uplink is not a cell (that being an entity with a cell identity that is transmitted on a downlink carrier frequency), and it is used only as an alternative to the normal uplink, never at the same time. A cell only implements the PUCCH on one of the normal or supplementary uplink, not both.

The *channel bandwidth* is the radio bandwidth that the carrier occupies. Frequency range 1 supports 13 different channel bandwidths of 5, 10, 15, 20, 25, 30, 40, 50, 60, 70, 80, 90 and 100 MHz. Frequency range 2 supports four channel bandwidths of 50, 100, 200 and 400 MHz.

Table 7.6 Frequency bands in frequency range 1.

Band	Initial release	First use by 5G	Uplink band (MHz)	Downlink band (MHz)	Duplex mode
n1	Rel-99	Rel-15	1920–1980	2110–2170	FDD
n2	Rel-99	Rel-15	1850–1910	1930–1990	FDD
n3	Rel-5	Rel-15	1710–1785	1805–1880	FDD
n5	Rel-6	Rel-15	824–849	869–894	FDD
n7	Rel-7	Rel-15	2500–2570	2620–2690	FDD
n8	Rel-7	Rel-15	880–915	925–960	FDD
n12	Rel-8	Rel-15	699–716	729–746	FDD
n14	Rel-8	Rel-16	788–798	758–768	FDD
n18	Rel-9	Rel-16	815–830	860–875	FDD
n20	Rel-9	Rel-15	832–862	791–821	FDD
n25	Rel-10	Rel-15	1850–1915	1930–1995	FDD
n28	Rel-11	Rel-15	703–748	758–803	FDD
n29	Rel-11	Rel-16	—	717–728	SDL
n30	Rel-12	Rel-16	2305–2315	2350–2360	FDD
n34	Rel-99	Rel-15	2010–2025	2010–2025	TDD
n38	Rel-7	Rel-15	2570–2620	2570–2620	TDD
n39	Rel-8	Rel-15	1880–1920	1880–1920	TDD
n40	Rel-8	Rel-15	2300–2400	2300–2400	TDD
n41	Rel-10	Rel-15	2496–2690	2496–2690	TDD
n48	Rel-14	Rel-16	3550–3700	3550–3700	TDD
n50	Rel-15	Rel-15	1432–1517	1432–1517	TDD
n51	Rel-15	Rel-15	1427–1432	1427–1432	TDD
n65	Rel-13	Rel-16	1920–2010	2110–2200	FDD
n66	Rel-13	Rel-15	1710–1780	2110–2200	FDD
n70	Rel-14	Rel-15	1695–1710	1995–2020	FDD
n71	Rel-15	Rel-15	663–698	617–652	FDD
n74	Rel-15	Rel-15	1427–1470	1475–1518	FDD
n75	Rel-15	Rel-15	—	1432–1517	SDL
n76	Rel-15	Rel-15	—	1427–1432	SDL
n77	Rel-15	Rel-15	3300–4200	3300–4200	TDD
n78	Rel-15	Rel-15	3300–3800	3300–3800	TDD
n79	Rel-15	Rel-15	4400–5000	4400–5000	TDD
n80	Rel-15	Rel-15	1710–1785	—	SUL
n81	Rel-15	Rel-15	880–915	—	SUL
n82	Rel-15	Rel-15	832–862	—	SUL

(continued)

Table 7.6 (Continued)

Band	Initial release	First use by 5G	Uplink band (MHz)	Downlink band (MHz)	Duplex mode
n83	Rel-15	Rel-15	703–748	—	SUL
n84	Rel-15	Rel-15	1920–1980	—	SUL
n86	Rel-15	Rel-15	1710–1780	—	SUL
n89	Rel-16	Rel-16	824–849	—	SUL
n90	Rel-16	Rel-16	2496–2690	2496–2690	TDD
n91	Rel-16	Rel-16	832–862	1427–1432	FDD
n92	Rel-16	Rel-16	832–862	1432–1517	FDD
n93	Rel-16	Rel-16	880–915	1427–1432	FDD
n94	Rel-16	Rel-16	880–915	1432–1517	FDD
n95	Rel-16	Rel-16	2010–2025	—	SUL

Source: Adapted from 3GPP TS 38.104.

Table 7.7 Frequency bands in frequency range 2.

Band	Initial release	First use by 5G	Frequency band (GHz)	Duplex mode
n257	Rel-15	Rel-15	26.5–29.5	TDD
n258	Rel-15	Rel-15	24.25–27.5	TDD
n260	Rel-15	Rel-15	37–40	TDD
n261	Rel-15	Rel-15	27.5–28.35	TDD

Source: Adapted from 3GPP TS 38.104.

7.2.2 Band Combinations

During carrier aggregation, a mobile communicates with a single base station using multiple cells, which operate on different carrier frequencies that are known as *component carriers*. There are three types of carrier aggregation: *contiguous intra-band aggregation*, in which the cells are in the same frequency band and on adjacent frequencies; *non-contiguous intra-band aggregation*, in which the cells are in the same frequency band but are separated by a frequency gap; and *inter-band aggregation*, in which the cells are in different frequency bands.

The specifications list a large number of *band combinations* in which carrier aggregation is allowed, and similar combinations for dual connectivity and the supplementary uplink. They are defined in response to requests from individual network operators, and are used for two reasons: to ensure that 3GPP defines test requirements for each individual combination, and to limit the implementation complexity of the mobile. As simple examples, the notation CA_n77 denotes 3GPP support for contiguous intra-band aggregation in band n77, while CA_n77_n257 denotes inter-band aggregation in bands n77 and n257. At the other extreme, DC_1-3-5-7_n78-n257 denotes inter-band EUTRA-NR dual connectivity (EN-DC), in which the master eNB uses inter-band aggregation in LTE bands 1, 3, 5 and 7, and the secondary gNB uses 5G bands n78 and n257 [28–30].

7.2.3 Bandwidth Classes

A mobile tells its master node about its own support for carrier aggregation, dual connectivity and the supplementary uplink, as part of the RRC signalling message *UE Capability Information*. The information is in two parts: the mobile first states which band combinations it supports, and then states how many contiguous intra-band carriers it supports for the individual bands within each combination. That second part is defined using a letter known as the *carrier aggregation bandwidth class*, which has a different meaning in the two frequency ranges [31, 32].

A mobile often supports fewer carriers in the uplink than it does in the downlink, so it makes separate declarations for the two directions. On the downlink, for example, the mobile might declare support for carrier aggregation using two adjacent carriers in band n77, and four adjacent carriers in band n257 that together span no more than 800 MHz, in a configuration denoted CA_n77C-n257F [33]. On the uplink, however, Release 15 restricts the mobile's capabilities in those bands to one component carrier each, in the configuration CA_n77A-n257A.

7.3 Frequency Domain Structure

7.3.1 Numerologies

5G uses five different subcarrier spacings, which are listed in Table 7.8 [34, 35]. Together with other related parameters, each of these is known as a *numerology*, and is identified by a *subcarrier spacing configuration* that is denoted by the Greek letter μ.

As shown in Table 7.8, frequency range 1 uses subcarrier spacings of 15, 30 and 60 kHz for most of its transmissions, while FR2 uses subcarrier spacings of 60 and 120 kHz. The PSS, SSS and PBCH use a different set of subcarrier spacings, for reasons that will become clearer when we discuss the synchronization procedure in Chapter 8. The subcarrier spacings for the PRACH are completely different, and will be introduced in Chapter 9.

Table 7.8 Numerologies.

μ	Subcarrier spacing (kHz)	RB bandwidth (kHz)	Data in FR1	Data in FR2	PSS, SSS and PBCH in FR1	PSS, SSS and PBCH in FR2
0	15	180	✓		✓	
1	30	360	✓		✓	
2	60	720	✓	✓		
3	120	1440		✓		✓
4	240	2880				✓

A mobile can use different numerologies on the uplink and downlink and, in cases of carrier aggregation or dual connectivity, in each of its serving cells. In any one cell, however, the mobile only uses one uplink numerology and one downlink numerology at a time. On the other hand, the cell can use multiple numerologies at the same time. To do so, the cell can assign different numerologies to individual parts of its channel bandwidth, and can then communicate on those different numerologies with different mobiles.

The 3GPP specifications state which combinations of frequency band, channel bandwidth and subcarrier spacing are allowed. In frequency range 1, for example, band n1 supports a bandwidth of 5 MHz using a subcarrier spacing of 15 kHz, and bandwidths of 10, 15 and 20 MHz using subcarrier spacings of 15, 30 and 60 kHz. Many combinations are omitted, typically because the channel bandwidth is too wide for the frequency band, or because they would result in too many subcarriers or too few.

Subcarriers are grouped into units known as *resource blocks* (RBs), which are used during scheduling. Each resource block contains 12 subcarriers, but its bandwidth in MHz depends on its subcarrier spacing. Unlike in LTE, a 5G resource block is only defined as a function of frequency: it has no particular time duration.

7.3.2 Transmission Bandwidth Configuration

The *transmission bandwidth configuration* is the set of resource blocks on which the base station transmits and receives. Tables 7.9 and 7.10 show the largest permitted values of the transmission bandwidth configuration in frequency ranges 1 and 2, for each combination of channel bandwidth and subcarrier spacing. There is an absolute maximum of 275 resource blocks (3300 subcarriers), so as to limit the implementation complexity of the mobile.

The upper and lower edges of the channel bandwidth are not occupied by resource blocks: instead, they are used as *guard bands*, to reduce the interference between the carrier of interest and the adjacent carriers at frequencies above and below. In LTE, the resource blocks usually occupied 90% of the channel bandwidth. In 5G, the specifications assume that the OFDMA signal is filtered more effectively before transmission, which reduces the amount of power that leaks outside the transmission bandwidth configuration. That in turn allows the resource blocks to occupy more of the channel bandwidth, with a typical value around 95% and a maximum of 98.28% for a channel bandwidth of 100 MHz and a subcarrier spacing of 30 kHz.

Table 7.9 Maximum transmission bandwidth configurations in frequency range 1.

Subcarrier spacing (kHz)	Channel bandwidth (MHz)												
	5	10	15	20	25	30	40	50	60	70	80	90	100
15	25	52	79	106	133	160	216	270	—	—	—	—	—
30	11	24	38	51	65	78	106	133	162	189	217	245	273
60	—	11	18	24	31	38	51	65	79	93	107	121	135

Source: Adapted from 3GPP TS 38.104.

Table 7.10 Maximum transmission bandwidth configurations in frequency range 2.

Subcarrier spacing (kHz)	Channel bandwidth (MHz)			
	50	100	200	400
60	66	132	264	—
120	32	66	132	264

Source: Adapted from 3GPP TS 38.104.

The specifications also state the minimum permitted values for the widths of the guard bands. Subject to a clarification that we will discuss later in this chapter, those values are consistent with the maximum transmission bandwidth configurations in Table 7.10, a constraint that fixes the location of the maximum transmission bandwidth configuration within the channel bandwidth. Unlike in LTE, however, the cell can use a transmission bandwidth configuration that is smaller than the permitted maximum. If it does so, then the location of the transmission bandwidth can be adjusted, provided that the guard bands remain sufficiently large.

7.3.3 Global and Channel Frequency Rasters

In the RRC signalling messages on the 5G air interface, radio frequencies are defined by means of an integer known as the *New Radio absolute radio frequency channel number* (NR-ARFCN). Each unit change of NR-ARFCN changes the corresponding *RF reference frequency* through a step that is known as the *global frequency raster*. As summarized in Table 7.11, the global raster has a value of 5, 15 or 60 kHz, depending on the frequency range.

In practice, however, a network operator can only adjust its reference frequencies using a larger step size, which is known as the *channel frequency raster*. The channel raster is 100 kHz in frequency bands that have been inherited from LTE, and 15, 30, 60 or 120 kHz otherwise, so each step of the channel raster might change the NR-ARFCN by more than one unit. For example, band n1 uses a global frequency raster of 5 kHz and a channel raster of 100 kHz, such that the network operator can only adjust the NR-ARFCN in steps of 20.

Table 7.11 Values of the global frequency raster and the channel raster.

Frequency range (GHz)	Global frequency raster (kHz)	Band-specific channel rasters (kHz)	Corresponding steps of NR-ARFCN
0–3	5	15, 30, 100	3, 6, 20
3–24.25	15	15, 30	1, 2
24.25–100	60	60, 120	1, 2

Source: Adapted from 3GPP TS 38.104.

7.3.4 Common Resource Blocks

Common resource blocks (CRBs) define a numbering scheme for the resource blocks in the transmission bandwidth configuration. A cell has one set of common resource blocks for each of its numerologies, in which the resource blocks are numbered upwards with increasing frequency, starting from zero.

The common resource blocks on the different numerologies are aligned at a frequency known as *point A*, which identifies subcarrier 0 on CRB 0 in each of the numerologies used. The exact location of point A is adjustable: in particular, it can lie below the base of a particular numerology's transmission bandwidth configuration, in which case the lowest-numbered CRBs remain unused. The base station signals the location of point A to the mobile by means of an NR-ARFCN.

Figure 7.5 illustrates the end result. In this example, the base station is transmitting in a channel bandwidth of 10 MHz, using the largest transmission bandwidth configurations that are allowed. The base station has defined point A to be the base of the transmission bandwidth configuration for a subcarrier spacing of 15 kHz. For subcarrier spacings of 30 and 60 kHz, CRB 0 is unused.

Figure 7.5 highlights two other issues. Firstly, the definition of point A implies that the subcarriers on different numerologies are aligned with each other, but the edges of the

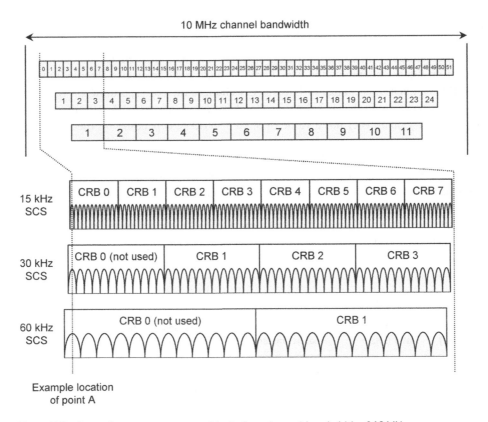

Figure 7.5 Example common resource blocks in a channel bandwidth of 10 MHz.

resource blocks are misaligned. Secondly, if the cell is using the maximum transmission bandwidth configuration, then one of the guard bands equals the minimum permitted value that we noted earlier in this chapter, while the other is wider by one subcarrier spacing. When using a 10 MHz channel bandwidth and a 15 kHz subcarrier spacing, for example, the transmission bandwidth configuration spans 52 resource blocks, in other words 9360 kHz. In that configuration, the minimum guard band spans 312.5 kHz, while the other guard band spans 327.5 kHz.

7.3.5 Bandwidth Parts

As in enhanced machine-type communication (eMTC) and the narrowband internet of things (NB-IoT), but unlike in Release 8 LTE, the mobile does not have to handle the whole of the base station's transmission bandwidth configuration. Instead, the base station can restrict a mobile to a smaller *bandwidth part* (BWP) [36–40]. That reduces the mobile's power consumption, and allows manufacturers to build inexpensive mobiles that are completely unable to communicate over the whole of the transmission band.

Every bandwidth part has a direction (uplink or downlink), and occupies a contiguous range of common resource blocks with a particular subcarrier spacing. In Figure 7.5, for example, the base station might define a bandwidth part with a subcarrier spacing of 15 kHz and a bandwidth of 24 resource blocks, beginning at CRB 28 and ending at CRB 51.

Every cell has an initial downlink bandwidth part and an initial uplink bandwidth part. Those are used for initial access, for example by mobiles that are accessing a primary cell from the states of RRC_INACTIVE and RRC_IDLE. Subsequently, the base station can configure a mobile with a maximum of four additional uplink bandwidth parts and four additional downlink bandwidth parts, by means of RRC signalling.

The mobile only communicates using one *active uplink BWP* and one *active downlink BWP* at a time. The base station changes a mobile's active bandwidth parts using downlink control information on the PDCCH. To supplement that procedure, the base station can also identify one of the mobile's downlink bandwidth parts as the *default downlink BWP*. The mobile reverts to the default downlink bandwidth part if it has not communicated with the base station for a time that is adjustable between 2 and 2560 ms. That ensures that the base station can find the mobile in the event of communication problems, for example if the mobile has moved to the wrong downlink bandwidth part due to a bit error in its downlink control information.

In FDD mode, the uplink and downlink bandwidth parts are completely independent, and the active uplink and downlink bandwidth parts can be independently changed. In TDD mode, the bandwidth parts are defined in {uplink, downlink} pairs in which the uplink and downlink have the same centre frequency, while the active uplink and downlink bandwidth parts always change together.

7.3.6 Virtual and Physical Resource Blocks

The resource blocks inside a bandwidth part are known as *physical resource blocks* (PRBs). They are numbered upwards from zero at the base of the bandwidth part, and map directly onto the common resource blocks that we defined earlier in this chapter. In practice, however, the base station schedules the mobile using *virtual resource blocks* (VRBs). Those are mapped onto the corresponding PRBs either directly, or by re-ordering them with an interleaver that gives the mobile some extra frequency diversity.

7.4 Time Domain Structure

7.4.1 Frame Structure

The timing of the 5G air interface is based on a timing unit T_c, which is defined as follows:

$$T_c = \frac{1}{4096 \times 480 \text{ kHz}} \approx 0.51 \text{ ns} \tag{7.1}$$

where T_c is the sample duration at the output of the transmitter's inverse fast Fourier transform (FFT), if it is using a subcarrier spacing of 480 kHz (twice the value that is actually supported), and a 4096 point FFT. It is 1/64 of the equivalent timing unit from LTE.

Figure 7.6 shows the resultant timing, while Table 7.12 lists the details for each of the numerologies [41]. The longest structure is a *frame*, which has a duration of 10 ms (19 660 800 T_c), and is used for a few slowly repeating signalling messages. Each frame is identified using a 10-bit *system frame number* (SFN), which runs from 0 to 1023 before repeating. In turn, a frame is divided into 10 *subframes*, each with a duration of 1 ms (1 966 080 T_c).

Depending on the choice of numerology, a subframe contains from 1 to 16 *slots*. The slot is the main time period for scheduling in 5G, as the air interface's uplink and downlink transmissions last either for one slot at a time, or for a duration that is closely related to the slot. Except at the smallest subcarrier spacing, those scheduling periods are shorter than the subframes that were used by LTE. That helps 5G to support low-latency communications, particularly when using a numerology with a wide subcarrier spacing and a short symbol duration.

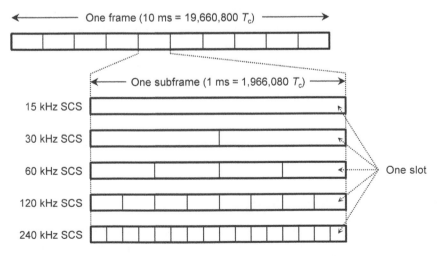

Figure 7.6 Frames, subframes and slots.

Table 7.12 Details of the frame structure for the different numerologies.

Subcarrier spacing configuration	0	1	2	3	4	2
Cyclic prefix	Normal	Normal	Normal	Normal	Normal	Extended
Slots per subframe	1	2	4	8	16	4
Symbols per slot	14	14	14	14	14	12
Symbol duration (units of T_c)	131 072	65 536	32 768	16 384	8 192	32 768
Cyclic prefix (units of T_c)	9 216	4 608	2 304	1 152	576	8 192
Symbol duration (µs)	66.66	33.33	16.66	8.33	4.16	16.66
Cyclic prefix (µs)	4.69	2.34	1.17	0.58	0.29	4.16

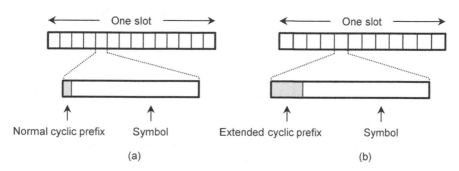

Figure 7.7 Slots and symbols, when using the (a) normal and (b) extended cyclic prefix.

A slot can be organized in two ways, which are shown in Figure 7.7. Normally, a slot contains 14 symbols. Each of those is preceded by a *normal cyclic prefix*, whose duration determines the largest delay spread that the receiver can handle without any inter-symbol interference. Ideally, the normal cyclic prefix would last for 1/14 of a symbol, but that would not be an integer multiple of T_c. Instead, most of the normal cyclic prefixes last for 9/128 of a symbol, while the two cyclic prefixes at the start of the subframe and halfway through are longer by 1024 T_c. At a subcarrier spacing of 15 kHz, for example, most of the cyclic prefixes last for 9216 T_c, but the first and eighth last for 10 240 T_c.

If the subcarrier spacing is 60 kHz, then the number of symbols per slot can be reduced to 12. If it is, then each symbol is preceded by an *extended cyclic prefix* which lasts for 1/4 of a symbol, and which has roughly the same duration as the normal cyclic prefix with a 15 kHz subcarrier spacing. Its main use in 5G is to help network operators carry out low-latency communications in an environment with a large delay spread by resolving the conflicting requirements for a short symbol duration and a long cyclic prefix [42].

7.4.2 Timing Advance

A mobile's uplink transmissions take place earlier than would otherwise be expected, by an amount known as the *timing advance* (TA). Timing advance ensures that the signals from different mobiles reach the base station at roughly the same time, even if the mobiles are at different distances from the base station, so as to prevent any inter-symbol interference between them. It can be estimated as follows:

$$\text{TA} \approx \frac{2r}{c} \tag{7.2}$$

where r is the distance from the base station to the mobile, and c is the speed of light.

In 5G, a mobile's timing advance is adjusted in steps, each of which is 1/128 of a symbol long [43, 44]. The base station initializes the timing advance as part of the random access procedure that we will discuss in Chapter 9. The maximum timing advance in that procedure is 3846 steps, which is about two slots: that limits the round-trip time to 2 ms at a subcarrier spacing of 15 kHz, and limits the cell radius to 300 km. Subsequently, the base station adjusts the timing advance using a MAC control element known as a timing advance command, with a maximum adjustment of 63 steps at a time.

In carrier aggregation, a base station might transmit different cells from different physical locations. To handle that situation, those cells can be grouped into *timing advance groups* (TAGs), each with a different value for the uplink timing advance.

7.4.3 TDD Configurations

In TDD mode, the uplink and downlink are on the same carrier frequency. To stop them from interfering, the base station assigns each individual symbol to either the uplink or downlink, but not both.

To see how this is done, let us first note that transmissions on the 5G air interface fall into three broad categories: periodic transmissions, which are configured using RRC signalling alone; semi-persistent transmissions, which are pre-configured by RRC signalling but switched on and off using MAC control elements; and dynamic transmissions, which are scheduled one at a time by means of the PDCCH. Dynamic scheduling is the usual mechanism for the PDSCH and PUSCH, while periodic and semi-persistent scheduling are more common for the CSI-RS and SRS.

The base station can then assign each symbol in three ways: as a downlink, uplink or flexible symbol. Although the details are complicated, the principles are as follows [45]. In downlink symbols, downlink transmissions are allowed, but uplink transmissions are suppressed. In uplink symbols, uplink transmissions are allowed, but downlink transmissions are suppressed. In flexible symbols, dynamic uplink and downlink transmissions are allowed, but periodic and semi-persistent transmissions (for example, a mobile's periodic transmissions of the SRS) are suppressed.

Usually, the base station assigns its symbols by means of a *TDD configuration*. That sets up a repeating pattern of slots, which begins with a number of downlink-only slots and ends with a number of uplink-only slots. Those are separated by a switching slot, which begins with downlink symbols, continues with flexible symbols and ends with uplink symbols. The pattern's duration lies between 0.5 and 10 ms, and the base station can optionally set up two patterns that alternate with each other.

The TDD configuration is defined on a reference subcarrier spacing, which is the smallest that the base station will ever use and which contains the longest symbols. If the mobile is assigned to a larger subcarrier spacing containing shorter symbols, then the pattern simply maps in time from the reference subcarrier spacing to the others. Figure 7.8 shows an

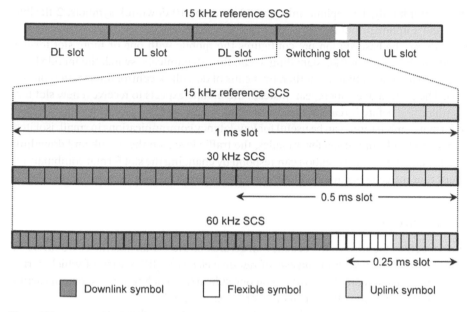

Figure 7.8 Example of a TDD configuration.

example of a single pattern containing three downlink slots, one switching slot and one uplink slot, which is drawn from the test specifications for 5G [46].

If the base station does not schedule any dynamic transmissions in its flexible symbols, then those symbols can be used as a *guard period*. The guard period allows the mobile to apply timing advance to the uplink symbols that follow, without any risk of them colliding with the downlink symbols that came before.

There are then two types of TDD configuration. Initially, the base station configures the mobile with a cell-specific *common TDD configuration*. That configuration might have a large number of flexible symbols to support a long guard period for mobiles that are close to the cell edge. Later, the base station can re-configure the mobile with a *dynamic TDD configuration*, using mobile-specific RRC signalling. In that configuration, the base station can re-assign any of the previous flexible symbols to the uplink or downlink, so as to shorten the guard period for mobiles that are nearby.

7.4.4 Slot Format Combinations

For most purposes, TDD configurations work well. For more dynamic control, the base station can re-assign any of the previous flexible symbols to the uplink and downlink using *slot format combinations* [47].

A slot format combination is a pattern of slots, with a maximum duration of 256 slots. Within that pattern, each slot is associated with a *slot format* that has a value of 0–255. In turn, the slot format acts as a pointer into a look-up table, which identifies each of its symbols as uplink, downlink or flexible symbols. As a simple example, the pattern in Figure 7.8 could be reproduced using a slot format combination of {0, 0, 0, 32, 1}, where a slot format of 0 is downlink only, 1 is uplink only, and 32 contains 10 downlink symbols, 2 flexible symbols and 2 uplink symbols.

To use them, the base station pre-configures the mobile with one or more slot format combinations by means of RRC signalling. It can then dynamically switch the mobile from one slot format combination to another, by means of downlink control information on the PDCCH. The pattern does not repeat: instead, the mobile expects to receive a new slot format combination before it reaches the end of an old one.

Slot format combinations can be useful for low-latency communications in small, isolated cells. In a small cell containing few mobiles, the traffic levels on the uplink and downlink can quickly change. The base station can respond by changing the slot format combination, confident that the cell's isolation will prevent any interference with its neighbours.

7.4.5 Resource Grid

Bringing together the time and frequency structures of the 5G air interface, we arrive at a *resource grid*. A resource grid consists of *resource elements* (REs), each of which corresponds to the transmission of one symbol on one subcarrier. Each resource element carries information from just one physical channel or physical signal.

In LTE, the specifications defined much of the mapping from channels and signals onto resource elements. In 5G that mapping is more flexible to help ensure that early design choices will not limit 3GPP's ability to enhance the system in the future. Instead, the base station assigns resource elements to channels and signals by means of RRC signalling, and provides the mobile with parameters that configure the way in which it will transmit or receive. Each of these assignments is known as a *resource*.

7.5 Multiple Antennas

7.5.1 Antenna Ports

An *antenna port* is not an antenna. Rather, an antenna port is a stream of information in the transmitter which carries one or more physical channels and physical signals, and which drives one or more antennas using digital precoding and/or analogue spatial filtering.

As shown in Tables 7.13 and 7.14, the specifications define four sets of antenna ports on the uplink and four on the downlink, each of which carries a different set of channels and signals [48]. The exact interpretation of the antenna ports differs from one set of channels and signals to another, while some of the antenna ports can be interpreted in more than one way, depending on how the channels and signals have been configured.

At least over short intervals of time and frequency, the transmitter processes different symbols on the same antenna port using the same precoder and the same spatial filter. In practical terms, that means that the receiver can carry out channel estimation by measuring the reference signals that arrive on a particular antenna port, and can then use the result to demodulate a physical channel that arrives on the same antenna port. On the downlink, for example, the base station can transmit up to 12 parallel PDSCH layers on antenna ports 1000 to 1011, using a mix of single-user and multiple-user MIMO. The mobile estimates the

Table 7.13 Uplink antenna ports.

Antenna ports	Channels and signals	Interpretation
0–11	DM-RS and PT-RS for PUSCH	One for each of the cell's MIMO layers
1000–1003	SRS, PUSCH	One for each of the UE's SU-MIMO layers, either one per digital antenna connector, or precoded, with one per digitally precoded beam
2000	PUCCH + DM-RS	One per cell
4000	PRACH	One per cell

SU-MIMO: single-user MIMO

Table 7.14 Downlink antenna ports.

Antenna ports	Channels and signals	Interpretation
1000–1011	PDSCH + DM-RS and PT-RS	One for each of the cell's MIMO layers, up to eight layers with SU-MIMO alone, or up to four layers per UE with MU-MIMO
2000	PDCCH + DM-RS	One per cell
3000–3031	CSI-RS	Either one per digital antenna connector, or precoded, with one per digitally precoded beam
4000	PSS, SSS, PBCH + DM-RS	One per cell

MU-MIMO: multiple-user MIMO; SU-MIMO: single-user MIMO

channel on those antenna ports by measuring the corresponding demodulation and phase tracking reference signals, and uses the result to demodulate the PDSCH.

7.5.2 Relationships Between Antenna Ports

The above discussion covered the relationships between channels and signals that are transmitted on the same antenna port. However, it did not tell us anything about the relationships between different antenna ports. Those relationships are especially important when configuring the spatial filters that are used for analogue beamforming, in both the transmitter and the receiver.

On the uplink, these issues are handled using *spatial relationship information*. That states that the mobile should transmit on one antenna port using the same spatially filtered beam that it has previously used to transmit or receive on a different antenna port. Spatial relationships can be configured in two ways: some are defined by the specifications, while others are signalled to the mobile by means of RRC signalling, MAC control elements or downlink control information.

On the downlink, these issues are handled using *quasi co-location* (QCL) relationships [49]. There are four types of quasi co-location. The most relevant is quasi co-location type D, which indicates that two sets of downlink antenna ports share the same spatially filtered transmission beam, and that the mobile should receive them using the same spatially filtered reception beam. Once again, these relationships can be configured in two ways: some are defined by the specifications, while others are signalled to the mobile using fields known as *transmission configuration indicator* (TCI) states.

In addition, quasi co-location types A, B and C indicate that two sets of downlink antenna ports share the same parameters for some combination of average delay, delay spread, average Doppler shift and Doppler spread. They are used to denote a CSI-RS antenna port as a *tracking reference signal* (TRS), which the mobile measures to maintain time and frequency synchronization with the base station.

7.6 Data Transmission

7.6.1 Transport Channel Processing

At this stage, it is useful to lay out a framework for the signal processing in the air interface's physical layer. While none of the channels and signals use this framework exactly, they all follow it closely, which will help us to focus on the individual details in the chapters that follow.

The transport channel processor is illustrated in Figure 7.9 and uses the procedures that we covered in Chapter 5 [50]. With variations, these steps are used by all the transport channels apart from the simpler RACH, and by the uplink and downlink control information.

The process can begin in two ways. In the case of transport channels, the physical layer receives one or two protocol data units from the MAC protocol that are known as *transport blocks*. Alternatively, the physical layer composes uplink or downlink control information of its own.

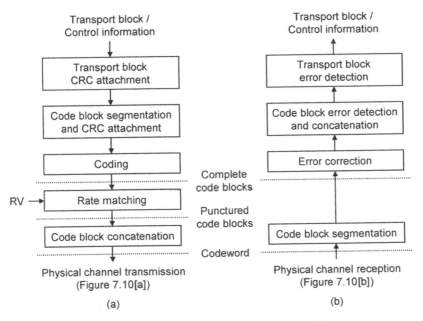

Figure 7.9 Transport channel processing. (a) Transmission. (b) Reception.

The transmitter begins by attaching a cyclic redundancy check (CRC) of up to 24 bits, which the receiver uses for error detection. In some cases, it segments the resulting information into a number of *code blocks*, and attaches a CRC to each one. It then applies error correction coding, using one of three algorithms: a *low-density parity check* (LDPC) code for large block lengths, a *polar code* for small block lengths or a *block code* for very small block lengths [51]. Finally, the transmitter selects the coded bits that it will actually transmit by means of rate matching, concatenates any code blocks together and sends the result onwards in the form of *codewords*.

7.6.2 Physical Channel Processing

Lower down, the physical layer supports a hybrid multiple-antenna architecture, in which the base station and mobile can both use a mix of digital beamforming and analogue beam selection. Figure 7.10 shows the most important steps in the physical channel processor, for a channel that supports multiple-layer transmission and hybrid ARQ, and for its associated demodulation reference signal [52, 53].

The transmitter begins by mixing the coded bits with a cell- or mobile-specific pseudo-random scrambling code, so as to randomize the interference between devices that are transmitting on the same radio frequency. It then converts the bits to symbols by means of modulation. The stage of layer mapping only applies to the PUSCH and PDSCH, and converts a stream of symbols for a single mobile into a set of parallel MIMO layers. At the same time, the transmitter generates one stream of demodulation reference signals for

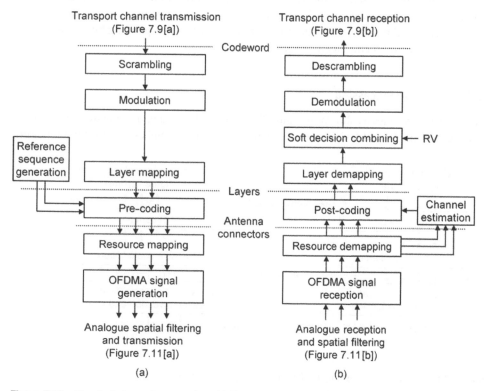

Figure 7.10 Physical channel processing. (a) Transmission. (b) Reception.

each MIMO layer. (We have deliberately avoided the term *antenna port* in that description, because its exact meaning varies from one channel or signal to another.)

The precoder is an implementation of digital beamforming, which maps the outgoing layers onto digital antenna connectors in the manner shown in Figure 6.12. It does so by applying a precoding matrix, in which each matrix element is an amplitude scaling factor and phase shift that maps one layer onto one antenna connector. The demodulation reference signals are precoded in the same way as the corresponding channels.

After precoding, the transmitter maps the symbols onto the underlying resource elements, each of which corresponds to the transmission of one symbol on one subcarrier and one antenna port. The transmitter then generates the time domain OFDMA signals by carrying out an inverse FFT and inserting a cyclic prefix, and sends them to the analogue processor for transmission.

The processes are reversed in the receiver. Using the demodulation reference signals, the channel estimation stage computes the amplitude scaling factors and phase shifts that mapped each of the transmitted layers onto each of the receiver's antenna connectors. The result is a composite channel matrix $H \cdot F$, in which F is the transmitter's precoding matrix, and H is the channel matrix that mapped the transmitter's antenna connectors onto the antenna connectors in the receiver. The receiver then computes the desired postcoding matrix G, in the manner described in Chapter 6. By applying the postcoding matrix, the receiver recovers the desired MIMO layers in the manner shown in Figure 6.10, and converts them back to a single stream.

The receiver then combines the symbols from the initial transmission and any hybrid ARQ re-transmissions, leaving gaps for any symbols that were omitted during rate matching. After demodulating the symbols and descrambling the corresponding bits, it passes the results to the transport channel receiver.

7.6.3 Analogue Processing

The process concludes as shown in Figure 7.11. In the first stage, the transmitter converts the time domain OFDMA signals to analogue form. It then applies a spatial filter, before

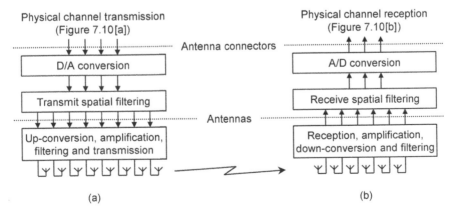

Figure 7.11   Analogue processing. (a) Transmission. (b) Reception.

mixing the signal up to radio frequency and carrying out amplification, filtering and transmission. The processes are reversed in the receiver.

The spatial filter is an implementation of analogue beamforming, which maps the antenna connectors onto the corresponding physical antennas. The base station typically selects one analogue beam for each antenna connector by means of a phase ramp, as shown in Figure 6.2. The mobile typically selects one antenna panel by means of a switch, as shown in Figure 6.21.

References

1 3GPP TS 38.300 (2019) NR; NR and NG-RAN overall description; Stage 2 (Release 15), December 2019.

2 3GPP TS 38.211 (2019) NR; Physical channels and modulation (Release 15), December 2019.

3 3GPP TS 38.212 (2019) NR; Multiplexing and channel coding (Release 15), December 2019.

4 3GPP TS 38.213 (2019) NR; Physical layer procedures for control (Release 15), December 2019.

5 3GPP TS 38.214 (2019) NR; Physical layer procedures for data (Release 15), December 2019.

6 Parkvall, S., Dahlman, E., Furuskär, A. et al. (2017). NR: the new 5G radio access technology. *IEEE Communications Standards Magazine* 1 (4): 24–30.

7 Bertenyi, B., Nagata, S., Kooropaty, H. et al. (2018). 5G NR radio interface. *Journal of ICT Standardization* 6 (1–2): 31–58.

8 3GPP TS 38.300 (2019) NR; NR and NG-RAN overall description; Stage 2 (Release 15), December 2019, Sections 4.4, 6.1.

9 3GPP TS 24.501 (2019) Non-access-stratum (NAS) protocol for 5G system (5GS); Stage 3 (Release 15), December 2019.

10 3GPP TS 38.331 (2019) NR; Radio resource control (RRC) protocol specification (Release 15), December 2019.

11 3GPP TS 37.324 (2018) E-UTRA and NR; Service Data Adaptation Protocol (SDAP) specification (Release 15), September 2018.

12 3GPP TS 38.323 (2019) NR; Packet Data Convergence Protocol (PDCP) specification (Release 15), June 2019.

13 3GPP TS 38.322 (2019) NR; Radio link control (RLC) protocol specification (Release 15), March 2019.

14 3GPP TS 38.321 (2019) NR; Medium access control (MAC) protocol specification (Release 15), December 2019.

15 3GPP TS 38.101-1 (2019) NR; User equipment (UE) radio transmission and reception; Part 1: Range 1 standalone (Release 15), December 2019.

16 3GPP TS 38.101-2 (2019) NR; User equipment (UE) radio transmission and reception; Part 2: Range 2 standalone (Release 15), December 2019.

17 3GPP TS 38.101-3 (2019) NR; User equipment (UE) radio transmission and reception; Part 3: Range 1 and range 2 interworking operation with other radios (Release 15), December 2019.

18 3GPP TS 38.101-4 (2019) NR; User equipment (UE) radio transmission and reception; Part 4: Performance requirements (Release 15), December 2019.

19 3GPP TS 38.104 (2019) NR; Base station (BS) radio transmission and reception (Release 15), December 2019.

20 3GPP TS 38.215 (2019) NR; Physical layer measurements (Release 15), December 2019.

21 3GPP TS 38.133 (2019) NR; Requirements for support of radio resource management (Release 15), December 2019.

22 3GPP TS 38.401 (2019) NG-RAN; Architecture description (Release 15), December 2019, Annex B.

23 3GPP TS 37.340 (2019) Evolved Universal Terrestrial Radio Access (E-UTRA) and NR; Multi-connectivity; Stage 2 (Release 15), December 2019, Section 4.2.

24 3GPP TS 38.321 (2019), NR; Medium access control (MAC) protocol specification (Release 15), December 2019, Section 4.5.

25 3GPP TS 38.212 (2019) NR; Multiplexing and channel coding (Release 15), December 2019, Sections 4, 6.3, 7.3.

26 3GPP TS 38.211 (2019) NR; Physical channels and modulation (Release 15), December 2019, Sections 6.1, 7.1.

27 3GPP TS 38.104 (2019) NR; Base station (BS) radio transmission and reception (Release 16), December 2019, Section 5.

28 3GPP TS 38.101-1 (2019) NR; User equipment (UE) radio transmission and reception; Part 1: Range 1 standalone (Release 15), December 2019, Section 5.2A.

29 3GPP TS 38.101-2 (2019) NR; User equipment (UE) radio transmission and reception; Part 2: Range 2 standalone (Release 15), December 2019, Section 5.2A.

30 3GPP TS 38.101-3 (2019) NR; User equipment (UE) radio transmission and reception; Part 3: Range 1 and range 2 interworking operation with other radios (Release 15), December 2019, Sections 5.2A, 5.5B.

31 3GPP TS 38.101-1 (2019) NR; User equipment (UE) radio transmission and reception; Part 1: Range 1 standalone (Release 15), December 2019, Section 5.3A.5.

32 3GPP TS 38.101-2 (2019) NR; User equipment (UE) radio transmission and reception; Part 2: Range 2 standalone (Release 15), December 2019, Section 5.3A.4.

33 3GPP TS 38.101-3 (2019) NR; User equipment (UE) radio transmission and reception; Part 3: Range 1 and range 2 interworking operation with other radios (Release 15), December 2019, Section 5.5A.1.

34 3GPP TS 38.211 (2019) NR; Physical channels and modulation (Release 15), December 2019, Sections 4.2, 4.4.

35 3GPP TS 38.104 (2019) NR; Base station (BS) radio transmission and reception (Release 15), December 2019, Sections 5.3, 5.4.

36 Jeon, J. (2018). NR wide bandwidth operations. *IEEE Communications Magazine* 56 (3): 42–46.

37 3GPP TS 28.541 (2019) Management and orchestration; 5G network resource model (NRM); Stage 2 and stage 3 (Release 15), December 2019, Section 4.3.7.

38 3GPP TS 38.211 (2019) NR; Physical channels and modulation (Release 15), December 2019, Section 4.4.5.

39 3GPP TS 38.213 (2019), NR; Physical layer procedures for control (Release 15) December 2019, Section 12.

40 3GPP TS 38.331 (2019) NR; Radio resource control (RRC) protocol specification (Release 15), December 2019, Annex B.2.

41 3GPP TS 38.211 (2019) NR; Physical channels and modulation (Release 15), December 2019, Sections 4.1, 4.3, 5.3.1.

42 Dahlman, E., Parkvall, S., and Sköld, J. (2018). *5G NR: The Next Generation Wireless Access Technology*. Academic Press, Section 7.2.

43 3GPP TS 38.211 (2019) NR; Physical channels and modulation (Release 15), December 2019, Section 4.3.1.

44 3GPP TS 38.213 (2019) NR; Physical layer procedures for control (Release 15), December 2019, Section 4.2.

45 3GPP TS 38.213 (2019) NR; Physical layer procedures for control (Release 15), December 2019, Section 11.1.

46 3GPP TS 38.508-1 (2019) 5GS; User equipment (UE) conformance specification; Part 1: Common test environment (Release 15), June 2019, Section 4.6.3 (TDD-UL-DL-Config).

47 3GPP TS 38.213 (2019) NR; Physical layer procedures for control (Release 15), December 2019, Section 11.1.1.

48 3GPP TS 38.211 (2019) NR; Physical channels and modulation (Release 15), December 2019, Section 4.4.1, 6.2, 7.2.

49 3GPP TS 38.214 (2019) NR; Physical layer procedures for data (Release 15), December 2019, 5.1.5.

50 3GPP TS 38.212 (2019) NR; Multiplexing and channel coding (Release 15), December 2019, Section 5.

51 Richardson, T. and Kudekar, S. (2018). Design of low-density parity check codes for 5G new radio. *IEEE Communications Magazine* 56 (3): 28–34.

52 3GPP TS 38.211 (2019) NR; Physical channels and modulation (Release 15), December 2019, Sections 5, 6.3, 7.3.

53 CPRI (2019) Common Public Radio Interface: eCPRI interface specification, Version 2.0 May 2019, Section 6.1.

8

Cell Acquisition

A mobile uses the acquisition procedure to discover a nearby 5G cell and acquire its main configuration parameters. Acquisition forms part of the higher-layer procedure for measurement reporting for mobiles in the state of RRC_CONNECTED. If a mobile supports standalone 5G communications with a primary 5G cell, then acquisition also forms part of the procedures for network and cell selection, and for cell reselection in the states of RRC_INACTIVE and RRC_IDLE.

This chapter explains the tasks that are carried out. The first three sections are relevant to both standalone and non-standalone operation, while the final section is relevant to standalone operation alone.

8.1 Acquisition Procedure

8.1.1 Introduction

The acquisition procedure involves the primary synchronization signal (PSS), the secondary synchronization signal (SSS), the physical broadcast channel (PBCH) and its demodulation reference signal (DM-RS). The base station transmits these using regions of time and frequency that are known as *synchronization signal / physical broadcast channel blocks* (SS/PBCH blocks or SSBs). A sequence of SSBs is known as an *SS/PBCH burst*. Within an SS/PBCH burst, each block is labelled using a distinct *SS/PBCH block identity* (SSB identity).

A key part of the procedure is its support for beamforming, with the beams created using digital precoding at low radio frequencies, analogue spatial filtering at high frequencies or some combination of the two [1]. During an SS/PBCH burst, the base station sweeps through all of its transmission beams, sending one or more SSBs per beam. The mobile waits for the best transmission beam to sweep past, perhaps trying out different reception beams in order to discover the best beam pair. Figure 8.1 shows an example in which the base station and mobile are using four and two beams respectively.

Once the procedure has completed, the mobile reports the best SSB identity back to the base station, in order to identify the best transmission beam that it found. Depending on the circumstances, the mobile can do so using the procedure for measurement reporting, channel state information reporting or random access.

An Introduction to 5G: The New Radio, 5G Network and Beyond, First Edition. Christopher Cox.
© 2021 John Wiley & Sons Ltd. Published 2021 by John Wiley & Sons Ltd.

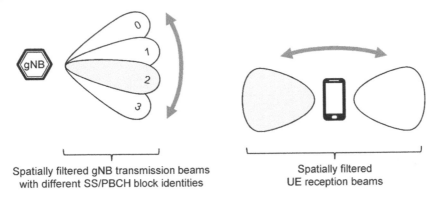

Spatially filtered gNB transmission beams with different SS/PBCH block identities

Spatially filtered UE reception beams

Figure 8.1 Analogue beam selection during the acquisition procedure.

Table 8.1 Summary of the acquisition procedure.

Step	Task	Information received
1	Receive PSS	Frequency synchronization
		Symbol timing
2	Receive SSS	Physical cell identity
3	Receive DM-RS and PBCH	Frame timing
		Best downlink transmit spatial filter
		Best downlink receive spatial filter
4	Read MIB from PBCH	Cell-barring flag
		System frame number
		Configuration of PDCCH that schedules SIB 1
5	Receive PDCCH	Scheduling command for reception of SIB 1
6	Read SIB 1 from PDSCH	Remaining minimum system information

8.1.2 Non-standalone Operation

To go further, let us investigate the case of non-standalone operation. Some 5G cells might only act as secondary cells, for use during carrier aggregation. Those cells do not have to broadcast any SSBs at all. A mobile can still discover them by measuring their channel state information reference signals, but the gNB central unit has to configure those signals in advance and tell the mobile about the relevant configuration parameters.

Other 5G cells might be able to act as primary SCG cells, for use during dual connectivity. Those cells broadcast SSBs in the manner described in the previous section. They can do so on any radio frequency, but the gNB central unit has to tell the mobile about that frequency in advance. By receiving the SSBs, the mobile discovers the information in the first three steps of Table 8.1.

Referring now to step 4, most of the information in the PBCH is a radio resource control (RRC) signalling message known as the master information block (MIB). If a cell can only act as a primary SCG cell, then the MIB only contains two important pieces of information. The first is the system frame number, which every gNB has to transmit in case it is not time-synchronized with other gNBs that are nearby. The second is a *cell-barring* flag, which indicates that the cell is unwilling to act as a primary cell and is unavailable for selection.

8.1.3 Standalone Operation

In standalone operation, some or all of the 5G cells can act as primary cells. In those cells, the SSBs have a few additional features. Firstly, they are only transmitted on frequencies defined by a *synchronization raster*, to ensure that the mobile can find them without prior signalling from the network. Secondly, the MIB indicates that the cell is not barred. Thirdly, the MIB tells the mobile how to receive system information block 1 (SIB 1; steps 5 and 6 of Table 8.1), which contains information such as the new radio cell identity and the PLMN-ID. An SSB meeting this last criterion is known as a *cell-defining SSB*.

The mobile can now read the base station's transmissions of SIB 1, which are scheduled on the physical downlink control channel (PDCCH) and delivered on the physical downlink shared channel (PDSCH) in the manner that we will describe in Chapter 11. It now has all the information required for the procedures of cell selection and reselection.

8.2 Resource Mapping

8.2.1 SS/PBCH Blocks

Figure 8.2 shows the internal structure of an SS/PBCH block [2, 3]. Each block has dimensions of 4 symbols by 240 subcarriers (20 resource blocks). The cell transmits the PSS on the first symbol, the SSS on the third symbol, and the PBCH on the second, third and fourth. The PBCH includes demodulation reference symbols on every fourth subcarrier, with the exact subcarrier offset depending on the physical cell identity.

The duration and bandwidth of an SSB, measured in μs and MHz, depend upon its subcarrier spacing. That can equal 15 or 30 kHz in frequency range 1, and 120 or 240 kHz in frequency range 2, and can be different from the subcarrier spacing used for data.

8.2.2 Transmission Frequency

The centre frequency of an SSB is the frequency used by subcarrier 0 of resource block 10. Unlike in LTE, that frequency is adjustable, with two main possibilities.

During the cell selection procedure, the mobile has to identify SSBs without any help from the network. To make that task easier, the centre of a cell-defining SSB can only lie on a few discrete frequencies, which are defined by the *global synchronization raster* in Table 8.2 [4] and are identified by a *global synchronization channel number* (GSCN). The mobile scans through most of the 5G frequency bands in unit steps of the global synchronization raster, pausing each time to search for transmissions of the SSBs. There are two complications: in a few of the frequency bands, the steps are integer multiples of the global synchronization raster; while at carrier frequencies below 3 GHz, there is an additional frequency offset of 50, 150, or 250 kHz.

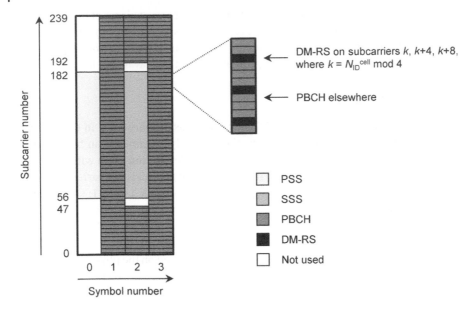

Figure 8.2 Internal structure of an SS/PBCH block, using a physical cell identity of 1. *Source:* Adapted from 3GPP TS 38.300.

Table 8.2 Values of the global synchronization raster.

Frequency range (GHz)	Global synchronization raster (MHz)	Possible offsets (kHz)
0–3	1.2	50, 150, 250
3–24.25	1.44	0
24.25–100	17.28	0

Source: Adapted from 3GPP TS 38.104.

Each frequency band supports either one or two possible values for the subcarrier spacing. If there are two possibilities, then the mobile has to try both. However, the mobile can store the frequencies and subcarrier spacings that it finds during one cell search to speed things up on a later occasion.

As an example, band n28 covers downlink frequencies from 758 to 803 MHz. In that band, the mobile takes unit steps through the global synchronization raster of 1.2 MHz, with possible offsets of 50, 150, or 250 kHz. The SSB subcarrier spacing is 15 kHz.

During the procedures for cell reselection and measurement reporting, the gNB central unit can direct the mobile to the SSBs of neighbouring cells by means of RRC signalling. In these cases, the centre frequency can lie on any possible value of the New Radio absolute radio frequency channel number (NR-ARFCN).

8.2.3 Transmission Timing

The maximum number of blocks in an SS/PBCH burst is 4 in frequency range 1 below a carrier frequency of 3000 MHz, 8 in FR1 above 3000 MHz, and 64 in FR2. The base station transmits each block within a single beam, so those figures limit the number of beams that it can use for its SSBs. (They do not necessarily limit the number of beams used by other channels and signals, because the base station might set up a small number of wide-angle beams for its SSBs, and then set up narrower beams for communication with individual mobiles.) If the base station is using fewer beams than the permitted maximum, then it can either map one beam onto multiple blocks or miss out some blocks altogether.

Within each burst, the base station transmits its SSBs in either the first or the second half of a frame, and labels each one using its SSB identity. That runs upwards from zero at the start of the half-frame, so it identifies not only the downlink beam but also the block's exact transmission time.

There are then five possible timing patterns for the SSBs within a burst, which are denoted by the letters A to E [5]. Each timing pattern is associated with one of the subcarrier spacings discussed earlier in this chapter. Each frequency band supports either one or two timing patterns, which the mobile has to discover during the cell selection procedure, and which the gNB central unit can signal otherwise.

Figure 8.3 shows the timing patterns that are used in frequency range 1. In each pattern, the SS/PBCH burst contains a maximum of four or eight blocks, depending on the carrier frequency, so there is a maximum of four or eight beams. There are two possible timings for a subcarrier spacing of 30 kHz, with the choice depending on the frequency band.

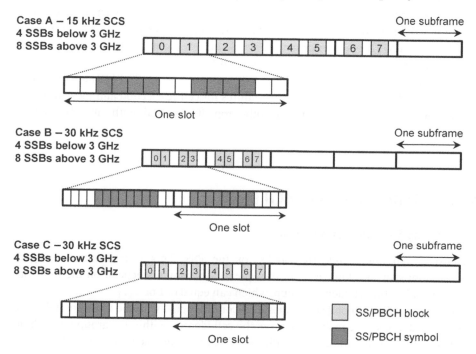

Figure 8.3 Timing patterns for the SS/PBCH blocks in frequency range 1.

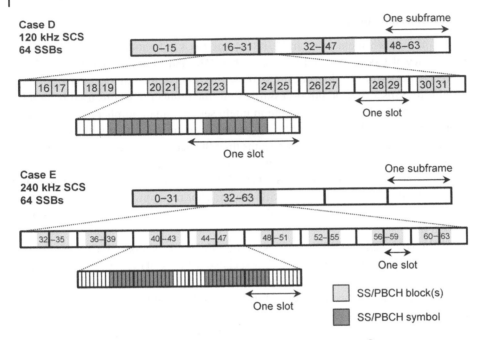

Figure 8.4 Timing patterns for the SS/PBCH blocks in frequency range 2.

In frequency range 2 (Figure 8.4), the SS/PBCH burst contains a maximum of 64 blocks, so there is a maximum of 64 beams. In case E, the use of a 240 kHz subcarrier spacing limits the amount of time that the burst occupies, at the expense of an increase in its transmission bandwidth [6]. (A 60 kHz subcarrier spacing would occupy too much time in FR2 and too much bandwidth in FR1, so it is not supported at all.)

During the cell selection procedure, the mobile assumes that the SS/PBCH bursts repeat with a period of 20 milliseconds. In the procedures for cell reselection and measurement reporting, the gNB central unit can indicate other repetition periods with values between 5 and 160 ms.

8.3 Acquisition of the SS/PBCH Block

8.3.1 Primary Synchronization Signal

To acquire a cell, the mobile first looks for the PSS. This is transmitted using a pseudo-random sequence known as an *m-sequence*. Its contents depend on the physical cell identity within the cell identity group, which can equal 0, 1 or 2 [7].

To receive the PSS, the mobile correlates its incoming signal with the three possible PSS sequences. In doing so, the mobile discovers the cell identity within the group for each of the nearby cells, measures the corresponding symbol timing and acquires initial frequency synchronization.

8.3.2 Secondary Synchronization Signal

The SSS is transmitted using another pseudo-random sequence known as a *Gold code*. Its contents depend on the cell identity group, which can take a value between 0 and 335 [8].

The SSBs only use the normal cyclic prefix, so the mobile knows the exact timing offset between the PSS and the SSS, and can immediately receive the SSS. By doing so, the mobile discovers the cell identity group for each of the nearby cells, and therefore its physical cell identity.

8.3.3 Demodulation Reference Signal for the PBCH

The mobile can now receive the demodulation reference signal for the PBCH. Its contents depend on two pieces of information, namely the half-frame number within the frame (0 or 1), and the two or three least significant bits (LSBs) of the SSB index [9].

In frequency range 1, these bits comprise the whole of the SSB index, so the mobile knows the beam on which the block was transmitted. The mobile also knows the block's transmission time from Figure 8.3, so it can calculate the time at which the frame began.

8.3.4 Physical Broadcast Channel

The mobile now has enough information to receive the PBCH. At a low level, the PBCH carries two pieces of information, namely the three most significant bits (MSBs) of the SSB index in frequency range 2, and the four least significant bits of the system frame number [10]. By measuring the first of these, the mobile completes its knowledge of the downlink transmission beam and the frame timing.

However, most of the information on the PBCH is an RRC signalling message known as the master information block. The most important information element in the MIB is the cell-barring flag, which indicates whether the cell is available for selection. In previous generations, network operators have barred cells before taking them down for maintenance. In 5G, the field acquires an additional meaning by indicating that the cell is unwilling to act as a primary cell. If the cell is barred, then another flag tells the mobile whether all of the 5G cells on that carrier frequency are barred as well.

In architectural options 3, 5 and 7, the network operator bars all of its 5G cells in this way. The mobile ignores them and instead uses the 4G acquisition procedure to select a nearby LTE cell. In options 2 and 4, however, at least some of the 5G cells are available for selection. The mobile can then read the rest of the MIB and can act upon it in the manner described in Section 8.4.

8.4 System Information

8.4.1 Master Information Block

The master information block contains three sets of information [3, 11, 12]. The first comprises the cell-barring flags that we identified in Section 8.3.4, while the second contains the six most significant bits of the system frame number.

The third set tells the mobile whether the cell is transmitting SIB 1, and, if so, how to receive the corresponding scheduling messages on the PDCCH. The simplest parameter is the downlink subcarrier spacing for the PDCCH. The next defines any subcarrier offset, from the base of the SSB to the subcarriers in the cell's common resource blocks. There are also pointers into two tables from which the mobile looks up two sets of information that we will cover in Chapter 11, namely a control resource set (CORESET) for the PDCCH, and an associated search space. The final parameter defines the symbols in which the base station will transmit demodulation reference signals for the PDSCH.

If the cell is not transmitting SIB 1, then the mobile treats it as barred. Otherwise, the mobile starts to listen for scheduling messages on the PDCCH, using the same reception beam on which it read the SSBs. In turn, those scheduling messages trigger reception of SIB 1 on the PDSCH.

8.4.2 System Information Block 1

SIB 1 contains all the remaining information that the mobile needs in order to listen to a cell and communicate with it. There are several sets of parameters. The first states the cell's tracking area code, radio access network (RAN) area code and New Radio cell identity, and identifies the networks that the cell belongs to by means of their PLMN-IDs. The second states the parameters that the mobile should use during the cell selection procedure. The third is a set of parameters that schedule broadcast transmissions of the other SIBs, in the manner described in Section 8.4.4.

The fourth set of information is a set of downlink configuration parameters. These include the resource block offset, from the base of the SSB to point A in the cell's downlink common resource blocks. By combining this quantity with the subcarrier offset from earlier in this chapter, the mobile can identify the exact frequency of point A. The downlink parameters also include the mobile's initial downlink bandwidth part (BWP). The fifth set of information is a similar set of parameters for the uplink and for any supplementary uplink, which include the absolute frequency of point A in the uplink common resource blocks, and an initial uplink bandwidth part. Amongst the remaining parameters, the most important is the common TDD configuration for the cell.

8.4.3 Other System Information Blocks

The other system information blocks are summarized in Table 8.3 [13]. SIBs 2–5 are used by mobiles in the states of RRC_IDLE and RRC_INACTIVE during the procedures for cell reselection. SIB 2 contains parameters that are common to all the cell reselection procedures, as well as cell-independent parameters for reselection on a single carrier frequency. SIB 3 contains the corresponding cell-specific parameters, while SIB 4 and SIB 5 cover reselection to a different 5G carrier frequency and to LTE.

The remaining SIBs are smaller and more specialized. SIBs 6 and 7 contain notifications from the *Earthquake and Tsunami Warning System* (ETWS), while SIB 8 contains notifications from the U.S. system for *Wireless Emergency Alerts* (WEA), also known as the *Commercial Mobile Alert System* (CMAS). Finally, SIB 9 contains the *Coordinated Universal Time* (UTC), as well as offsets to *Global Positioning System* (GPS) time and local time.

Table 8.3   Organization of the system information.

Block	Information	Examples
MIB	Master information block	Cell-barring flag
		Intra-frequency reselection from barred cell?
		6 MSBs of system frame number
		Common downlink subcarrier spacing
		Offset to common resource grid
		Initial configuration of PDCCH
SIB 1	Cell access information	PLMN-ID
		Tracking area code
		RAN area code
		Global cell ID
	Common serving cell configuration	Downlink frequency and initial BWP
		Uplink frequency and initial BWP
		TDD configuration
	Scheduling of system information	SIB mapping, period and window size
	Cell selection parameters	Q_{rxlevmin}, Q_{qualmin}
SIB 2	Common cell reselection parameters	Q_{hyst}
	Cell-independent intra-frequency reselection	$S_{\text{IntraSearchP}}$, $S_{\text{IntraSearchQ}}$, $T_{\text{Reselection,NR}}$
SIB 3	Cell-specific intra-frequency reselection	$Q_{\text{offset,s,n}}$
SIB 4	Inter-frequency reselection	Frequency, priority, $\text{Thresh}_{\text{x,HighP}}$
SIB 5	Reselection to LTE	Frequency, priority, $\text{Thresh}_{\text{x,HighP}}$
SIB 6	ETWS primary notification	Urgent earthquake or tsumani alert
SIB 7	ETWS secondary notification	Supplementary ETWS information
SIB 8	CMAS notification	US Presidential or public safety alert
SIB 9	Timing information	Universal, GPS and local time

8.4.4   Transmission and Reception of the System Information

The base station can transmit its system information in a few different ways [14]. As we saw in this chapter, it broadcasts the MIB in each of its downlink transmission beams as part of the corresponding SS/PBCH burst. It also broadcasts SIB 1 in each beam with a period of 160 ms, and can repeat the transmission at other times within that period.

Optionally, the base station can also broadcast any or all of the other SIBs, for use by mobiles in RRC_INACTIVE and RRC_IDLE. To do so, it collects the SIBs into *scheduling groups*, and gives each group a *scheduling window* that has a duration of 5–1280 slots and a period of 8–512 frames. It then lists those scheduling parameters as part of SIB 1, and states whether or not each group is actually being broadcast.

If the base station is broadcasting a particular scheduling group, then it does so using each of its downlink transmission beams, at some time during the scheduling window and at some frequency in the initial downlink bandwidth part. The mobile can then receive the system information in the same way as any other downlink transmission.

If the base station is not broadcasting a particular scheduling group, then the mobile can request it using the random access procedure. Mobiles in RRC_CONNECTED receive their system information using mobile-specific signalling.

References

1 Liu, J., Au, K., Maaref, A. et al. (2018). Initial access, mobility, and user-centric multi-beam operation in 5G New Radio. *IEEE Communications Magazine* 56 (3): 35–41.

2 3GPP TS 38.300 (2019) NR; NR and NG-RAN overall description; Stage 2 (Release 15), December 2019, Section 5.2.4.

3 3GPP TS 38.211 (2019) NR; Physical channels and modulation (Release 15), December 2019, Section 7.4.3.

4 3GPP TS 38.104 (2019) NR; base station (BS) radio transmission and reception (Release 15), December 2019, Section 5.4.3.

5 3GPP TS 38.213 (2019), NR; Physical layer procedures for control (Release 15), December 2019, Section 4.1.

6 Dahlman, E., Parkvall, S., and Sköld, J. (2018). *5G NR: The Next Generation Wireless Access Technology*. Academic Press, Section 16.1.1.

7 3GPP TS 38.211 (2019) NR; Physical channels and modulation (Release 15), December 2019, Section 7.4.2.2.

8 3GPP TS 38.211 (2019) NR; Physical channels and modulation (Release 15), December 2019, Section 7.4.2.3.

9 3GPP TS 38.211 (2019) NR; Physical channels and modulation (Release 15), December 2019, Section 7.4.1.4.

10 3GPP TS 38.212 (2019) NR; Multiplexing and channel coding (Release 15), December 2019, Section 7.1.

11 3GPP TS 38.213 (2019), NR; Physical layer procedures for control (Release 15), December 2019, Sections 4.1, 13.

12 3GPP TS 38.331 (2019) NR; radio resource control (RRC) protocol specification (Release 15), December 2019, Section 6.2.2 (MIB).

13 3GPP TS 38.331 (2019) NR; radio resource control (RRC) protocol specification (Release 15), December 2019, Section 6.3.1.

14 3GPP TS 38.331 (2019) NR; radio resource control (RRC) protocol specification (Release 15), December 2019, Section 5.2.

9

Random Access

The mobile uses the random access procedure if it needs to contact a 5G cell but has no means of doing so using the physical uplink control channel (PUCCH) or the physical uplink shared channel (PUSCH). There are several triggers. In the state of RRC_CONNECTED, the mobile uses random access during the procedures for handover and dual connectivity to initialize its timing advance. Other applications include recovery from the interruption of a spatially filtered beam. During standalone operation, the mobile also uses random access for initial access to a primary cell from the 5G states of RRC_INACTIVE and RRC_IDLE, while other applications include requests for the transmission of system information.

The procedure begins with an uplink transmission on the physical random access channel (PRACH), and continues with the ensuing signalling. The later parts of the procedure anticipate issues that we will cover in Chapter 11, so we will highlight those towards the end of the chapter.

9.1 Physical Random Access Channel

9.1.1 PRACH Formats

The mobile begins the random access procedure by transmitting one or more *preambles* on the physical random access channel [1]. A PRACH preamble has its own subcarrier spacing, which can be different from that of the other uplink channels, and which is the reciprocal of the PRACH symbol duration.

There are two types of preamble. Long PRACH preambles occupy 839 subcarriers with a spacing of 1.25 or 5 kHz, and are only used in frequency range 1 (FR1). Short PRACH preambles occupy 139 subcarriers, with a spacing of 15 or 30 kHz in FR1 and 60 or 120 kHz in FR2. Taken together, the preamble length and subcarrier spacing determine the bandwidth of the PRACH transmission. By combining that quantity with the PUSCH subcarrier spacing, we can calculate the number of PUSCH resource blocks that it occupies. Table 9.1 summarizes the results.

The preamble sequence is preceded by a cyclic prefix, which prevents inter-symbol interference in the base station receiver. The sequence is transmitted without any timing advance, so it is usually followed by a guard period, which prevents the preambles from

An Introduction to 5G: The New Radio, 5G Network and Beyond, First Edition. Christopher Cox.
© 2021 John Wiley & Sons Ltd. Published 2021 by John Wiley & Sons Ltd.

Table 9.1 Number of PUSCH resource blocks occupied by the PRACH.

PRACH sequence length	PRACH subcarrier spacing (kHz)	Number of PUSCH resource blocks occupied by the PRACH			
		15 kHz PUSCH	30 kHz PUSCH	60 kHz PUSCH	120 kHz PUSCH
839	1.25	6	3	2	—
839	5	24	12	6	—
139	15	12	6	3	—
139	30	24	12	6	—
139	60	—	—	12	6
139	120	—	—	24	12

Source: Adapted from 3GPP TS 38.211.

Table 9.2 Durations of the long PRACH formats.

PRACH format	Subcarrier spacing (kHz)	Duration (subframes)			
		Cyclic prefix	Symbols	Guard period	Total
0	1.25	0.103	0.8	0.097	1
1	1.25	0.684	2×0.8	0.716	3
2	1.25	0.153	4×0.8	0.647	4
3	5	0.103	4×0.2	0.097	1

Source: Adapted from 3GPP TS 38.211.

colliding with any timing-advanced uplink transmissions that follow. The PRACH can then be transmitted in a variety of different formats, which define the duration of the cyclic prefix, the number of symbols in the preamble sequence and the shortest possible guard period. (The base station can extend the guard period if it wishes to, simply by not scheduling anything immediately after the PRACH.)

Table 9.2 lists the long PRACH formats, whose timing is best related to that of the sub-frame. The usual choice is format 0. Format 1 uses a long cyclic prefix and guard period, and increases the received signal energy by the use of two PRACH symbols, so it is suitable for large cells with weak received signals. Format 2 uses four symbols, so it is suitable if the received signal is even weaker. Format 3 is an alternative to format 2, with a shorter duration that is intended for fast-moving mobiles [2].

Table 9.3 lists the short PRACH formats, whose timing is best related to that of the PRACH symbol. The distinction between the numbers 0 to 4 lies in the number of symbols that the mobile transmits, and therefore the received signal energy, while the distinction between the letters A, B and C lies in the durations of the cyclic prefix and the guard period. (The guard period is zero for formats A1, A2 and A3, but that does not cause any problems because those formats are always followed by another PRACH transmission.)

Table 9.3 Durations of the short PRACH formats.

PRACH format	Duration (PRACH symbols)			
	Cyclic prefix	Symbols	Guard period	Total
A1	0.141	2	0	2.141
A2	0.281	4	0	4.281
A3	0.422	6	0	6.422
B1	0.105	2	0.035	2.141
B2	0.176	4	0.105	4.281
B3	0.246	6	0.176	6.422
B4	0.457	12	0.387	12.844
C0	0.605	1	0.535	2.141
C2	1	4	1.422	6.422

Source: Adapted from 3GPP TS 38.211.

Table 9.4 Number of PUSCH symbols occupied by the short PRACH

PRACH subcarrier spacing (kHz)	Number of PUSCH symbols occupied by the short PRACH with formats {A1, B1, C0 / A2, B2 / A3, B3, C2 / B4}			
	15 kHz PUSCH	30 kHz PUSCH	60 kHz PUSCH	120 kHz PUSCH
15	2 / 4 / 6 / 12	4 / 8 / 12 / 24	8 / 16 / 24 / 48	—
30	1 / 2 / 3 / 6	2 / 4 / 6 / 12	4 / 8 / 12 / 24	—
60	—	—	2 / 4 / 6 / 12	4 / 8 / 12 / 24
120	—	—	1 / 2 / 3 / 6	2 / 4 / 6 / 12

At first glance, the durations of the short PRACH formats look rather arbitrary. However, each one is an exact multiple of 137/128 PRACH symbols, which is the combined duration of a PUSCH symbol at the same subcarrier spacing and its normal cyclic prefix (which itself is 9/128 symbols long). Noting that the PRACH and PUSCH can have different subcarrier spacings and symbol durations, Table 9.4 lists the number of PUSCH symbols that the PRACH transmission occupies for each combination of PUSCH subcarrier spacing, PRACH subcarrier spacing and PRACH format.

The specifications also speak about *PRACH slots*. When using the long PRACH formats, each subframe contains one PRACH slot. When using the short formats, each subframe contains one, two, four or eight PRACH slots for PRACH subcarrier spacings of 15, 30, 60 or 120 kHz, in the usual way.

9.1.2 Generation of the PRACH Preamble

Each cell has 64 PRACH preambles, which the mobile creates using parameters supplied by the base station. The preambles are based on *Zadoff–Chu sequences* [3, 4]. To create a preamble, the mobile begins by computing a Zadoff–Chu *root sequence*. There are 138 of these when using a short PRACH preamble, and 838 when using a long preamble. The mobile then applies a *cyclic shift* by shifting the data points along the sequence and wrapping them around the ends.

If two mobiles transmit at the same time and frequency using different cyclic shifts of the same root sequence, then the base station can decode one transmission without any interference from the other. If they use different root sequences, then there is a small amount of interference between the two. (Mathematically, the different cyclic shifts are orthogonal, while the different root sequences are uncorrelated.) Because of this issue, we prefer where possible to distinguish different mobiles in the same cell by assigning them different cyclic shifts.

However, those cyclic shifts look very like time delays, so there is a more serious risk that the base station might mistake one cyclic shift from a nearby mobile with a different cyclic shift from a mobile that is far away. To deal with that problem, the base station advertises a parameter called the *zero correlation zone configuration*, which depends on the expected cell size, and determines the number of cyclic shifts that a mobile can generate from a single root sequence. In a large cell, the mobile only uses a few widely spaced cyclic shifts, while, in a small cell, the number of cyclic shifts is large and their spacing is narrow.

The base station then defines a logical *root sequence index* for each cell. The mobile converts that to a physical root sequence index using a look-up table, and generates as many preambles as it can from that root sequence by means of cyclic shifts. If it has not yet reached 64 preambles, then it increments the logical root sequence index by one, and continues. The network operator should assign different sets of logical root sequences to nearby cells during radio network planning to ensure that their PRACH transmissions do not interfere.

9.1.3 Resource Mapping

A mobile can transmit the PRACH using reserved parts of the uplink resource grid that are known as *PRACH occasions*. These are defined using *PRACH frequency occasions* and *PRACH time occasions*, and are the places where those frequency and time occasions intersect. The base station reserves these resources as part of an uplink bandwidth part.

The PRACH frequency occasions occupy a single contiguous set of resource blocks within the bandwidth part. The base station defines that allocation using two parameters, which state the physical resource block at which it begins, and the number of PRACH frequency occasions (one to eight) that it contains.

The PRACH time occasions are defined using a parameter called the *PRACH configuration index*. That serves as a pointer into a look-up table, which starts by stating the PRACH format to use and the system frame numbers in which PRACH transmissions are allowed. The table then states the subframes in which PRACH transmissions can begin in FR1, or the 60 kHz slots (four per subframe) when they can begin in FR2. The short PRACH formats can be shorter than a subframe or a slot, so the table also defines the times at which short PRACH transmissions can begin within the subframe or the 60 kHz slot.

9.2 Random Access Procedure

9.2.1 Random Access Preamble

The PRACH forms the basis of the random access procedure [5, 6]. During that procedure, the base station provides the mobile with three quantities, namely an identity known
as a *cell radio network temporary identifier* (C-RNTI), an initial value for the uplink timing advance, and a scheduling grant for the transmission of an uplink message. The two
devices also initialize their choices of transmission and reception beams. We will discuss
the *contention-based* version of the procedure, which a mobile uses to access a primary cell
from the state of RRC_INACTIVE or RRC_IDLE, and which is illustrated in Figure 9.1.

As part of its random access preamble, the mobile reports back a *synchronization sig-
nal/physical broadcast channel* (SS/PBCH) block index to the base station, in other words
the identity of a downlink beam. Usually, this is the best SS/PBCH block index that the
mobile found during the acquisition procedure. However, the only requirement is that the
received signal power on that beam should exceed a threshold defined by the base station.

To configure that reporting process, the base station reserves some of the random access
preambles for the contention-free procedure that we will introduce towards the end of this
chapter, and for special cases such as system information requests. It then sets up a mapping from the SS/PBCH block indexes to the preambles and PRACH occasions, in which
successive indexes are mapped first to the unreserved PRACH preambles, then to PRACH
frequency occasions, then to the different PRACH time occasions within a PRACH slot, and
finally to different PRACH slots.

The base station also divides the unreserved preambles into two groups, namely group
B, which is used for large packets in good radio conditions, and group A, which is used

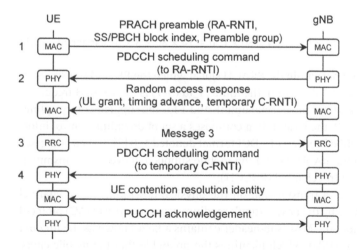

Figure 9.1 Contention-based random access procedure.

otherwise. A preamble in group A eventually triggers a small scheduling grant that can accommodate a small transmission, while a preamble in group B triggers a larger grant.

Based on these configuration parameters, the mobile identifies its preamble group, looks up the SS/PBCH block index that it wishes to report, and chooses a suitable PRACH occasion and preamble. If more than one PRACH occasion or preamble is suitable, then the mobile chooses one of them at random. The mobile then computes the initial uplink transmit power for the PRACH, which is based on the power that the base station wishes to receive and the mobile's estimate of the downlink path loss. Having done so, the mobile can send its PRACH transmission (step 1 of Figure 9.1).

The mobile then waits for a response from the base station, for a response window whose duration lies between one slot and the smaller of 80 slots and one frame. If there is no response, then the mobile increases its transmit power by a value between 2 and 6 dB, and repeats the transmission. That process continues until the base station responds as described in Section 9.2.2, or until the mobile reaches a maximum number of transmissions.

9.2.2 Random Access Response

From this point onwards, the random access procedure makes use of the underlying procedures for data transmission and reception, which we will discuss in Chapter 11. The resulting process is very like the one from LTE, but readers who are not familiar with LTE may prefer to read that chapter before they tackle the details.

The mobile's PRACH transmission is implicitly associated with an identity, known as the *random access radio network temporary identifier* (RA-RNTI). That is computed as follows:

$$RA\text{-}RNTI = 1 + s\_id + 14 \times t\_id + 14 \times 80 \times f\_id + 14 \times 80 \times 8 \times ul\_carrier\_id \quad (9.1)$$

where s_id (0 to 13) is the symbol within the PRACH slot at which the PRACH transmission began, t_id (0–79) is the PRACH slot within the frame, f_id (0–7) is the PRACH frequency occasion, and ul_carrier_id is 0 for the normal uplink and 1 for the supplementary uplink.

To start its response (step 2), the base station composes a set of downlink control information (DCI), using a simplified version of a DCI format denoted 1_0 [7]. It then addresses the control information to the RA-RNTI that the mobile supplied above, and transmits it on the physical downlink control channel (PDCCH).

That information schedules a transmission on the physical downlink shared channel (PDSCH), which is composed by the base station's medium access control (MAC) protocol [8]. Within that transmission, the MAC sub-header contains a field known as the *random access preamble identifier* (RAPID), which identifies the preamble that the mobile chose and ensures that the base station can respond to multiple preambles at the same time. The MAC payload contains three fields, namely an initial value for the uplink timing advance, a *temporary cell radio network temporary identifier* (TC-RNTI) and an uplink scheduling grant. There is no uplink acknowledgement, so no opportunity for a re-transmission.

9.2.3 Message 3

On receiving the random access response, the mobile initializes its uplink timing advance. Using the uplink scheduling grant, the mobile then sends the message that originally triggered the random access procedure, for example the *RRC Setup Request* that we will discuss in Chapter 13 (step 3). That message contains a unique identity for the mobile, which is either a MAC C-RNTI control element if the mobile already has a C-RNTI, or a 5G S temporary mobile subscriber identity (5G-S-TMSI), or a 39-bit random number. If the base station does not hear the mobile, then it can ask for a re-transmission using a further uplink scheduling grant.

9.2.4 Contention Resolution

There is a risk that two or more mobiles might send a random access request using the same PRACH occasion and the same preamble. If that happens, then the base station cannot tell them apart, so it replies with a single random access response. All of the mobiles receive the response, so they all transmit their uplink messages in step 3.

The uplink messages contain different mobile identities, so the base station either decodes one message in the presence of interference from the others, or fails to decode any of them. If the base station does decode one of the messages, then it replies with a *contention resolution* response (step 4). That response begins with a downlink scheduling command on the PDCCH, which is addressed either to the temporary C-RNTI from earlier, or to the C-RNTI if the mobile already has one. The response then continues with a transmission on the PDSCH. That contains a MAC control element known as the *UE contention resolution identity*, which echoes back the uplink message, and therefore the unique identity, that the base station received.

If the mobile hears its own identity in the contention resolution response, then it concludes that the procedure has succeeded, sends an uplink acknowledgement on the PUCCH and adopts the temporary C-RNTI as a full C-RNTI if it did not already have one. Otherwise, the mobile backs off and starts the procedure again.

9.2.5 Contention-free Procedure

Sometimes, the radio access network can allocate random access resources to the mobile before the procedure begins. There are several examples amongst the procedures for mobility management in RRC_CONNECTED, for example a change of primary cell (in other words, a handover), and the addition or change of a primary SCG cell. In each of these cases, the mobile uses the random access procedure for two main reasons: to initialize its uplink timing advance with the new cell, and to report the identity of the best downlink beam.

As part of the surrounding radio resource control (RRC) signalling, the radio access network can associate each of the new cell's downlink beams with a PRACH preamble and a set of PRACH occasions. After acquiring the new cell, the mobile contacts it in a *contention-free* version of the random access procedure, in which it uses the resources that the network has assigned. There is no risk of contention with another mobile, so the final contention resolution step is omitted.

A mobile can also use the contention-free procedure in the states of RRC_INACTIVE and RRC_IDLE, to ask a serving cell for a group of system information blocks that it is not currently broadcasting. To configure that, the base station associates each scheduling group with a PRACH preamble and a set of PRACH occasions, using the scheduling information in system information block 1. The mobile then transmits its preamble, as before. The base station omits the MAC payload from its random access response, as there is no need for any uplink transmission. Instead, the base station broadcasts the requested system information blocks in the usual way.

References

1 3GPP TS 38.211 (2019) NR; Physical channels and modulation (Release 15), December 2019, Section 6.3.3.

2 Liu, J., Au, K., Maaref, A. et al. (2018). Initial access, mobility, and user-centric multi-beam operation in 5G new radio. *IEEE Communications Magazine* 56 (3): 35–41.

3 Frank, R., Zadoff, S., and Heimiller, R. (1962). Phase shift pulse codes with good periodic correlation properties. *IEEE Transactions on Information Theory* 8: 381–382.

4 Chu, D. (1972). Polyphase codes with good periodic correlation properties. *IEEE Transactions on Information Theory* 18: 531–532.

5 3GPP TS 38.321 (2019) NR; medium access control (MAC) protocol specification (Release 15), December 2019, Section 5.1.

6 3GPP TS 38.213 (2019) NR; Physical layer procedures for control (Release 15), December 2019, Sections 7.4, 8.

7 3GPP TS 38.212 (2019) NR; Multiplexing and channel coding (Release 15), December 2019, Section 7.3.1.2.1.

8 3GPP TS 38.321 (2019) NR; medium access control (MAC) protocol specification (Release 15), December 2019, Sections 6.1.5, 6.2.2.

10

Link Adaptation

As radio signals travel from the transmitter to the receiver, they are subjected to attenuations and phase shifts. The base station and mobile measure those attenuations and phase shifts, so as to compensate for them. If the system is using analogue beam selection, then the base station and mobile use related measurements to identify the best spatially filtered transmission and reception beams. We will address those issues in this chapter.

We will begin with the downlink, in which the processes are more transparent. The base station transmits channel state information reference signals (CSI-RSs), which the mobile uses to compute channel state information (CSI) that describes the state of the downlink radio channel. The mobile returns that information to the base station for use during scheduling, using the physical uplink shared channel (PUSCH) if it is transmitting uplink data in the same slot, and the physical uplink control channel (PUCCH) otherwise. On the uplink, the base station tells the mobile to transmit sounding reference signals (SRSs), calculates the state of the uplink channel by inspecting the signals that arrive and uses thee information internally.

In the absence of channel reciprocity the two processes are independent, but if channel reciprocity applies then shortcuts are possible. For example, the base station might use the SRS to adapt to most aspects of the downlink radio channel, but might assist that process using interference measurements that are derived from the CSI.

10.1 CSI Reference Signals

10.1.1 Transmission and Reception

Figure 10.1 shows how the CSI reference signals are transmitted and received. The base station transmits the signals using a maximum of 32 antenna ports, starting at downlink port number 3000, and optionally precodes them. The mobile measures the signals that arrive, and compares them with the sequences defined in the specifications. By doing so, it estimates the attenuations and phase shifts that mapped the base station's CSI-RS antenna ports onto the mobile's antenna connectors. The mobile can then compute the corresponding channel state information and return it to the base station.

An Introduction to 5G: The New Radio, 5G Network and Beyond, First Edition. Christopher Cox.
© 2021 John Wiley & Sons Ltd. Published 2021 by John Wiley & Sons Ltd.

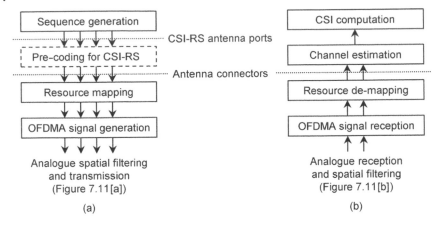

Figure 10.1 Physical channel processing for the CSI reference signal. (a) Transmission. (b) Reception.

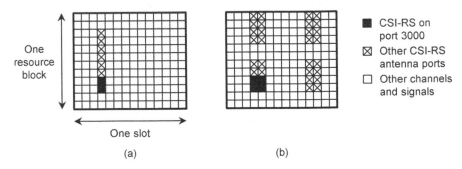

Figure 10.2 Example resource mappings for the CSI reference signal on port 3000. (a) Eight antenna ports in four CDM groups of two, using subcarriers 2 and 3 of symbol 3. (b) 32 antenna ports in eight CDM groups of four, using subcarriers 2 and 3 of symbols 3 and 4.

10.1.2 Resource Mapping

The base station can map the CSI-RS to the underlying resources using several different patterns [1]. Depending on the number of ports that it is using and the chosen pattern, the base station collects the ports into groups containing one, two, four or eight ports each, and spreads each group over the same number of resource elements using *code division multiplexing* (CDM). Separate parameters determine the precise choices of symbols in a slot and of subcarriers in a resource block. The reference signals are then transmitted over part or all of a downlink bandwidth part, usually on every resource block or every alternate one. There is also a special pattern that is intended for use as a tracking reference signal, in which the reference signal occupies three subcarriers in every resource block.

Figure 10.2 shows some examples. The first is taken from the 5G test specifications for frequency range 1, and uses eight antenna ports that are arranged into four CDM groups containing two ports each [2]. The second is a more extreme case, with 32 antenna ports in eight CDM groups of four.

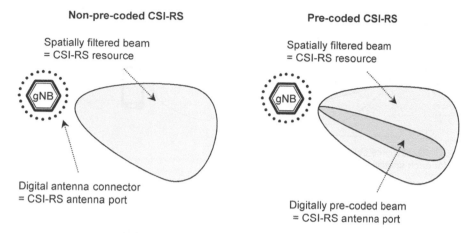

Figure 10.3 Interpretations of the CSI reference signal.

The CSI-RS antenna ports can be interpreted in two ways, which are illustrated in Figure 10.3. If the CSI-RS is not precoded, then each port corresponds to a digital antenna connector, which might drive a single antenna in a purely digital system, or an antenna panel in a hybrid system. If it is precoded, then each port corresponds to a digitally precoded beam.

10.1.3 CSI-RS Resources

To configure the CSI-RS, the base station supplies the mobile with a number of *CSI reference signal resources* by means of mobile-specific radio resource control (RRC) signalling. Each resource defines a set of antenna ports and resource elements on which the base station is transmitting the CSI-RS, and the way in which the mobile should receive it [3].

The antenna ports in a CSI-RS resource all use the same spatial filter, so the resource is associated with a particular choice of spatially filtered beam (Figure 10.3). Although those are similar to the beams used by the synchronization signal/physical broadcast channel (SS/PBCH) blocks, they might not be exactly the same. For initial acquisition, the base station transmits the SS/PBCH blocks using wide-angle beams which collectively span the entire cell. However, it might use narrower beams for the CSI-RS, which might only be directed in the vicinity of individual mobiles, and which do not have to cover the cell.

There are three types of resource. The most important is a *non-zero power* (NZP) *CSI-RS resource*, which consists of a pseudo-random sequence that is mapped and transmitted in the manner described here. The mobile uses the resource to measure the reference signals that arrive on the corresponding beam, as well as any interference from neighbouring cells. A *zero power* (ZP) *CSI-RS resource* has a similar resource mapping but is intended for use by other mobiles within the cell, and its contents are undefined. If a mobile is configured with a ZP CSI-RS resource, then it simply skips the corresponding resource elements when receiving the physical downlink shared channel (PDSCH). Finally, a *CSI interference measurement* (CSI-IM) *resource* has a simpler resource mapping and is used to measure the interference from neighbouring cells.

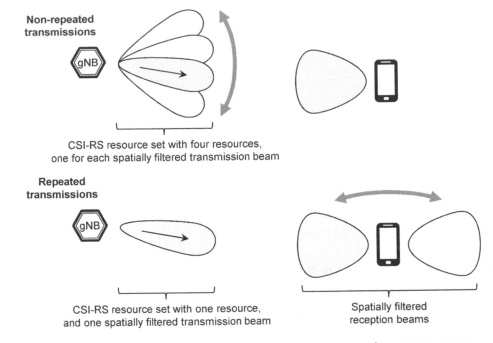

Figure 10.4 Non-repeated and repeated transmissions of the resources in a non-zero power CSI-RS resource set.

10.1.4 CSI-RS Resource Sets

CSI-RS resources are grouped into *CSI-RS resource sets*, each of which has a defined behaviour in the time domain. A *periodic* CSI-RS resource set is fully configured by RRC signalling. It has a period of 4–640 slots, with each of its constituent resources having a different slot offset within that period. A *semi-persistent* CSI-RS resource set is similar, but is switched on and off by a control element that is composed by the medium access control (MAC) protocol. An *aperiodic* CSI-RS resource set is a one-off transmission whose measurement is pre-configured by a MAC control element, and then triggered by means of downlink control information.

As before, the most important is an *NZP CSI-RS resource set*, in other words a set of NZP CSI-RS resources. Usually, the base station transmits each resource within that set using a different spatially filtered beam, switching from one to another by means of beam sweeping. The mobile can then measure all the resources within the set and can establish its preferred choice. However, the base station can also configure the resource set for *repetition*, in which it transmits all the constituent resources using the same spatial filter. The mobile can then receive the different resources using different spatially filtered reception beams and can determine the best one to use. Figure 10.4 illustrates the distinction.

The base station can also define a *CSI SSB resource set*. This is simply a list of SS/PBCH block indexes, each of which corresponds to a spatially filtered transmission beam, and which the mobile can measure in the same way as a non-repeated NZP CSI-RS resource set. The measurements are less sophisticated (for example, there is only one antenna port), but there is also an advantage, as the SS/PBCH blocks do not have to be explicitly configured.

Finally, a *zero power CSI-RS resource set* is a set of ZP CSI-RS resources, while a *CSI-IM resource set* is a set of CSI-IM resources.

10.2 Channel State Information

10.2.1 Introduction

By measuring the CSI-RS, the mobile computes several types of channel state information, which it feeds back to the base station as part of its uplink control information [4]. The computations are made over a set of resource blocks known as the *CSI reporting band*. The base station configures this by dividing a downlink bandwidth part into sub-bands, and defining a reporting band that encompasses some or all of those sub-bands.

10.2.2 CSI-RS and SS/PBCH Block Resource Indicators

The first quantity of interest is the *CSI-RS resource indicator* (CRI). That identifies an NZP CSI-RS resource that the mobile has found, in other words a spatially filtered transmission beam. An alternative is the *SS/PBCH block resource indicator* (SSBRI), which identifies an SS/PBCH block index.

10.2.3 Layer 1 RSRP

After identifying the incoming beams, the mobile can calculate the *layer 1 reference signal received power* (L1-RSRP) for each one. The L1-RSRP is the physical layer's average received signal power per resource element, measured on resource elements that are carrying either the secondary synchronization signal or the CSI-RS [5]. By filtering it, the mobile can calculate the layer 3 *reference signal received power* (RSRP) that it uses during the RRC procedures for cell selection, cell reselection and measurement reporting [6].

Using RRC signalling, the base station can tell the mobile to report the received signal power on either one, two or four spatially filtered beams. The strongest beam is reported using a 7-bit quantity with a step size of 1 dBm, while any others are reported using 4-bit offsets with step sizes of 2 dB.

10.2.4 Rank Indication

The remaining quantities can only be derived from the CSI-RS. The first is the *rank indication* (RI), which is the mobile's estimate of the number of MIMO layers that it can successfully receive. The RI has a maximum value of 8. It is also limited by the number of CSI-RS antenna ports that the base station is using, the number of receive antenna connectors at the mobile and the nature of the precoding matrix indicator (Section 10.2.5).

10.2.5 Precoding Matrix Indicator

The *precoding matrix indicator* (PMI) signals the additional precoding that the mobile would like applied to the CSI-RS antenna ports, when the base station directs a digitally precoded beam towards it. The mobile derives the PMI alongside the rank indication, for example by estimating the combination of RI and PMI that yields the highest data rate.

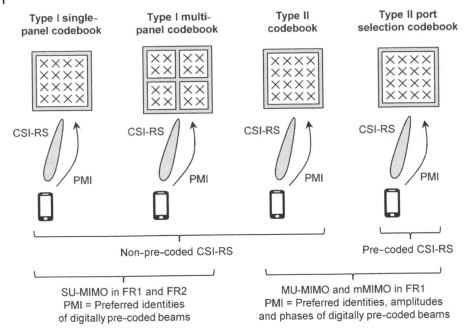

Figure 10.5 Types of PMI codebook.

To limit the amount of information that the mobile has to return, the specifications make a number of simplifications. The most important concerns the layout of the base station's antenna array. In a purely digital system, the base station has a maximum of 32 antennas, arranged in a two-dimensional array of 16 cross-polarized pairs. In a hybrid system, each pair is replaced by a cross-polarized antenna panel whose constituent antennas are controlled using analogue spatial filtering. Usually, there are two further assumptions. Firstly, the antennas or panels are arranged in a uniform linear array. Secondly, each antenna or panel is associated with a CSI-RS antenna port, which is not itself precoded.

The mobile then compresses its PMI feedback using two pairs of *PMI codebooks*, which are illustrated in Figure 10.5 [7, 8]. The *type I single panel codebook* is designed for single user MIMO in frequency ranges 1 and 2, and supports a maximum of eight SU-MIMO layers. Adding some more detail, the codebook defines a limited set of polarized, digitally precoded transmission beams. The mobile selects the beams that it would like to receive, with one polarized beam per layer, and feeds that selection back as the PMI. The base station then implements the mobile's selection as the downlink precoding matrix F, with each polarized beam approximating one of the independent propagation paths that we introduced in Section 6.3.8.

The *type I multi-panel codebook* is similar. Even in a purely digital system, however, it assumes that the antennas are arranged into two or four panels which are spaced more widely than the antennas within each panel. To construct its beams, the base station has to apply additional phase shifts between those panels, so the mobile reports back the desired phase shifts as part of the PMI. The maximum number of SU-MIMO layers is four.

The *type II codebook* provides richer feedback for use by multiple-user or massive MIMO in frequency range 1, with a maximum of two layers per mobile. As before, the codebook defines a limited set of polarized, digitally precoded beams, and the mobile selects the beams that it would like to receive. This time, however, the mobile reports back a compressed version of its entire downlink channel sub-matrix by expressing that sub-matrix as a weighted sum of two, three or four beams, and reporting back the identity, amplitude and phase of each one. By combining the reports from all the mobiles, the base station can construct the full downlink channel matrix H_{DL}, and can compute the desired downlink precoding matrix F in the manner described in Section 6.3.5.

The *type II port selection codebook* is similar. However, it reduces the amount of feedback by assuming that the CSI-RS antenna ports are themselves precoded, such that each antenna port corresponds to a single beam. In all four codebooks, the mobile reports the PMI in two parts: a wideband portion denoted i_1 for the CSI reporting band as a whole, and portions for each individual sub-band that are denoted i_2.

10.2.6 Channel Quality Indicator

After computing the RI and PMI, the mobile can derive the *channel quality indicator* (CQI). The CQI indicates the fastest modulation scheme and coding rate that the mobile can handle with a certain *block error ratio* (BLER), in other words the percentage of transport blocks that still contain errors after forward error correction, but before any hybrid ARQ retransmissions.

The most important part is a four-bit *wideband CQI*. Depending on its configuration, the mobile can report the wideband CQI in one of three ways, which are shown in Table 10.1. CQI table 2 (see Table 10.1, top row) assumes an error ratio of 10%, and is for mobiles that support downlink 256-QAM and are configured to use it. Table 1 is similar, but does not include 256-QAM. Table 3 is intended for ultra-reliable low-latency communications: it gives more weight to slower modulation schemes and coding rates, and it assumes an error ratio of 0.001%.

Sometimes the mobile can receive two downlink codewords, in which case it reports separate values of the wideband CQI for each one. The mobile can also report a CQI for each individual sub-band by means of a two-bit *sub-band differential CQI*.

10.2.7 Layer Indicator

The *layer indicator* (LI) identifies the strongest MIMO layer that the mobile would receive if the base station precoded its transmissions by means of the PMI. The base station uses the layer indicator when transmitting the physical downlink control channel (PDCCH) and when transmitting the phase tracking reference signals for the PDSCH.

10.2.8 CSI Reporting

To configure its CSI reports, the base station provides the mobile with one or more *CSI reporting configurations* [9]. Each one defines the cell on which the mobile should make its measurements, and the corresponding CSI-RS resource sets. It also states the cell on which

Table 10.1 Interpretation of the channel quality indicator.

CQI	Table 1 (BLER = 10%)			Table 2 (BLER = 10%)			Table 3 (BLER = 0.001%)		
	Modulation	Code rate × 1024	Bits per resource element	Modulation	Code rate × 1024	Bits per resource element	Modulation	Code rate × 1024	Bits per resource element
0	—	—	—	—	—	—	—	—	—
1	QPSK	78	0.15	QPSK	78	0.15	QPSK	30	0.06
2	QPSK	120	0.23	QPSK	193	0.38	QPSK	50	0.10
3	QPSK	193	0.38	QPSK	449	0.88	QPSK	78	0.15
4	QPSK	308	0.60	16-QAM	378	1.48	QPSK	120	0.23
5	QPSK	449	0.88	16-QAM	490	1.91	QPSK	193	0.38
6	QPSK	602	1.18	16-QAM	616	2.41	QPSK	308	0.60
7	16-QAM	378	1.48	64-QAM	466	2.73	QPSK	449	0.88
8	16-QAM	490	1.91	64-QAM	567	3.32	QPSK	602	1.18
9	16-QAM	616	2.41	64-QAM	666	3.90	16-QAM	378	1.48
10	64-QAM	466	2.73	64-QAM	772	4.52	16-QAM	490	1.91
11	64-QAM	567	3.32	64-QAM	873	5.12	16-QAM	616	2.41
12	64-QAM	666	3.90	256-QAM	711	5.55	64-QAM	466	2.73
13	64-QAM	772	4.52	256-QAM	797	6.23	64-QAM	567	3.32
14	64-QAM	873	5.12	256-QAM	885	6.91	64-QAM	666	3.90
15	64-QAM	948	5.55	256-QAM	948	7.41	64-QAM	772	4.52

Source: Adapted from 3GPP TS 38.214.

the mobile should send its report and the quantities that should be reported: either SSBRI and L1-RSRP; or CRI and L1-RSRP; or CRI, RI and some combination of PMI, CQI and LI.

Each CSI reporting configuration has a time domain behaviour, and a channel on which to send the report. There are four combinations. Periodic CSI reports on the PUCCH are configured by RRC signalling alone, and have a period of 4–320 slots. Semi-persistent reports on the PUCCH are similar, but are switched on and off by means of a MAC control element. Semi-persistent reports on the PUSCH have a period of 5–320 slots, and are switched on and off by means of downlink control information (DCI). Finally, aperiodic reports on the PUSCH are triggered solely by means of DCI.

We now have two types of time domain behaviour: one for the base station's CSI-RS resource sets, and one for the mobile's CSI reporting configuration. Only some combinations are allowed, which are listed in Table 10.2.

The specifications limit the amount of information that the mobile has to transmit, in two ways. Firstly, there are restrictions on the use of the PUCCH: for example, the mobile can only use periodic reporting on the PUCCH for wideband reports that are based on type I PMI codebooks. Secondly, the mobile can omit the least important parts of the CSI from both the PUCCH and the PUSCH, if there is not enough capacity available.

Table 10.2 Valid combinations of time domain behaviour for the measurement and reporting of channel state information.

Time domain behaviour of the CSI-RS resource set	Time domain behaviour for CSI reporting			
	Periodic CSI reporting on PUCCH	Semi-persistent CSI reporting on PUCCH	Semi-persistent CSI reporting on PUSCH	Aperiodic CSI reporting on PUSCH
Periodic	✓	✓	✓	✓
Semi-persistent	✕	✓	✓	✓
Aperiodic	✕	✕	✕	✓

Source: Adapted from 3GPP TS 38.214.

10.3 Physical Uplink Control Channel

10.3.1 Introduction

Uplink control information (UCI) comprises the channel state information that we have just been discussing, as well as hybrid ARQ acknowledgements (HARQ-ACKs) of the base station's downlink transmissions, and uplink scheduling requests (SRs). If the mobile wishes to transmit CSI or HARQ-ACK bits within a particular slot, and has received a scheduling grant for uplink data transmission on the PUSCH, then it embeds the UCI into the data. Otherwise, the mobile sends the information on the PUCCH.

10.3.2 PUCCH Formats

The PUCCH can be transmitted using five formats, which are summarized in Table 10.3. There are two main distinctions. Firstly, PUCCH formats 0 and 1 can handle either one or two HARQ-ACK bits or a scheduling request, with format 0 also able to handle both. Formats 2, 3 and 4 can handle more information and are essential for the transmission of CSI. Secondly, PUCCH formats 0 and 2 confine the transmission to one or two symbols. Formats 1, 3 and 4 spread the transmission over four to 14 symbols, so they are suitable when there is a need to limit the mobile's transmit power, for example at the cell edge. In each case, an individual PUCCH transmission is confined to a single slot, although it can be repeated over two, four, or eight slots so as to boost the received signal energy.

Table 10.3 PUCCH formats.

	Format 0	Format 1	Format 2	Format 3	Format 4
Number of UCI bits	1–2	1–2	> 2	> 2	> 2
Number of symbols	1–2	4–14	1–2	4–14	4–14
Number of resource blocks	1	1	1–16	1–16	1–16

Source: Adapted from 3GPP TS 38.211.

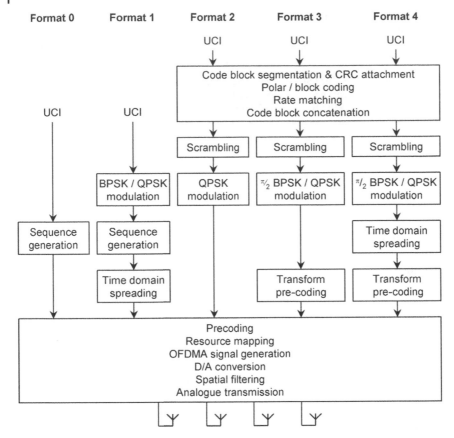

Figure 10.6 Transmission of uplink control information on the PUCCH.

The different formats are processed as shown in Figure 10.6 [10, 11]. In format 0, the control bits are spread over a single resource block, using a Zadoff–Chu sequence that depends on the control bits to be transmitted. In format 1, the bits are modulated using BPSK if there is one control bit and QPSK if there are two. The resulting symbols are spread over frequency using a Zadoff–Chu sequence, and over time by code division multiplexing with a mobile-specific *orthogonal cover code*. The base station can share the resources amongst different mobiles by assigning them different Zadoff–Chu sequences, as well as different cover codes in format 1.

In formats 2, 3 and 4, the control bits are initially processed for error correction, using two code blocks if the number of control bits is large and one otherwise. Formats 3 and 4 are intended for low-power transmission, so the transmit power variations are limited by the use of transform precoding and by the option of $\pi/2$ BPSK modulation. In format 4, the transmission is confined to one resource block and is spread over time using a mobile-specific orthogonal cover code.

In formats 1, 2, 3 and 4, the PUCCH is accompanied by a demodulation reference signal. That occupies every third subcarrier when using format 2, and designated symbols within the transmission when using formats 1, 3 and 4.

Using a MAC control element, the base station configures the mobile with an uplink spatial relationship, which tells it to transmit the PUCCH with the same spatial filter that it used either to receive a particular SS/PBCH block index or CSI-RS resource, or to transmit one of the SRS resources that we will discuss in Section 10.4. If the SRS resource in that last case was itself precoded, then the PUCCH is precoded as well.

10.3.3 PUCCH Resources

To configure the PUCCH, the base station supplies the mobile with a number of *PUCCH resources*, each of which defines the resource elements that the mobile can use for the PUCCH, and the parameters that it should apply [12]. Each resource is associated with a PUCCH format and a physical resource block at which to begin. To support frequency hopping, the base station can optionally tell the mobile to change to a different resource block halfway through. There are also parameters specific to each format, for example the number of resource blocks that are occupied in the case of formats 2 and 3.

PUCCH resources are collected into *PUCCH resource sets*, each of which can handle a certain maximum number of UCI bits. In turn, PUCCH resource sets form the basis for a mobile's *PUCCH configuration*.

Initially, the base station supplies the mobile with a cell-specific PUCCH configuration in its initial uplink bandwidth part. The mobile uses that configuration for one purpose only: to send uplink HARQ-ACK bits during the procedure for RRC connection establishment, which takes it into the state of RRC_CONNECTED.

As part of that procedure, the base station supplies the mobile with a dedicated PUCCH configuration in each of its subsequent uplink bandwidth parts. Each configuration includes a maximum of four PUCCH resource sets, which have different capabilities. The first set contains a maximum of 32 PUCCH resources, which can handle two UCI bits at a time by means of formats 0 and 1. The others each contain a maximum of 8 PUCCH resources, which can handle more UCI bits by means of formats 2, 3 and 4.

When the mobile has to send uplink control information on the PUCCH, it uses the number of UCI bits to determine its choice of PUCCH resource set. Meanwhile, the base station tells the mobile which resource to use within the chosen set, using either RRC signalling or downlink control information. Together, these fields tell the mobile which PUCCH resource to use. The only exception is a stand-alone scheduling request, which is handled using PUCCH resources that are set up for that purpose alone.

10.4 Sounding

10.4.1 Transmission and Reception

The sounding reference signals are transmitted and received in a similar way to the CSI reference signals, as shown in Figure 10.7 [13]. The mobile transmits the signals using one, two or four antenna ports, starting at uplink port number 1000, and optionally precodes them. The base station measures the incoming signals and compares them with the sequences defined in the specifications. By doing so, it estimates the attenuations and phase shifts that mapped the mobile's SRS antenna ports onto the base station's antenna connectors.

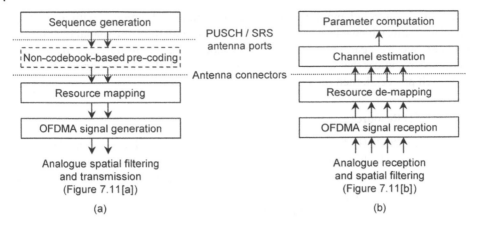

Figure 10.7 Physical channel processing for the sounding reference signal. (a) Transmission. (b) Reception.

It uses the information to support its uplink scheduling decisions and, in cases of channel reciprocity, its downlink scheduling decisions as well.

10.4.2 Resource Mapping

In the time domain, the SRS is mapped to one, two or four consecutive symbols, which are chosen from the last six symbols in a slot, and which take place after any PUSCH transmissions within that slot. In the frequency domain, the signal is mapped to every second or fourth subcarrier by means of a *transmission comb*. Figure 10.8 shows an example.

The contents are computed using a Zadoff–Chu sequence. The cyclic shift depends on the antenna port, so the mobile's transmissions on the same resource elements from different antenna ports are orthogonal. The sequence number is cell- and mobile-specific, and supports the possibility of sequence hopping from one symbol to the next so as to randomize the interference between different transmissions.

10.4.3 SRS Resources

The base station configures a mobile's sounding reference signals using *SRS resources*, each of which defines the resource elements that the mobile should use and the parameters that it

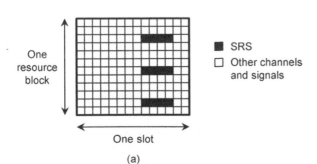

■ SRS
□ Other channels and signals

Figure 10.8 Example resource mapping for the sounding reference signal, using symbols 8–11, and a transmission comb of four subcarriers starting at subcarrier 1.

Non-pre-coded SRS   **Pre-coded SRS**

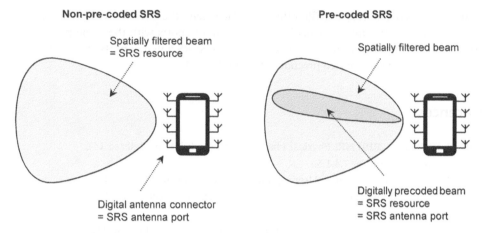

Figure 10.9   Interpretations of the sounding reference signal.

should apply [14]. In the simplest frequency domain configuration, each resource spans the whole of an uplink bandwidth part. However, the base station can also reduce the required transmit power by limiting each individual transmission to a narrower bandwidth, which changes from one slot to the next by means of frequency hopping.

In turn, SRS resources are grouped into *SRS resource sets*, each of which has a defined behaviour in the time domain. A periodic SRS resource set is fully configured by RRC signalling. It has a period of 1–2560 slots, with each of its constituent resources having a different slot offset within that period. A semi-persistent SRS resource set is similar, but is switched on and off by a MAC control element. An aperiodic SRS resource set is associated with a one-off transmission, which is triggered by means of downlink control information.

There are now two possible interpretations for the sounding reference signal, which are illustrated in Figure 10.9. If the signal is not precoded, then each antenna port corresponds to a digital antenna connector. In turn, each SRS resource corresponds to a spatial filter, in other words an antenna panel, and supports a maximum of four ports, in other words the connectors to that panel. Within an SRS resource set, different resources can use different spatial filters, but the mobile only transmits using one resource at a time.

The base station configures that option by sending uplink spatial relationship information to the mobile. That tells the mobile to transmit an SRS resource with the same spatial filter that it used either to receive an SS/PBCH block index or a CSI-RS resource, or to transmit another SRS resource. This option is the basis of codebook based precoding for the PUSCH, which we will examine in Chapter 11.

In conditions of channel reciprocity, the SRS can be precoded. Each antenna port corresponds to a digitally precoded beam. Similarly, each SRS resource corresponds to one antenna port, in other words one beam. Within an SRS resource set, different resources can use different sets of digital precoding weights, but they use the same spatial filter. The mobile can transmit using multiple resources within the set at the same time.

The base station configures that option by associating an SRS resource set with a CSI-RS resource, which tells the mobile two things. Firstly, the mobile should transmit the SRS resource set with the same spatial filter that it used to receive the CSI-RS resource. Secondly, the mobile should transmit the constituent SRS antenna ports with the same precoding matrix that it constructed to receive the constituent CSI-RS antenna ports. This option is the basis of non-codebook-based precoding for the PUSCH.

References

1 3GPP TS 38.211 (2019) NR; Physical channels and modulation (Release 15), December 2019, Section 7.4.1.5.

2 3GPP TS 38.508-1 (2019) 5GS; User equipment (UE) conformance specification; Part 1: Common test environment (Release 15), June 2019, Section 4.6.3 (CSI-RS-ResourceMapping).

3 3GPP TS 38.214 (2019) NR; Physical layer procedures for data (Release 15), December 2019, Sections 5.1.4.2, 5.1.6.1, 5.2.1.2, 5.2.2.3, 5.2.2.4.

4 3GPP TS 38.214 (2019) NR; Physical layer procedures for data (Release 15), December 2019, Sections 5.2.1.4.3, 5.2.2.1, 5.2.2.2.

5 3GPP TS 38.215 (2019) NR; Physical layer measurements (Release 15), December 2019, Sections 5.1.1, 5.1.2.

6 3GPP TS 38.331 (2019) NR; Radio resource control (RRC) protocol specification (Release 15), December 2019, Section 5.5.3.2.

7 Onggosanusi, E., Saifur Rahman, M., Guo, L. et al. (2018). Modular and high-resolution channel state information and beam management for 5G new radio. *IEEE Communications Magazine* 56 (3): 48–55.

8 Huang, Y., Li, Y., Ren, H. et al. (2018). Multi-panel MIMO in 5G. *IEEE Communications Magazine* 56 (3): 56–61.

9 3GPP TS 38.214 (2019) NR; Physical layer procedures for data (Release 15), December 2019, Sections 5.2.1.1, 5.2.1.4, 5.2.1.5, 5.2.1.6, 5.2.3, 5.2.4, 5.2.5.

10 3GPP TS 38.212 (2019) NR; Multiplexing and channel coding (Release 15), December 2019, Section 6.3.1.

11 3GPP TS 38.211 (2019) NR; Physical channels and modulation (Release 15), December 2019, Sections 6.3.2, 6.4.1.3.

12 3GPP TS 38.213 (2019) NR; Physical layer procedures for control (Release 15), December 2019, Section 9.2.

13 3GPP TS 38.211 (2019) NR; Physical channels and modulation (Release 15), December 2019, Sections 6.4.1.4.

14 3GPP TS 38.214 (2019) NR; Physical layer procedures for data (Release 15), December 2019, Sections 6.1.1, 6.2.1.

11

Data Transmission and Reception

The procedures for data transmission and reception are at the core of the 5G air interface. In this chapter, we will describe how data transmissions are scheduled by means of downlink control information (DCI) on the physical downlink control channel (PDCCH), and how the information is actually delivered using the physical uplink and downlink shared channels. We will also address the transmission of reference signals and uplink hybrid ARQ acknowledgements, and discuss the data rates that the 5G air interface can achieve. The procedures mainly lie in the physical layer, but the higher-level aspects are controlled by the medium access control (MAC) protocol.

11.1 Introduction

11.1.1 Data Transmission Procedure

Most of the data transmissions in 5G take place by means of *dynamic scheduling*, in which the base station schedules each individual transmission by delivering downlink control information on the PDCCH. Figure 11.1 shows the procedures that are used [1].

Downlink transmissions are triggered by the arrival of downlink data at the gNB distributed unit. That reacts by sending the mobile a downlink scheduling command on the PDCCH (Figure 11.1, step 1), which alerts the mobile to a forthcoming data transmission on the physical downlink shared channel (PDSCH) (2). The mobile processes the PDSCH, and generates a positive acknowledgement if the data arrive successfully or a negative acknowledgement if there are any errors. It then transmits that acknowledgement as part of its uplink control information (3). Depending on the nature of the acknowledgement, the base station either re-transmits the previous data or moves to a new data transmission, and the process continues.

Uplink transmissions are triggered when the mobile's application delivers data to its air interface protocol stack. If the mobile has no resources available on the physical uplink shared channel (PUSCH), then it asks for resources by sending the base station a scheduling request (SR) on the physical uplink control channel (PUCCH). That triggers the delivery of a scheduling grant on the PDCCH (1), which gives the mobile permission for an uplink data transmission on the PUSCH (2). The base station processes the PUSCH and sends an implicit acknowledgement in its next scheduling grant (3), by telling the mobile either

An Introduction to 5G: The New Radio, 5G Network and Beyond, First Edition. Christopher Cox.

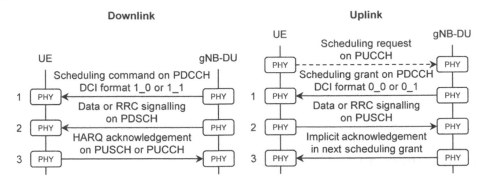

Figure 11.1 Data transmission procedures for the downlink and uplink.

to re-transmit the previous data or to move to a new transmission. As the process continues, the mobile keeps the base station informed about how much data are waiting for transmission. When the mobile reaches the end of the data stream, it simply ignores any more scheduling grants that arrive.

11.1.2 Downlink Control Information

The base station can write its downlink control information using a number of different formats. Table 11.1 lists the DCI formats that are defined as part of Release 15 [2]. The first four trigger data transmission on the uplink and downlink. In particular, formats 0_0 and 1_0 carry short scheduling messages that have limited capabilities (for example, they do not support MIMO transmissions), while formats 0_1 and 1_1 are longer and more powerful. Formats 2_0 to 2_3 control other aspects of the mobile's operation, and are covered later in the chapter.

11.1.3 Radio Network Temporary Identifiers

When transmitting its downlink control information, the base station identifies one or more target mobiles by means of a 16-bit *radio network temporary identifier* (RNTI) [3]. There are

Table 11.1 DCI formats.

DCI format	Contents
0_0	Simple uplink scheduling grant for the PUSCH
0_1	Complex uplink scheduling grant for the PUSCH
1_0	Simple downlink scheduling command for the PDSCH
1_1	Complex downlink scheduling command for the PDSCH
2_0	Slot format indicators
2_1	Pre-emption indications
2_2	Power control commands for the PUCCH and PUSCH
2_3	SRS requests and power control commands

Table 11.2   Radio network temporary identifiers.

Type of RNTI	Application	DCI formats used
P-RNTI	Paging message	1_0
SI-RNTI	System information broadcast	1_0
RA-RNTI	Random access responses	1_0
TC-RNTI	Random access contention resolution	0_0, 1_0
C-RNTI	Dynamic scheduling	0_0, 0_1, 1_0, 1_1
MCS-C-RNTI	Dynamic scheduling for high-reliability data	0_0, 0_1, 1_0, 1_1
CS-RNTI	Configured and semi-persistent scheduling	0_0, 0_1, 1_0, 1_1
SP-CSI-RNTI	Activation of semi-persistent CSI reporting	0_1
SFI-RNTI	Slot format indicators	2_0
INT-RNTI	Pre-emption indications	2_1
TPC-PUCCH-RNTI	Power control commands for the PUCCH	2_2
TPC-PUSCH-RNTI	Power control commands for the PUSCH	2_2
TPC-SRS-RNTI	SRS requests and power control commands	2_3

Source: Adapted from 3GPP TS 38.321.

several types of RNTI, which are listed in Table 11.2 along with the DCI formats that they support.

The most important is the *cell radio network temporary identifier* (C-RNTI), which is assigned during the random access procedure, and which identifies the target mobile in most of the dynamic scheduling messages that we introduced at the start of this chapter. The *modulation and coding scheme C-RNTI* (MCS-C-RNTI) is a variant of the C-RNTI, which is used for dynamic changes to the modulation and coding scheme. The *configured scheduling RNTI* (CS-RNTI) schedules repeated data transmissions by means of procedures known as configured and semi-persistent scheduling, while the *semi-persistent CSI-RNTI* (SP-CSI-RNTI) schedules repeated transmissions of channel state information on the PUSCH.

Of the remainder, the random access radio network temporary identifier (RA-RNTI) and temporary cell radio network temporary identifier (TC-RNTI) both appeared during the random access procedure in Chapter 9. The *paging RNTI* (P-RNTI) and *system information RNTI* (SI-RNTI) have fixed hexadecimal values of FFFE and FFFF respectively, which the base station uses to schedule paging and system information transmissions that can be read by any of the mobiles in the cell. The other RNTIs are all shared amongst groups of mobiles, and are used alongside DCI formats 2_0 to 2_3.

11.2   Transmission and Reception of the PDCCH

11.2.1   Transmission of the PDCCH

The base station transmits the PDCCH by following the underlying scheme that we introduced in Chapter 7. Figure 11.2 shows the details [4, 5]. After composing its DCI bits, the

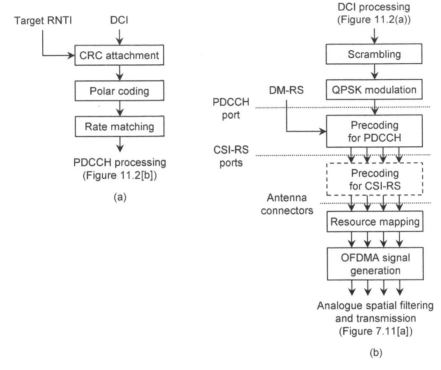

Figure 11.2 Transmission of downlink control information (DCI) on the PDCCH. (a) DCI processing. (b) Physical channel processing.

base station computes a 24-bit cyclic redundancy check (CRC), mixes the CRC bits with an RNTI that identifies the target mobile(s) and appends the result to the DCI. It then applies error correction coding and rate matching to compute a PDCCH codeword.

The physical channel processor begins by scrambling the codeword. It then applies a slow modulation scheme, specifically QPSK, to ensure that the message reaches distant mobiles reliably.

Optionally, the base station can precode the PDCCH in an implementation of digital beamforming. In turn, that allows the base station to transmit different sets of DCI bits to different mobiles by means of downlink multiple-user MIMO. The different mobiles share the same PDCCH antenna port, so they share the same sequences for the associated demodulation reference signal (DM-RS), and they cannot distinguish the different transmissions of the PDCCH. However, that does not matter, because the precoding process from Section 6.3.5 ensures that each mobile only receives a small amount of interference from those other transmissions.

It is easiest to understand downlink precoding by breaking it down into two stages. The first stage maps the PDCCH and DM-RS onto the base station's channel state information reference signal (CSI-RS) antenna ports [6]. The second stage is only used if the CSI-RS is itself precoded. It applies the CSI-RS precoding matrix to the PDCCH and DM-RS as well, so as to map the outgoing symbols onto the digital antenna connectors.

The base station can calculate the first precoding matrix in two ways: either from the sounding reference signals (SRSs) in cases of channel reciprocity, or from the mobile's precoding matrix indicator (PMI). If the rank indication is greater than one, then the PMI contains enough precoding weights for the transmission of multiple MIMO layers. If it does, then the base station can pick out the precoding weights for the best layer using the mobile's layer indication, and can use those weights to transmit the PDCCH.

In both stages, the base station precodes the DM-RS in exactly the same way as the PDCCH. From the mobile's point of view, the effect of downlink precoding is indistinguishable from the effect of the channel matrix, so the base station can apply any precoding matrix that it likes. As a result, downlink precoding does not actually appear in the 5G specifications, except for a few isolated remarks such as Reference [6]. The same issue applies in the case of the PDSCH (Section 11.4.3).

Precoding is applied in the frequency domain, so the base station can use different precoding matrices for different sets of subcarriers. To help the demodulation process, however, the base station supplies the mobile with a radio resource control (RRC) information element known as the *precoder granularity*, which indicates the number of resource blocks across which the precoding matrix remains the same.

The base station can also apply analogue spatial filtering to the PDCCH. It discovers the best spatial filter either from the mobile's CSI-RS resource indicator or, in cases of channel reciprocity, from its measurements of the sounding reference signals. It then signals its choice to the mobile using a MAC control element known as the *TCI (transmission configuration indicator) state indication*. That defines a quasi co-location relationship between the PDCCH and either a synchronization signal/physical broadcast channel (SS/PBCH) block index or a CSI-RS resource, which indicates that the base station is transmitting the PDCCH on the same spatially filtered beam that it used for the quasi co-located signal. The mobile can then receive the PDCCH using the corresponding spatially filtered reception beam.

11.2.2 Control Resource Sets

The PDCCH is mapped to the resource grid using regions of time and frequency known as *control resource sets* (CORESETs), which are illustrated in Figure 11.3. Each CORESET has a duration of one, two or three symbols, and occupies groups of six consecutive resource blocks that are defined using a bitmap. There are one or more CORESETs in each of the cell's downlink bandwidth parts.

At the finest level of detail, a CORESET consists of *resource element groups* (REGs). Each group spans one symbol by one resource block, and contains nine resource elements assigned to the PDCCH and three to the DM-RS. In turn, a *control channel element* (CCE) comprises six REGs. Finally, the PDCCH is transmitted using an *aggregation level* of one, two, four, eight or 16 CCEs. A large aggregation level is appropriate in the case of a weak received signal, which implies the need for a low coding rate, or if the DCI format contains a large number of bits.

11.2.3 Search Spaces

The control resource set defines the duration of a PDCCH transmission and the resource blocks that it occupies, but says nothing about its timing. To handle that, each CORESET is

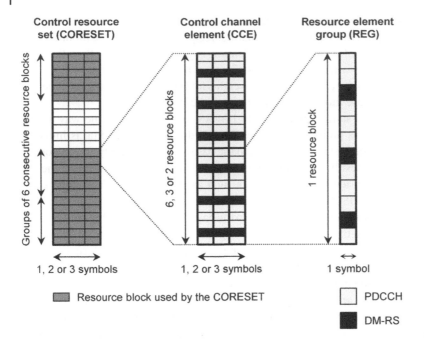

Figure 11.3 Example of a control resource set.

associated with one or more *search spaces* [7]. Each search space has a duration of 1–2559 slots, which repeats with a period of 1–2560 slots, and which defines the slots in which the CORESET might be used. The search space also identifies one or more symbols within the slot at which the CORESET might begin, usually just the first, but potentially anywhere for the case of low-latency applications. The search space also defines the number of *PDCCH candidates* that might be transmitted at each aggregation level, from which the mobile can calculate the exact locations within the CORESET at which the PDCCH might arrive.

Each search space is either mobile-specific or common to all the mobiles in the cell. It also has a type, which defines the cells in which it can be used, and the DCI formats and RNTIs that it might contain. Table 11.3 lists the types of search space that 5G supports.

11.2.4 Reception of the PDCCH

To receive the PDCCH, the mobile monitors every search space in the active downlink bandwidth part of each of its serving cells. In each of those search spaces, the mobile examines every location in which the PDCCH might arrive, and processes it using each of its configured DCI formats. At the end of the procedure, the mobile computes the received CRC bits, mixes them with each of its configured RNTIs and checks whether the message passes the CRC.

If it does pass, then the mobile concludes that the base station has sent DCI to the corresponding RNTI, and interprets the information as described later in this chapter. Otherwise, the mobile concludes either that it has not received any DCI or that there has been a bit error. In both cases, the mobile can safely ignore the message.

Table 11.3 Types of PDCCH search space.

Search space type	Cell	RNTIs used	Purpose
Common type 0	PCell	SI-RNTI	SIB 1
Common type 0A	PCell	SI-RNTI	Other SIBs
Common type 1	SpCell	RA-RNTI, TC-RNTI	Random access
Common type 2	PCell	P-RNTI	Paging
Common type 3	Any	SFI-RNTI, INT-RNTI, TPC-SRS-RNTI, TPC-PUCCH-RNTI, TPC-PUSCH-RNTI	DCI formats 2_0 to 2_3
	SpCell	C-RNTI, MCS-C-RNTI, CS-RNTI, SP-CSI-RNTI	Data
UE-specific	Any	C-RNTI, MCS-C-RNTI, CS-RNTI, SP-CSI-RNTI	Data

PCell: Primary cell; SpCell: special cell.

11.3 Scheduling Messages

11.3.1 DCI Formats 0_0 and 1_0

DCI formats 0_0 and 1_0 carry short scheduling messages which trigger simple uplink and downlink transmissions on the PUSCH and PDSCH. Their exact contents depend on the type of RNTI being used, so Table 11.4 lists those contents when the message is directed to a C-RNTI, MCS-C-RNTI or CS-RNTI. We will discuss the individual fields in this chapter.

11.3.2 Time Domain Resource Assignment

The *time domain resource assignment* is a pointer into a look-up table that has previously been configured using RRC signalling [8]. The table indicates the time when the data transmission begins and ends, using three separate fields.

The *slot offset* (0–32) is the number of slots between the scheduling message and the data transmission, with a value of zero indicating that they are in the same slot. Sometimes the numerologies used for data and scheduling can be different, for example when changing the active bandwidth part, so the slot durations can be different as well. In that situation, the slot offset uses the same numerology as the data. As examples, the 5G test specifications use offsets of two, four, six or eight slots before the PUSCH, depending on the frequency range and subcarrier spacing, and zero before the PDSCH [9].

The *mapping type* is A for a *slot-based transmission*, in which the data span most or all of a slot, and B for a *mini-slot-based transmission*, in which the duration of the data is shorter. Mapping type A is for normal data transmissions, while mapping type B is suitable for low-latency communications. In practical terms, the mapping type determines the locations of the demodulation reference signals for the PDSCH and PUSCH.

Finally, the *start and length indicator value* (SLIV) encodes the symbol within the chosen slot where the data transmission begins, and the number of symbols that it contains. The valid combinations are constrained by the choice of mapping type, as indicated in

Table 11.4 Contents of DCI formats 0_0 and 1_0.

Field	Bits	Format 0_0	Format 1_0
Identifier for DCI formats	1	✓	✓
Frequency domain resource assignment	See [a]	✓	✓
Time domain resource assignment	4	✓	✓
Frequency-hopping flag	1	✓	
VRB-to-PRB mapping	1		✓
Modulation and coding scheme	5	✓	✓
New data indicator	1	✓	✓
Redundancy version	2	✓	✓
HARQ process number	4	✓	✓
Downlink assignment index	2		✓
TPC command for PUSCH	2	✓	
TPC command for PUCCH	2		✓
PUCCH resource indicator	3		✓
PDSCH-to-HARQ feedback timing indicator	3		✓
Padding bits	As required	✓	
UL/SUL indicator	1 if configured	✓	

a) Number of bits $= \log_2 [N_{RB} (N_{RB} + 1)/2]$, for a bandwidth part containing N_{RB} resource blocks.

Table 11.5 Valid choices for the start and length indicator value.

Mapping type	Cyclic prefix	Valid uplink combinations			Valid downlink combinations		
		Start	Length	Start + length	Start	Length	Start + length
A	Normal	0	4–14	4–14	0–3	3–14	3–14
A	Extended	0	4–12	4–12	0–3	3–12	3–12
B	Normal	0–13	1–14	1–14	0–12	2, 4, or 7	2–14
B	Extended	0– 12	1–12	1–12	0– 10	2, 4, or 6	2–12

Source: Adapted from 3GPP TS 38.214.

Table 11.5. Downlink slot-based transmissions start in the first four symbols of a downlink or switching slot, usually immediately after the corresponding scheduling command. Uplink slot-based transmissions start at the very beginning of an uplink slot, while mini-slot-based transmissions are more flexible.

As implied by Table 11.5, each individual data transmission is confined to a single slot. However, the transmission can be repeated over an *aggregation level* of 2, 4 or 8 consecutive slots, to increase the received signal energy if the radio conditions are poor.

Table 11.6 Minimum PUSCH preparation time.

Subcarrier spacing (kHz)	Minimum number of symbols	
	Normal UE	Low-latency UE
15	10	5
30	12	5.5
60	23	11 (FR1 only)
120	36	Not supported

Source: Adapted from 3GPP TS 38.214.

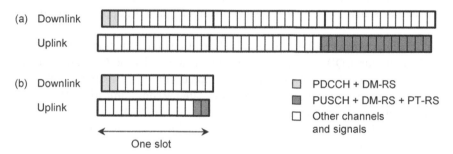

Figure 11.4 Example time domain resource assignments for the uplink. (a) Slot-based transmission, with a 26-symbol delay and a length of 14 symbols. (b) Mini-slot-based transmission, with a 10-symbol delay and a length of two symbols.

On the uplink, the mobile expects a minimum time delay between the end of the scheduling grant and the beginning of the data transmission. That delay depends on the chosen numerology, using whichever of the scheduling and data numerologies leads to the longer delay, and on whether the mobile is capable of low-latency communications. Table 11.6 lists the minimum symbol delays. An extra symbol is added if the time domain allocation begins with the PUSCH itself, rather than the DM-RS, while more symbols are required to account for the uplink timing advance.

Figure 11.4 shows some examples for the case of the uplink. The first is a slot-based transmission, while the second is a low-latency mini-slot-based transmission in which the scheduling grant and data transmission are in the same slot. (In TDD mode, that slot must be configured as a switching slot.)

11.3.3 Frequency Domain Resource Assignment

The *frequency domain resource assignment* indicates the physical resource blocks (PRBs) that the data will occupy, within the mobile's active downlink bandwidth part. There are two ways to specify it [10].

All four DCI formats support assignments of *type 1*, in which the base station gives the mobile a contiguous range of virtual resource blocks (VRBs). On both the uplink and downlink, those resource blocks can be mapped directly to the corresponding physical resource blocks. To provide additional frequency diversity, the downlink also supports an interleaved mapping, while the uplink also supports a frequency-hopping procedure in which the physical resource blocks can change either within a slot or between slots.

In addition, DCI formats 0_1 and 1_1 support more flexible assignments of *type 0*, in which the base station sends the mobile a bitmap indicating the *resource block groups* (RBGs) that it has assigned. Each resource block group has a bandwidth of 2–16 physical resource blocks, with the exact value constrained by the size of the active downlink bandwidth part and configured by RRC signalling.

11.3.4 Modulation and Coding Scheme

The *modulation and coding scheme* (MCS) is a 5-bit pointer into a look-up table, which indicates the modulation scheme to use for the data transmission and the approximate coding rate [11].

The downlink supports three look-up tables, which are used in conjunction with the three channel quality indicator (CQI) tables from Table 10.1. Table 1 omits 256-QAM, while table 2 supports it. Table 3 omits 256-QAM and places the emphasis on slow modulation schemes and coding rates, so it is intended for ultra-reliable low-latency communications (URLLC). The base station tells the mobile which table to use by means of RRC signalling. It can also switch dynamically to the use of table 3, by sending a downlink scheduling command towards an MCS-C-RNTI that it has previously supplied.

The uplink uses those tables as well, but also supports two others. Uplink table 1 supports the use of $\pi/2$ BPSK, and is used alongside the PUSCH processing step of transform precoding. Those steps limit the mobile's average transmit power, so the table is suitable for applications such as machine-type communications in which we wish to preserve the mobile's battery life. Uplink table 2 is similar, but also emphasizes slow modulation schemes and coding rates for use by URLLC.

To determine the size of the transport block, the mobile first uses the time and frequency domain resource assignments to calculate the total number of resource elements that are available, and subtracts an overhead to account for the reference signals. It then uses the MCS to calculate the approximate number of bits to be delivered, and rounds the result. In the case of downlink paging messages and random access responses, the number of bits can optionally be reduced by using a scaling factor of 0.25 or 0.5. That reduces the coding rate by the same scaling factor, which increases the reliability when communicating with distant mobiles.

11.3.5 Other Fields

DCI formats 0_0 and 1_0 contain a few other fields. Amongst them, the *identifier for DCI formats* indicates which of those DCI formats is being used. The *new data indicator* is toggled for a new transmission and stays unchanged for a retransmission. The *redundancy version* indicates precisely which coded bits are to be transmitted, while the *HARQ process number* indicates the HARQ process being used. *Transmit power control* (TPC) commands indicate how the mobile should adjust the power of its PUSCH transmission or PUCCH acknowledgement, as appropriate.

On the uplink, the *UL/SUL flag* indicates whether the mobile should use the normal or supplementary uplink (SUL), while the *frequency-hopping flag* indicates whether it should use frequency hopping. Padding bits are then added if required, to make DCI formats 0_0 and 1_0 the same length. On the downlink, the *VRB-to-PRB mapping* indicates whether the mapping of virtual to physical resource blocks is direct or interleaved. Finally, the *downlink assignment index*, *PUCCH resource indicator* and *PDSCH-to-HARQ feedback timing indicator* configure the mobile's HARQ acknowledgements on the uplink.

11.3.6   DCI Formats 0_1 and 1_1

DCI formats 0_1 and 1_1 are more powerful and use several other fields, but we will limit ourselves to the most important ones. When using carrier aggregation, the *carrier indicator* states which cell will be used for the forthcoming data transmission. It implements a technique known as *cross-carrier scheduling*, in which the scheduling message is sent in one cell (for example, a macrocell with a large coverage area), while the data are sent in another (for example, a microcell with a large capacity). The *bandwidth part indicator* changes the mobile's active bandwidth parts within the target cell, for example to increase the downlink bandwidth before reception of the PDSCH.

Other fields configure the use of single- and multiple-user MIMO. These are the *SRS resource indicator* and *precoding information and number of layers* on the uplink, the *transmission configuration indication* on the downlink, and the *antenna ports* on both. We will discuss them later in this chapter.

Release 16 introduces two new DCI formats, denoted 0_2 and 1_2, which are designed for URLLC [12]. In those formats, the sizes of several fields can be adjusted so as to limit the signalling overhead, while some of the additional fields from formats 0_1 and 1_1 are supported as well.

11.4   Transmission and Reception of the PUSCH and PDSCH

11.4.1   Transport Channel Processing

Now let us consider how the physical layer carries out a data transmission. Scheduled by the DCI, the MAC protocol creates either one or two protocol data units in the manner described in Chapter 12. It passes them to the physical layer in the form of *transport blocks*, and that layer starts to process them as shown in Figure 11.5 [13].

On the uplink, the mobile transmits one transport block at a time. On the downlink, the mobile receives one transport block at a time if there are one to four single-user MIMO layers, and two if there are five to eight layers. (Different MIMO layers can reach the receiver with different values of the signal-to-interference plus noise ratio (SINR), so it can be beneficial to transmit, acknowledge and re-transmit them individually. However, that increases the amount of signalling required, so the use of one or two transport blocks is a compromise.)

The transmitter begins by attaching a CRC to each transport block, using 24 CRC bits if the payload is larger than 3824 bits, and 16 bits otherwise. If the resulting payload is larger than

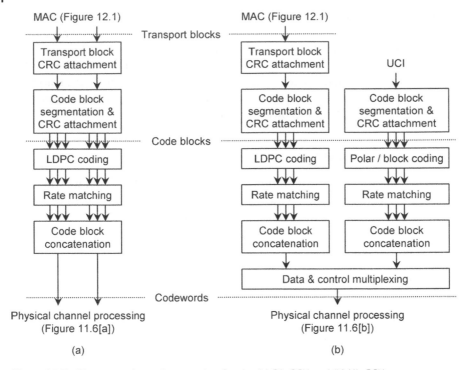

Figure 11.5 Transport channel processing for the (a) DL-SCH and (b) UL-SCH.

a maximum code block size (either 3840 or 8448 bits, depending on the coding rate), then it is segmented into *code blocks*, each of which receives a 24-bit CRC of its own to support re-transmission of individual code block groups (Section 11.6). The transmitter then applies error correction coding and rate matching, joins the code blocks back together, and sends the result onwards as one or more *codewords*.

On the uplink, the mobile may need to send uplink control information in the same slot as an uplink data transmission. If it does, then the control information is multiplexed into the PUSCH, and the PUCCH is not used.

11.4.2 Physical Channel Processing

The process continues as shown in Figure 11.6 [14, 15]. The transmitter first scrambles and modulates the outgoing codewords, and then maps the resulting symbols onto a set of parallel layers for use by single-user MIMO.

If there is just one single-user MIMO layer in the uplink, then the base station can instruct the mobile to apply an additional step of transform precoding. That applies a forward fast Fourier transform (FFT) to the PUSCH symbols, which reduces the power variations in the transmitted signal, and in turn reduces the average transmit power. The transmitter then applies digital precoding to the outgoing symbols, followed eventually by analogue spatial filtering and transmission. Those steps are rather different in the uplink and downlink, so we will address the two links in turn.

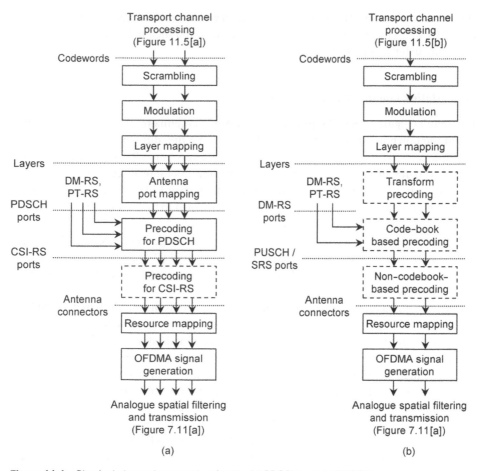

Figure 11.6 Physical channel processing for the (a) PDSCH and (b) PUSCH.

11.4.3 Downlink MIMO

In the downlink (Figure 11.6a), the stage of antenna port mapping associates each MIMO layer with a single PDSCH antenna port. In DCI format 1_1, the field *antenna ports* defines the number of layers that are assigned to an individual mobile, and the precise set of PDSCH antenna ports that it should use. In the case of single-user MIMO, the maximum number of PDSCH antenna ports is eight. In the case of multiple-user MIMO, there is a maximum of 12 PDSCH antenna ports in each cell, with up to four assigned to each individual mobile.

As in the case of the PDCCH (Section 11.2.1), downlink precoding is best considered in two stages. The first precoding matrix is calculated either from the PMI or, in cases of channel reciprocity, from the SRS. The precoding matrices can again be frequency specific, but this time the base station signals the precoder granularity either by RRC signalling, or by a field in DCI format 1_1 known as the *PRB bundling size indicator*.

As before, the base station discovers the best spatial filter either from the CSI-RS resource indicator or from the SRS, and indicates its choice to the mobile by means of a transmission configuration indication. This time, however, the base station first sets up a shortlist of up to eight TCI states by means of a MAC control element, and then indicates its choice more dynamically using a field in DCI format 1_1.

11.4.4 Uplink Codebook-based MIMO

In the uplink (Figure 11.6b), the principles of multiple antenna transmission are very like those on the downlink, but the details are different. The first difference concerns the use of antenna ports. There is a maximum of 12 MIMO layers per cell, each associated with an uplink DM-RS antenna port. There is also a maximum of four layers per mobile, each associated with a PUSCH/SRS antenna port. In DCI format 0_1, the field *antenna ports* identifies the DM-RS antenna ports that have been assigned to the mobile, and maps them onto the mobile's PUSCH/SRS antenna ports.

As before, the uplink precoder is shown in two separate stages. The first stage is applied to the PUSCH, DM-RS, and phase-tracking reference signal (PT-RS), while the second is applied to the SRS as well. This time, however, the two stages correspond to two alternative implementations.

In *codebook-based precoding*, the base station explicitly tells the mobile which precoding matrix to use. To implement this technique, the base station measures the sounding reference signals, determines the number of layers that the mobile should use and determines the best precoding matrix. It then sends that information to the mobile as part of DCI format 0_1, in a field known as *precoding information and number of layers*. The mobile applies that matrix in the first precoder, which maps its layers onto the corresponding PUSCH/SRS antenna ports. The SRS is not precoded, so the second precoder is not used.

The mobile can also apply a spatial filter, which directs the outgoing information to a spatially filtered transmission beam. When the SRS is not precoded, each SRS resource corresponds to a particular choice of spatial filter. The base station determines the best spatial filter from its measurements of the SRS. It then indicates its choice as part of DCI format 0_1, in a field known as the *SRS resource indicator*.

11.4.5 Uplink Non-codebook-based MIMO

Non-codebook-based precoding is only suitable in conditions of channel reciprocity. The SRS is itself precoded, using a precoder and a spatial filter that the mobile has previously derived from measurements of the downlink CSI-RS. The first precoder is omitted.

When the SRS is precoded, each SRS resource corresponds to a digitally precoded beam. As in the case of codebook-based precoding, the base station measures the mobile's SRS, and determines the number of layers that the mobile should use for the forthcoming transmission. This time, however, the base station also determines the best set of SRS resources on which to transmit those layers, in other words the best set of digitally precoded beams.

The base station sends all the information to the mobile using the SRS resource indicator field in DCI format 0_1. That field now indicates multiple SRS resources, not just one, such that the total number of SRS resources equals the total number of layers. The other DCI field, namely the precoding information and number of layers, is omitted.

11.5   Reference Signals

11.5.1   Demodulation Reference Signals

Like most of the other physical channels, the PUSCH and PDSCH are associated with demodulation reference signals, which help the receiver demodulate the incoming data [16, 17]. Several parameters determine how the reference signal is mapped to the underlying resource elements. We have already introduced the mapping type, which the base station signalled as part of the DCI. Mapping type A refers to slot-based transmission, in which the reference signal starts at the third or fourth symbol of the slot. Mapping type B refers to mini-slot-based transmission, in which it starts at the first symbol of the time domain assignment.

The other parameters are configured using RRC signalling. The *configuration type* is either 1 for single-user MIMO with a maximum of eight antenna ports, or 2 for multiple-user MIMO with a maximum of 12 ports. The reference signal is configured either for *single-symbol* transmission, which supports half the maximum number of ports, or for *double-symbol* transmission, which supports all of them. It can also occupy *additional symbols* later in the slot, which are suitable for fast-moving mobiles with a rapidly changing radio channel.

The antenna ports are collected into groups of two or four, for single- or double-symbol transmission respectively, and the corresponding reference signals are spread over two or four resource elements using code division multiplexing (CDM). If a resource element is assigned to one CDM group, then the remaining antenna ports leave that resource element unused, to ensure that the receiver can make an unambiguous measurement of the incoming signal.

Figure 11.7 shows some examples. Figure 11.7a is drawn from the 5G test specifications [18], and supports four antenna ports using mapping type A, configuration type 1 and a single-symbol DM-RS. The others are variations, using, in Figure 11.7b, a double-symbol DM-RS (eight ports); in 11.7c, configuration type 2 (six ports); and, in 11.7d, mapping type B.

11.5.2   Phase-tracking Reference Signals

In frequency range 2 (FR2), the base station can also associate the PUSCH and PDSCH with a phase-tracking reference signal [19, 20]. This has a similar role to the DM-RS, but with two main differences. Firstly, it has a higher density in the time domain (every one, two or four symbols), to help the receiver compensate for rapid phase fluctuations caused by timing jitter in the local oscillators. Secondly, it has a lower density in the frequency domain (every two or four resource blocks), because the phase fluctuations in nearby subcarriers are nearly the same.

On the downlink, the signal is only transmitted on one of the mobile's MIMO layers, because the phase fluctuations on different layers are nearly the same. The best layer is the one with the highest received SINR, which the base station discovers from the mobile's layer indicator. The base station then maps that layer to the one with the lowest port number. On the uplink, the PT-RS can be transmitted on multiple layers, as the mobile may be unable to maintain phase coherence between them. The base station tells the mobile which layers to use as part of its DCI.

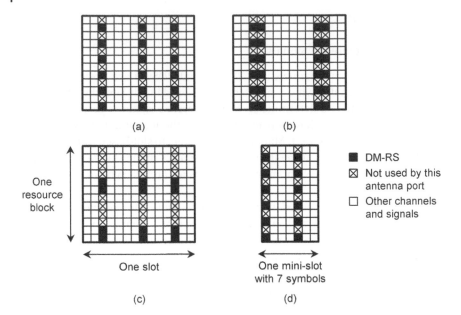

Figure 11.7 Example resource mappings for the DM-RS, using downlink port 1000 and uplink port 0. (a) Mapping type A, configuration type 1, single symbol DM-RS and two additional symbols. (b) Mapping type A, configuration type 1, double symbol DM-RS and one additional symbol. (c) Mapping type A, configuration type 2, single symbol DM-RS and two additional symbols. (d) Mapping type B over seven symbols, configuration type 1, single symbol DM-RS and one additional symbol.

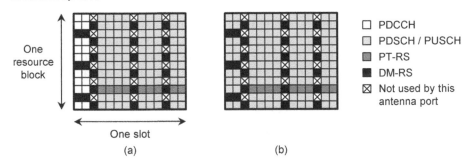

Figure 11.8 Example resource mappings for the PDSCH and PUSCH, using mapping type A, configuration type 1, single symbol DM-RS and two additional symbols, and PT-RS resource element offset 01. (a) Downlink port 1000. (b) Uplink port 0.

Figure 11.8 illustrates the end result. In these examples, the DM-RS uses the same parameters as in Figure 11.7a. The first two downlink symbols are occupied by the PDCCH, while the PTRS is transmitted once per symbol, and the remaining resource elements are assigned to the PDSCH or PUSCH.

11.6 Hybrid ARQ Acknowledgements

11.6.1 Downlink Acknowledgements of Uplink Data

After receiving an uplink data transmission on the PUSCH, the base station compares the code blocks and transport blocks with their respective CRC bits, and determines which ones have arrived successfully. It then sends an implicit acknowledgement using a further scheduling grant, by requesting either a new transmission or a re-transmission.

In the case of a re-transmission, the base station can ask the mobile to send either the whole transport block or individual *code block groups* (CBGs), each of which is a group of adjacent code blocks [21]. In the latter case, the base station first configures the maximum number of CBGs in a transport block (2, 4, 6 or 8) using RRC signalling. It then tells the mobile which CBGs to re-transmit as part of DCI format 0_1, using a bitmap known as the *CBG transmission information*. If the transport block only contains a few errors, then the acknowledgement and re-transmission of individual code block groups can greatly reduce the impact of HARQ re-transmissions on the air interface.

11.6.2 Uplink Acknowledgements of Downlink Data

A similar procedure applies for uplink acknowledgements of downlink data. The mobile determines whether the code blocks and transport blocks have arrived correctly, and generates a set of HARQ acknowledgement (HARQ-ACK) bits. It then returns those acknowledgements to the base station as part of its uplink control information, through a cell that is defined in the configuration parameters for the PDSCH. The base station then re-transmits either the whole transport block, or individual code block groups that are identified using CBG transmission information in DCI format 1_1.

The mobile may have to send several acknowledgements in the same uplink slot. For example, the mobile may be acknowledging individual code block groups; it may be using carrier aggregation and acknowledging several downlink transmissions in a single uplink cell; or it may be using TDD mode, with the uplink acknowledgements scheduled for a time when an uplink slot is available. To handle those situations, the mobile can compress its acknowledgements into a smaller number of HARQ-ACK bits by means of a codebook [22, 23].

11.6.3 Timing of Uplink Acknowledgements

The base station indicates when a mobile should send an uplink acknowledgement by means of the DCI field *PDSCH-to-HARQ feedback timing indicator* [24]. In DCI format 1_0, that field indicates the number of slots delay (1–8) from the end of the data transmission to the beginning of the uplink acknowledgement. In DCI format 1_1, it acts as a pointer into a look-up table that has previously been configured using RRC signalling, which in turn indicates a delay of 0–15 slots. If the uplink and downlink numerologies are different, then the delay is measured using the uplink slot duration.

The mobile expects a minimum time delay between the end of the data transmission and the beginning of the acknowledgement [25]. That delay depends on the chosen numerology, using whichever of the scheduling, data and acknowledgement numerologies leads to the longest delay; on whether there are any additional DM-RSs; and on whether the mobile is capable of low-latency communications. Table 11.7 lists the minimum symbol delays that result. Longer delays are needed in practice to account for the uplink timing advance, and also if the time domain assignment has fewer than seven symbols.

Figure 11.9 shows some examples. The first is for a slot-based transmission, while the second is for a low-latency mini-slot-based transmission in which the scheduling command,

Table 11.7 Minimum PDSCH processing time.

Subcarrier spacing (kHz)	Minimum number of symbols		
	Additional DM-RS Any UE	No additional DM-RS Normal UE	No additional DM-RS Low-latency UE
15	13	8	3
30	13	10	4.5
60	20	17	9 (FR1 only)
120	24	20	Not supported

Source: Adapted from 3GPP TS 38.214.

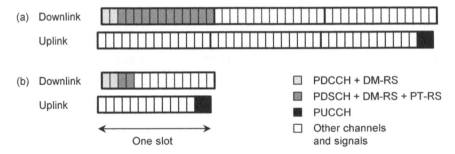

Figure 11.9 Example time domain resource assignments for the downlink. (a) Slot-based transmission, with a length of 12 symbols and a 26-symbol delay before the acknowledgement. (b) Mini-slot-based transmission, with a length of two symbols and an eight-symbol delay before the acknowledgement.

data transmission and uplink acknowledgement are in the same slot. (In TDD mode, that slot must be configured as a switching slot.)

11.7 Other DCI Formats

11.7.1 Introduction

DCI formats 2_0 to 2_3 deliver various kinds of downlink notification [26, 27]. They do not schedule any data transmissions: instead, the base station supplies all the relevant information within the downlink control information itself.

In each case, the DCI carries a set of notifications, each one for a different mobile within a group. Using RRC signalling, the base station tells each mobile the location of its own notification within the DCI, and sets up a single RNTI that is recognized by all the mobiles within the group. It can then notify all of those mobiles using a single DCI transmission.

11.7.2   Slot Format Indications

Slot format indications (SFIs) are written using DCI format 2_0, and are delivered to a group of mobiles identified by a *slot format indication RNTI* (SFI-RNTI). As discussed in Chapter 7, each indicator directs the mobile to a slot format combination, which defines the allocation of slots and symbols to the uplink and downlink over a maximum duration of 256 slots. That in turn supports dynamic changes to the slot format. Those can be valuable if the balance of traffic between the uplink and downlink is rapidly changing, provided that the cell is sufficiently isolated from its neighbours that there are no interference problems.

11.7.3   Pre-emption Indications

Pre-emption indications are written using DCI format 2_1, and are delivered to an *interruption RNTI* (INT-RNTI). They indicate that a previously scheduled downlink transmission has not in fact taken place, because it has been over-written by a low-latency transmission that was scheduled a moment later. Pre-emption indications are useful in congested cells that contain a mix of normal and low-latency applications, precisely because the base station may have to cancel a normal transmission so as to send low-latency information through.

Each pre-emption indication contains 14 bits. If a mobile receives one, then it either divides the time since the previous monitoring occasion for DCI format 2_1 into 14, or divides the time into seven and the active downlink bandwidth part into two. It then interprets the pre-emption indication as a bitmap, which shows the regions of time and frequency that have been over-written, and discards the affected symbols.

The mobile may be able to reconstruct a received code block by means of error correction, even if some of the received symbols have been over-written. If it cannot do so, then it retains the unaffected symbols in its HARQ reception buffer, for use during the forthcoming re-transmission.

Release 16 introduces a similar procedure to cancel a previously scheduled transmission on the uplink. The *cancellation indications* are written using a new DCI format denoted 2_4, and are delivered to a *cancellation indication RNTI* (CI-RNTI).

11.7.4   Transmit Power Control Commands

In common with other wireless communication systems, the mobile's transmit power should be the lowest that is consistent with satisfactory reception, both to preserve the mobile's battery and to minimize the interference that it delivers into nearby cells. In 5G, the mobile estimates its transmit power as best it can, using the base station's request for the received signal power per resource element, the mobile's estimate of the downlink propagation loss, and other parameters such as the transmission bandwidth [28].

The mobile's estimate may be inaccurate, for example if the uplink is subject to high levels of interference or, in FDD mode, if the uplink and downlink propagation losses are different. The base station therefore adjusts the mobile's estimate using *transmit power control* (TPC) commands. It can deliver those for the PUSCH using DCI formats 0_0 and 0_1, for the PUCCH using formats 1_0 and 1_1, and for either channel by sending DCI format 2_2 to the TPC-PUCCH-RNTI or TPC-PUSCH-RNTI. Each power control command is a two-bit field, which defines absolute power offsets of $-4, -1, +1$ or $+4$ dB, or accumulated offsets of $-1, 0,$

+1 or +3 dB. In the latter case, the mobile keeps separate records of the accumulated power offset for each cell, for each uplink bandwidth part and for the normal and supplementary uplinks.

Finally, DCI format 2_3 is delivered to the TPC-SRS-RNTI. It has two roles: it instructs mobiles to transmit aperiodic sounding reference signals by means of *SRS requests*, and it carries TPC commands that act in a similar way to the others.

11.8   Related Procedures

11.8.1   Scheduling Requests

Whenever a mobile is transmitting uplink data on the PUSCH, it keeps the base station informed about the amount of data in its transmit buffers by sending MAC control elements known as buffer status reports. If new uplink data arrive, and the mobile wishes to send a buffer status report but does not have the resources to do so, then it asks the base station for a scheduling grant by sending it a *scheduling request* (SR) on the PUCCH [29, 30].

To configure the use of scheduling requests, the base station provides the mobile with one or more *scheduling request resources* in each of its uplink bandwidth parts. Each resource has a period of two symbols to 640 slots and an offset within that period, which defines the times at which the mobile can send a scheduling request. The resource is associated with a PUCCH resource within the same bandwidth part, which defines how the request should be transmitted.

On receiving the scheduling request, the base station provides the mobile with an uplink scheduling grant. The mobile can then send its buffer status report, and can start to transmit the data that triggered the report.

11.8.2   Semi-persistent and Configured Scheduling

Dynamic scheduling is suitable for most types of data transmission, but there are two exceptions. The first is voice over IP. In IP voice communications, each transport block is small, no more than a few hundred bits. If each one were scheduled dynamically, then the overhead of the PDCCH would be large, limiting the number of voice calls that the cell could handle. On the other hand, the data rate of a voice call is roughly constant, so the resource allocations on the PDSCH and PUSCH are predictable.

Voice calls are handled using a technique known as *semi-persistent scheduling* (SPS) on the downlink, and as a *type 2 configured grant* on the uplink [31, 32]. In both cases, the RRC protocol configures the mobile with a transmission period, typically 20 ms, and a configured scheduling RNTI.

To start transmission, the base station sends downlink control information to the CS-RNTI, but otherwise writes the information in the usual way. The resource assignment repeats with the defined transmission period, while any re-transmissions are explicitly scheduled. To stop transmission, the base station uses a specially defined set of DCI fields that do not otherwise occur together. In the case of an uplink configured grant, the mobile confirms that it has received these DCI messages by means of a MAC control element.

The second exception is uplink low-latency communications. Sometimes, the mobile may have to send uplink packets at unpredictable times, but with a low delay. Using dynamic scheduling, the mobile would have to send a scheduling request, wait for the scheduling grant and only then transmit the packet. Each step would incur an unwanted delay.

It is better to handle these applications by means of *type 1 configured grants*. Using RRC signalling alone, the base station configures the mobile with a set of PUSCH resources that repeat with a period of two symbols to 640 ms. When a transmission opportunity occurs, the mobile sends any uplink data that are waiting for transmission, without the need for a scheduling request. Otherwise, it leaves the opportunity unused. To avoid wasting resources, type 1 configured grants should only be used sparingly.

11.8.3 Discontinuous Reception

In *discontinuous reception* (DRX), the mobile preserves its battery by spending most of its time in a low-power state in which it cannot be scheduled. There are two implementations.

In RRC_IDLE and RRC_INACTIVE, the mobile enters a *discontinuous reception cycle* with a length of 32, 64, 128 or 256 frames [33]. The mobile wakes once per DRX cycle, in a *paging frame* (PF) whose system frame number depends on the mobile's 5G S temporary mobile subscriber identity (5G-S-TMSI), and then stays awake throughout a *paging occasion* (PO), which starts at a time that also depends on the 5G-S-TMSI. (Several mobiles might share the same paging occasion, depending on their exact temporary identities.) In turn, the paging occasion contains one occurrence of the paging search space for each of the spatially filtered downlink beams within the serving cell.

To page a mobile, the base station first sends downlink control information to the P-RNTI within the mobile's paging occasion, using all the occurrences of the paging search space, and therefore using all of its spatially filtered beams. The DCI is written using a special version of DCI format 1_0 that includes either a *short messages* field, a scheduling command or both. The short messages field tells the mobile either to read an earthquake and tsunami warning system (ETWS) or commercial mobile alert system (CMAS) notification on system information blocks (SIBs) 6, 7 and 8, or to update the rest of its system information. The scheduling command tells the mobile to receive a downlink transmission on the PDSCH, which carries an RRC *Paging* message that uniquely identifies the target mobile. If the mobile does receive a paging message, then it contacts the base station using first the random access procedure, and then the procedure for establishment or resumption of the RRC connection. The resulting signalling is covered in Chapters 17 and 18.

Discontinuous reception in RRC_CONNECTED is suitable for low data rate communications [34]. The mobile wakes at the start of a *long DRX cycle*, whose duration is between 10 and 10 240 ms. It then stays awake, initially for an *on duration* of 1–1600 ms, and then for an *inactivity time* of 0–2560 ms after the arrival of any scheduling messages. If the on duration and inactivity time both expire, or the mobile receives a MAC control element known as a *DRX command*, then the mobile goes back to sleep.

The base station can also provide the mobile with a *short DRX cycle*, whose duration is between 2 and 640 ms. The mobile enters the short DRX cycle if it is scheduled, but reverts to the long cycle if a maximum number of short cycles (1–16) expires without any scheduling, or if it receives a MAC control element known as a *long DRX command*. The short DRX cycle is useful for voice communications, as it allows the mobile to take short bursts of sleep between the arrival of IP voice packets.

11.9 Performance of 5G

11.9.1 Peak Data Rate

In this final section, we will address the peak and typical data rates of 5G. The material draws upon 3GPP's submission to the International Telecommunication Union (ITU) [35], which demonstrates that the 3GPP 5G system complies with the requirements of IMT-2020. Amongst other parameters, the submission also evaluates the latency, energy efficiency and maximum mobile speed, as well as the user plane reliability for the case of URLLC applications. The appendices include several example link budgets for 5G.

The peak spectral efficiency of 5G is the maximum data rate that the air interface can handle per unit bandwidth, measured in units of bits s^{-1} Hz^{-1}. It is calculated as follows [36, 37]:

$$ SE = \frac{1}{BW} \left(v_{\text{layers}} Q_{\text{m}} f R_{\text{max}} \frac{12 N_{PRB}^{BW,\mu}}{T_s^\mu} (1 - OH) \right) \tag{11.1} $$

Here, BW is the maximum cell bandwidth, namely 100 MHz in frequency range 1 (FR1) and 400 MHz in FR2. v_{layers} is the maximum number of layers, which equals four in the uplink and eight in the FR1 downlink, but often falls to six in the FR2 downlink due to the impact of timing jitter [36]. Q_{m} is the maximum modulation order, which is eight for the case of 256-QAM. f is the scaling factor which we used earlier when calculating the transport block size, and which in this case equals one. R_{max} is the fastest coding rate that is used by the MCS, which equals 948/1024. $N_{PRB}^{BW,\mu}$ is the maximum resource block allocation in the chosen numerology, which equals 273 in FR1 and 264 in FR2. T_s^μ is the average duration of a symbol plus its cyclic prefix – in other words, the slot duration divided by the number of symbols per slot. Finally, OH is an overhead that accounts for other channels and signals. When calculating the data rate that it should be able to handle, the mobile assumes overheads of 8% and 14% respectively for the uplink and downlink of FR1, and 10% and 18% for the uplink and downlink of FR2.

Using that last assumption, we can calculate the peak spectral efficiencies that are listed in the fourth column of Table 11.8. (Reference [36] estimates the overhead from first principles, so its results are slightly different.)

Table 11.8 Peak spectral efficiency of 5G.

Direction	Frequency range	Duplex mode	Spectral efficiency (bits s^{-1} Hz^{-1})	Maximum bandwidth (MHz)	Example duty cycle (%)	Example data rate per cell (Gbps)
UL	FR1	FDD	25.0	100	100	2.50
UL	FR1	TDD	25.0	100	43	1.07
UL	FR2	TDD	23.7	400	43	4.05
DL	FR1	FDD	46.7	100	100	4.67
DL	FR1	TDD	46.7	100	56	2.60
DL	FR2	TDD	32.3	400	56	7.20

The final column of Table 11.8 contains some example figures for the peak data rate per cell. To calculate this quantity, we multiply the spectral efficiency first by the cell bandwidth, and then by the uplink or downlink duty cycle in the case of TDD. In estimating that last figure, we have assumed the use of two uplink slots, two downlink slots and one switching slot, which in turn contains 11 downlink symbols, one guard symbol and two uplink symbols. Other TDD configurations lead to different results.

Figure 11.10 compares the peak spectral efficiency of 5G with the values achieved using different releases of LTE. In LTE, Release 8 supports one uplink layer and four downlink layers, with those values increasing to four and eight respectively in Release 10. Release 8 supports 64-QAM, Release 12 introduces downlink 256-QAM, Release 14 introduces uplink 256-QAM and Release 15 introduces downlink 1024-QAM. In Release 15, the peak spectral efficiencies of the two systems are much the same. Even so, the use of a wider cell bandwidth means that the peak data rate of a 5G cell is much greater than it is in LTE.

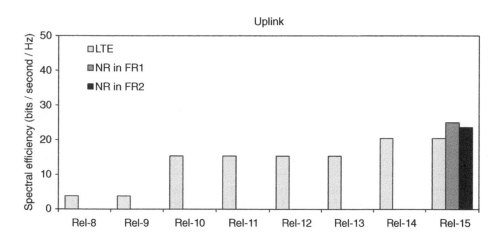

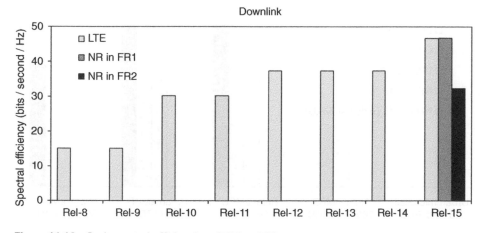

Figure 11.10 Peak spectral efficiencies of LTE and 5G.

We can then compute the actual data rate by adding the data rates of the mobile's serving cells. In FR2, for example, carrier aggregation bandwidth class C denotes the use of three contiguous component carriers spanning 400 MHz each, making a total bandwidth of 1200 MHz. In the Release 16 specifications for band n258, that bandwidth class can be used in both the downlink and uplink [38]. If a mobile supported that configuration, and used the TDD duty cycle quoted here, then its peak downlink and uplink data rates would be 21.6 and 12.2 Gbps respectively.

It cannot be stressed too strongly, however, that these data rates can only be achieved in ideal conditions. Amongst those conditions, the cell must be transmitting and receiving using its maximum radio bandwidth, namely 100 MHz in FR1 and 400 MHz in FR2. The mobile must be close to the base station, so that the received SINR is high, which allows the

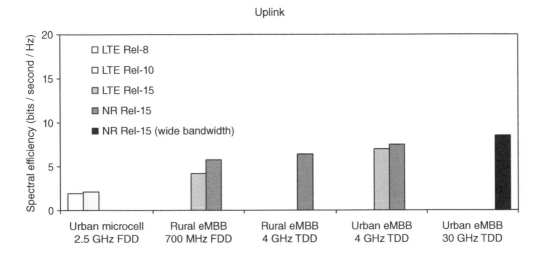

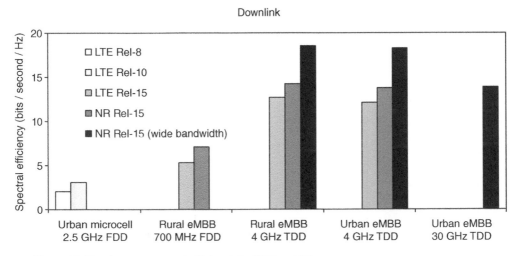

Figure 11.11 Average spectral efficiencies of LTE and 5G.

transmitter to use a fast modulation scheme and a high coding rate. Despite that proximity, there must be enough multipath for single-user MIMO with four, six or eight layers, as appropriate. There must be only one active mobile in the cell, so that it does not have to share the cell's resources with other mobiles, and the mobile and base station must support all the relevant optional features of the air interface. In any realistic scenario, these conditions are completely unachievable.

11.9.2 Typical Cell Capacity

For more realistic estimates of the performance of 5G, we have to resort to simulations. To illustrate these, Figure 11.11 shows the average spectral efficiencies that are measured in 3GPP simulations of the 5G New Radio and Release 15 LTE [39], and compares them with some representative results for LTE Releases 8 and 10 [40]. To estimate the typical cell capacity, we simply have to multiply these results by the bandwidth of each cell.

The results quoted for Release 15 are in the ITU's dense urban and rural environments for enhanced mobile broadband (eMBB) in IMT-2020 [41], which allows a like-for-like comparison between 5G and Release 15 LTE. The earlier results are in the ITU's urban microcell environment for IMT-Advanced [42], which is different from the later environments, but still gives an impression of how the capabilities of LTE have evolved. The wideband results are for a 100 MHz carrier in TDD mode, while the others are for a 20 MHz carrier in TDD mode or for a pair of 10 MHz carriers in FDD. Table 11.9 lists some of the more important test conditions.

In each environment, the 3GPP technical reports quote several different values for the average spectral efficiency, using different antenna configurations. Each of the results in Figure 11.11 is an average over those values, weighted by the number of samples quoted for each one. One result (for the LTE Release 15 uplink in the 4 GHz rural environment) is omitted, because it comes from a single sample with an anomalous antenna configuration.

Table 11.9 Test conditions for the 3GPP simulations of 5G and LTE.

	IMT-Advanced urban microcell	IMT-2020 rural eMBB	IMT-2020 dense urban eMBB
Number of sites	19	19	19
Number of cells per site	3	3	3
Inter-site distance	200 m	1732 m	200 m
Antenna height	10 m	35 m	25 m
Users per cell	10	10	10
Distribution of users	50% indoor at 3 km h^{-1}	50% indoor at 3 km h^{-1}	80% indoor at 3 km h^{-1}
	50% outdoor at 3 km h^{-1}	50% in cars at 120 km h^{-1}	20% in cars at 30 km h^{-1}
Carrier frequencies	2.5 GHz	700 MHz, 4 GHz	4 GHz, 30 GHz
Cell bandwidths in Figure 11.11	10+10 MHz FDD	10+10 MHz FDD/ 20 MHz TDD	20 MHz TDD/ 100 MHz TDD

In other cases, the configurations for 5G and Release 15 LTE are similar: for example, a typical configuration for dense urban eMBB at 4 GHz is 32×4 multiple-user MIMO. Full details are in References [40–42].

If the two systems occupy the same cell bandwidth, then the spectral efficiency of 5G in these simulations is around 10–30% greater than that of Release 15 LTE, and much greater than that of earlier releases. If the 5G system occupies a bandwidth of 100 MHz, then its spectral efficiency is even larger, around 50% greater than that of Release 15 LTE. Nevertheless, the quoted figures should be treated with caution for two main reasons. Firstly, the test conditions are still optimistic, as the mobiles and base stations all have powerful capabilities, and the load on each cell is high. Secondly, the results are sensitive to quantities such as the cell geometry, propagation model and antenna configuration, and might not translate to a realistic network.

References

1 3GPP TS 38.321 (2019) NR; Medium access control (MAC) protocol specification (Release 15), December 2019, Sections 5.3, 5.4.

2 3GPP TS 38.212 (2019) NR; Multiplexing and channel coding (Release 15), December 2019, Section 7.3.1.

3 3GPP TS 38.321 (2019), NR; Medium access control (MAC) protocol specification (Release 15), December 2019, Section 7.1.

4 3GPP TS 38.212 (2019) NR; Multiplexing and channel coding (Release 15), December 2019, Sections 7.3.2, 7.3.3, 7.3.4.

5 3GPP TS 38.211 (2019) NR; Physical channels and modulation (Release 15), December 2019, Section 7.3.2.

6 3GPP TS 38.214 (2019) NR; Physical layer procedures for data (Release 15), December 2019, Section 5.2.2.1.1.

7 3GPP TS 38.213 (2019) NR; Physical layer procedures for control (Release 15), December 2019, Section 10.1.

8 3GPP TS 38.214 (2019) NR; Physical layer procedures for data (Release 15), December 2019, Sections 5.1.2.1, 6.1.2.1, 6.4.

9 3GPP TS 38.508-1 (2019) 5GS; User equipment (UE) conformance specification; Part 1: Common test environment (Release 15), June 2019, Section 4.6.3 (PDSCH-TimeDomainResourceAllocationList, PUSCH-TimeDomainResourceAllocationList).

10 3GPP TS 38.214 (2019) NR; Physical layer procedures for data (Release 15), December 2019, Sections 5.1.2.2, 6.1.2.2, 6.3.

11 3GPP TS 38.214 (2019) NR; Physical layer procedures for data (Release 15), December 2019, Sections 5.1.3, 6.1.4.

12 3GPP TR 38.824 (2019) Study on physical layer enhancements for NR ultra-reliable and low latency case (URLLC) (Release 16), March 2019.

13 3GPP TS 38.212 (2019) NR; Multiplexing and channel coding (Release 15), December 2019, Sections 5, 6.2, 6.3.2, 7.2.

14 3GPP TS 38.211 (2019) NR; Physical channels and modulation (Release 15), December 2019, Sections 6.3.1, 7.3.1.

15 CPRI (2019) Common Public Radio Interface: eCPRI interface specification, Version 2.0, May 2019, Section 6.1.

16 3GPP TS 38.211 (2019) NR; Physical channels and modulation (Release 15), December 2019, Sections 6.4.1.1, 7.4.1.1.

17 3GPP TS 38.214 (2019) NR; Physical layer procedures for data (Release 15), December 2019, Sections 5.1.6.2, 6.2.2.

18 3GPP TS 38.101-1 (2019) NR; User equipment (UE) radio transmission and reception; Part 1: Range 1 standalone (Release 15), December 2019, Annex A.2.3, A.3.1.

19 3GPP TS 38.211 (2019) NR; Physical channels and modulation (Release 15), December 2019, Sections 6.4.1.2, 7.4.1.2.

20 3GPP TS 38.214 (2019) NR; Physical layer procedures for data (Release 15), December 2019, Sections 5.1.6.3, 6.2.3.

21 3GPP TS 38.214 (2019) NR; Physical layer procedures for data (Release 15), December 2019, Sections 5.1.7, 6.1.5.

22 Dahlman, E., Parkvall, S., and Sköld, J. (2018). *5G NR: The Next Generation Wireless Access Technology*. London: Academic Press Section 13.1.5.

23 3GPP TS 38.213 (2019) NR; Physical layer procedures for control (Release 15), December 2019, Section 9.1.

24 3GPP TS 38.213 (2019) NR; Physical layer procedures for control (Release 15), December 2019, Section 9.2.3.

25 3GPP TS 38.214 (2019) NR; Physical layer procedures for data (Release 15), December 2019, Section 5.3.

26 3GPP TS 38.212 (2019) NR; Multiplexing and channel coding (Release 15), December 2019, Section 7.3.1.3.

27 3GPP TS 38.213 (2019) NR; Physical layer procedures for control (Release 15), December 2019, Section 11.

28 3GPP TS 38.213 (2019) NR; Physical layer procedures for control (Release 15), December 2019, Section 7.

29 3GPP TS 38.321 (2019) NR; Medium access control (MAC) protocol specification (Release 15), December 2019, Section 5.4.4.

30 3GPP TS 38.213 (2019) NR; Physical layer procedures for control (Release 15), December 2019, Section 9.2.4.

31 3GPP TS 38.321 (2019) NR; Medium access control (MAC) protocol specification (Release 15), December 2019, Section 5.8.

32 3GPP TS 38.213 (2019) NR; Physical layer procedures for control (Release 15), December 2019, Section 10.2.

33 3GPP TS 38.304 (2019) NR; User equipment (UE) procedures in idle mode and RRC inactive state (Release 15), December 2019, Section 7.

34 3GPP TS 38.321 (2019) NR; Medium access control (MAC) protocol specification (Release 15), December 2019, Section 5.7.

35 3GPP TR 37.910 (2019) Study on self evaluation towards IMT-2020 submission (Release 16), September 2019.

36 3GPP TR 37.910 (2019) Study on self evaluation towards IMT-2020 submission (Release 16), September 2019, Section 5.1.1.

37 3GPP TS 38.306 (2019) NR; User equipment (UE) radio access capabilities (Release 15), December 2019, Section 4.1.2.

38 3GPP TS 38.101-2 (2019) NR; User equipment (UE) radio transmission and reception; Part 2: Range 2 standalone (Release 16), December 2019, Section 5.5A.1.

39 3GPP TR 37.910 (2019) Study on self evaluation towards IMT-2020 submission (Release 16), September 2019, Section 5.4.

40 3GPP TR 36.814 (2017) Evolved Universal Terrestrial Radio Access (E-UTRA); Further advancements for E-UTRA physical layer aspects (Release 9), March 2017, Section 10.2.

41 ITU-R M.2412-0 (2017) Guidelines for evaluation of radio interface technologies for IMT-2020, Sections 8.3, 8.4.

42 ITU-R M.2135 (2008) Guidelines for evaluation of radio interface technologies for IMT-Advanced, Sections 8.3, 8.4.

12

Air Interface Layer 2

This chapter addresses the air interface's higher-level transport protocols, which form layer 2 of an OSI protocol stack. We will work our way upwards from the bottom of the protocol stack, beginning with the medium access control (MAC) protocol, and progressing through radio link control (RLC), the packet data convergence protocol (PDCP) and the service data adaptation protocol (SDAP). The protocols are simpler than the air interface's physical layer: the first three have not changed much in comparison with LTE, while the last one is new to 5G.

12.1 Medium Access Control

12.1.1 Protocol Architecture

The medium access control protocol schedules transmissions on the air interface and controls the low-level operation of the physical layer [1]. There is one MAC entity per node, so a mobile in 5G dual connectivity has two 5G MAC entities, one each for the master and secondary. Figure 12.1 shows the protocol architecture from the viewpoint of the mobile.

Above the MAC protocol, each logical channel carries data from a single radio bearer. In the transmitter, the radio link control protocol stores outgoing data in its transmit buffers, and delivers them to the MAC protocol on receiving a request from the MAC. Inside the MAC protocol, the control function generates control elements (CEs), which convey MAC-related signalling information over to the receiver. The multiplexing function merges the data and control elements together, and the protocol sends the resulting data on a transport channel to the underlying physical layer. The processes are reversed in the receiver.

The MAC protocol also controls several procedures within the physical layer, notably the ones for random access and hybrid ARQ. We discussed those procedures in earlier chapters and have little more to say about them here.

12.1.2 Scheduling

The base station's MAC protocol is associated with a scheduling function, which decides the contents of its downlink scheduling commands and its uplink scheduling grants. In dual connectivity, the master node (MN) contains one scheduling function that controls the

An Introduction to 5G: The New Radio, 5G Network and Beyond, First Edition. Christopher Cox.
© 2021 John Wiley & Sons Ltd. Published 2021 by John Wiley & Sons Ltd.

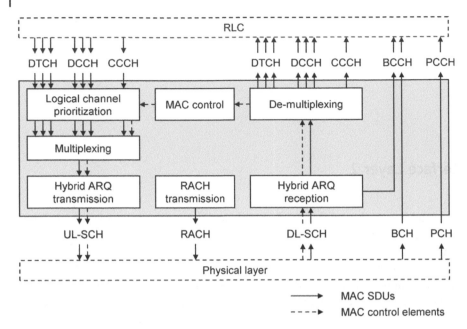

Figure 12.1 Architecture of the mobile's medium access control protocol. BCH: broadcast channel; DL-SCH: downlink shared channel; PCH: paging channel; RACH: random access channel; UL-SCH: uplink shared channel. Source: Adapted from 3GPP TS 38.321.

whole of the master cell group (MCG), while the secondary node (SN) contains a separate scheduling function that controls the secondary cell group (SCG).

The scheduling function uses three main sets of information. The first covers the quality of service (QoS) targets of the individual radio bearers, for example their priorities, maximum error rates and maximum delays. The second covers the amount of data in the transmit buffers and the time that the data have been there. The third covers the quality of the radio channel, using the mobiles' channel state information (CSI) reports and the base station's measurements of the sounding reference signals. Other information includes the mobiles' hybrid ARQ acknowledgements and scheduling requests, and any knowledge of the load in nearby cells.

Using that information, the base station decides which mobiles to schedule on the uplink and downlink at every scheduling opportunity. For each mobile, it then has to decide the transport block size, the number of resource blocks to allocate, and parameters such as the modulation scheme, the coding rate and the number of MIMO layers. The algorithm is a proprietary one, written by the equipment vendor, with parameters that can be adjusted by the network operator.

12.1.3 Logical Channel Prioritization

An uplink scheduling grant only defines the size of a mobile's transport block: it does not say what the transport block should contain. That issue is handled using a separate algorithm for uplink logical channel prioritization.

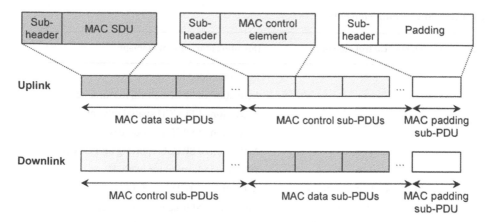

Figure 12.2 Structure of a MAC PDU. Source: Adapted from 3GPP TS 38.321.

To configure the algorithm, the base station associates each dedicated logical channel with several parameters. The most important are a priority from 1 to 16, where 1 is the most important, and a *prioritized bit rate* from 0 to 65 536 kbps, which is a target for the long-term average data rate. (The specifications also support an infinite bit rate, with the interpretation 'as fast as possible'.) Ultimately, those quantities are derived from the QoS targets that we will discuss in Chapter 15. Other parameters include restrictions on the serving cells and subcarrier spacings that an individual logical channel can use.

The specifications also prioritize the control elements and the common logical channels. The greatest importance is attached to signalling associated with message 3 of the random access procedure, namely MAC cell radio network temporary identifier (C-RNTI) control elements and transmissions on the common control channel.

On receiving a scheduling grant, the MAC protocol runs through the logical channels in priority order, and asks the RLC protocol to deliver enough data on each one to maintain its prioritized bit rate. If any resources remain, then it runs through the logical channels in priority order once again, and asks for as much data from each one as it can. The downlink uses a similar procedure, but the details are a matter for the base station.

12.1.4 Multiplexing and De-multiplexing

In the transmitter, the multiplexing function joins MAC service data units (SDUs) and control elements together, and creates MAC protocol data units (PDUs) that have the internal structure shown in Figure 12.2. Each MAC SDU is associated with a sub-header, which states the number of bytes in the SDU and which identifies the logical channel using a field known as the *logical channel identity* (LCID). Each control element is also associated with a sub-header, which identifies the type of control element by means of a reserved LCID. If there are not enough data available to fill the resource allocation, then the PDU ends with a padding SDU, which is identified by its own sub-header. The resulting PDUs are identical to the transport blocks from Chapter 11.

The control elements lie before the downlink SDUs but after the uplink SDUs, and each of these lies immediately after its corresponding sub-header. That layout helps to reduce

Table 12.1 MAC control elements.

LCID	Uplink MAC control element	Downlink MAC control element
47	—	Recommended bit rate
48	—	Activation of SP ZP CSI-RS resource set
49	—	Activation of PUCCH spatial relation
50	—	Activation of SP SRS resource set
51	—	Activation of SP CSI reporting on PUCCH
52	—	TCI state indication for PDCCH
53	Recommended bit rate query	Activation of TCI states for PDSCH
54	Multiple-entry PHR (four octets)	Aperiodic CSI trigger state sub-selection
55	Configured grant confirmation	Activation of SP CSI-IM / NZP CSI-RS resource set
56	Multiple-entry PHR (one octet)	Activation of PDCP duplication
57	Single-entry PHR	Activation of SCell (four octets)
58	C-RNTI	Activation of SCell (one octet)
59	Short truncated BSR	Long DRX command
60	Long truncated BSR	DRX command
61	Short BSR	Timing advance command
62	Long BSR	UE contention resolution identity
63	Padding	Padding

CSI-IM: channel state information interference measurement; CSI-RS: channel state information reference signal; NZP: non-zero power; PDCCH: physical downlink control channel; PDSCH: physical downlink shared channel; SP: semi-persistent; SRS: sounding reference signal; TCI: transmission configuration indicator; ZP: zero power.
Source: Adapted from 3GPP TS 38.321.

the latency in the mobile [2]: for example, a mobile can start to assemble a PDU from one or more high-priority SDUs even before a scheduling grant is available.

12.1.5 MAC Control Elements

The MAC protocols in the mobile and base station communicate using MAC control elements that are listed in Table 12.1 [3]. Each control element has a defined format and interpretation, and is associated with a reserved value for the LCID. Most of the control elements are covered elsewhere, so we will briefly describe the ones that remain.

Using *buffer status reports* (BSRs), the mobile keeps the base station informed about the amount of data in its RLC transmit buffers. These are triggered in three situations. The mobile sends a *regular BSR* if new data arrive on a logical channel that has a higher priority than the buffers were previously storing. If the mobile does not have any resources on the physical uplink shared channel (PUSCH), then it instead sends a scheduling request on the physical uplink control channel (PUCCH). The mobile also sends a *periodic BSR* at regular intervals during uplink data transmission, and a *padding BSR* if it has enough spare room during a normal PUSCH transmission.

The mobile keeps the base station informed about its uplink transmit power by means of *power headroom reports* (PHRs). Each report indicates the difference between the mobile's maximum power and the power required for an uplink transmission, with negative values used if the required power exceeds the amount available [4]. It can be triggered in several ways, for example if the downlink path loss has changed significantly since the last report or if a periodic reporting timer expires.

When using the multimedia telephony service for voice over IP, the mobile can request a *recommended bit rate* from the base station by means of an uplink control element, and the base station can provide one on the downlink. The mobile uses that recommendation in the multimedia telephony application, when negotiating the bit rate of its voice codec with its peer [5]. If, for example, the radio conditions are poor, then the base station can recommend a lower bit rate to the mobile. If the air interface's resource allocation remains the same, then that recommendation lowers the coding rate and therefore reduces the number of packet errors.

Using an *SCell activation/deactivation* control element, the base station can activate and deactivate the mobile's use of individual secondary cells (SCells). Deactivation can reduce the mobile's power consumption if the data rate is low.

12.2 Radio Link Control

12.2.1 Protocol Architecture

The RLC protocol maintains the layer 2 data link between the mobile and the base station [6]. As shown in Figure 12.3, it has three modes of operation, which are known as *transparent mode* (TM), *unacknowledged mode* (UM) and *acknowledged mode* (AM). Transmission and reception are handled by separate RLC entities in transparent and unacknowledged modes, and by a single bi-directional entity in acknowledged mode. These entities are set up on a bearer-by-bearer basis.

In transparent mode, dual connectivity is not supported, so the radio bearers are all MCG bearers. Each bearer is associated with one RLC transmit entity and/or one RLC receive entity in the mobile, and likewise in the master node.

In unacknowledged mode, dual connectivity is supported, so the radio bearers can be MCG, SCG or split bearers. Furthermore, each radio bearer can optionally be associated with two RLC transmit and/or two RLC receive entities in the mobile, and likewise in the

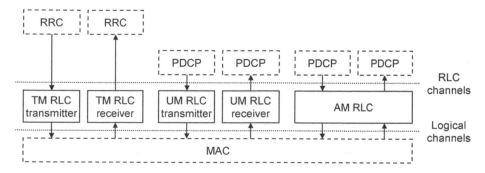

Figure 12.3 Radio link control protocol. Source: Adapted from 3GPP TS 38.322.

network. The second RLC entity can be used in two ways. Firstly, it is essential for the implementation of split bearers, in which the network has one RLC entity in the master node and one in the secondary. Secondly, it supports a technique known as PDCP duplication, in which the network's RLC entities can be either in different nodes or in the same one, and the transmitter delivers packets through both to increase its reliability.

The same applies in acknowledged mode, except that the RLC entities are bi-directional. In both modes, the maximum number of RLC entities increases in Release 16 from two to four [7].

12.2.2 Transparent Mode

Transparent mode handles three types of signalling message: system information messages on the broadcast control channel (BCCH), paging messages on the paging control channel (PCCH), and messages in RRC_INACTIVE and RRC_IDLE on the common control channel (CCCH). Its architecture (Figure 12.4) is very simple.

The transmit entity receives signalling messages from the radio resource control (RRC) protocol in the form of RLC SDUs and stores them in a buffer. The MAC protocol grabs those messages from the buffer in the form of RLC PDUs without any modification. (The messages are short enough to fit into a single transport block without segmentation.) The receive entity reverses the process by passing the incoming messages directly up to the RRC.

12.2.3 Unacknowledged Mode

Unacknowledged mode handles data radio bearers on the dedicated traffic channel (DTCH) that require a low latency and do not require any further error correction. There are two main types. The first is traditional real-time traffic such as conversational IP voice and video, for which the physical layer delivers a packet error ratio of around 0.1–1% by means of hybrid ARQ. The second is traffic for ultra-reliable low-latency communication (URLLC), for which the physical layer delivers a packet error ratio of around 0.001% to 0.01% using a slow modulation scheme and a low coding rate, perhaps with the assistance of PDCP duplication.

Figure 12.5 shows the architecture. The transmit entity receives an RLC SDU from the PDCP, generates a header and stores the SDU in a buffer. Later, the MAC protocol asks

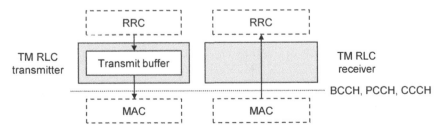

Figure 12.4 Architecture of the RLC protocol in transparent mode. Source: Adapted from 3GPP TS 38.322.

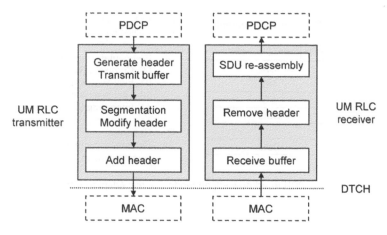

Figure 12.5 Architecture of the RLC protocol in unacknowledged mode. Source: Adapted from 3GPP TS 38.322.

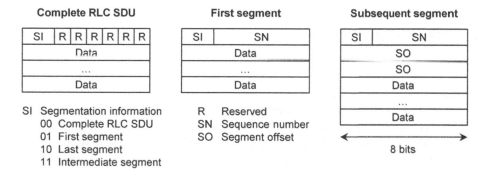

Figure 12.6 Structure of an RLC data PDU in unacknowledged mode. Source: Adapted from 3GPP TS 38.322.

the RLC for one or more PDUs with a specific total size. In response, the transmit entity attaches the headers to their respective SDUs and sends them on.

If an SDU is too large, then the transmit entity can segment it into smaller PDUs. In doing so, it modifies *segmentation information* (SI) in the first PDU's header to indicate that other segments follow, and includes an SDU *sequence number* (SN). It then generates a header for each subsequent PDU that indicates the *segment offset* (SO) within the original SDU. Figure 12.6 shows the structures of the resulting PDUs, for the case where the sequence number contains 6 bits.

This process allows the transmit entity to create the header of an unsegmented SDU before receiving any requests from the MAC protocol, which reduces its latency [8]. Unlike in LTE, there is no opportunity to concatenate the end of one SDU with the start of the next one, as the latency would then be higher.

When a PDU reaches the receiver, there are two possibilities. If the PDU contains a complete SDU, then the receiver passes it directly to the PDCP. Otherwise, the receiver stores the

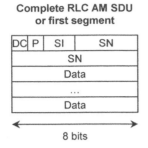

Complete RLC AM SDU or first segment

DC	P	SI	SN
SN			
Data			
...			
Data			

←——————————→
8 bits

Subsequent segment

DC	P	SI	SN
SN			
SO			
SO			
Data			
...			
Data			

DC   Data / control
P   Polling bit

SI   Segmentation information
    00 Complete RLC SDU
    01 First segment
    10 Last segment
    11 Intermediate segment

R   Reserved
SN   Sequence number
SO   Segment offset

Figure 12.7 Structure of an RLC data PDU in acknowledged mode. Source: Adapted from 3GPP TS 38.322.

PDU in a buffer, and uses its sequence number and segment offset to help reconstruct the corresponding SDU. In doing so, the RLC begins the process of hybrid ARQ re-ordering.

12.2.4 Acknowledged Mode

Acknowledged mode handles radio bearers that require a high reliability and can tolerate a long delay. There are two types: data radio bearers on the DTCH that are carrying non-real-time traffic such as web pages and emails, and signalling radio bearers on the dedicated control channel (DCCH) that are carrying mobile-specific signalling. To avoid the use of a very low coding rate, the physical layer is configured to deliver a relatively high packet error ratio, perhaps around 10%. The RLC protocol reduces that error rate by re-transmitting packets that have not reached the receiver correctly, at the expense of adding additional time delays. The typical result is a packet error ratio of around 10^{-6}, with the remaining errors corrected by means of end-to-end re-transmissions using TCP.

There are two types of PDU. *Data PDUs* carry higher-layer traffic and signalling messages. As shown in Figure 12.7, they are similar to PDUs in unacknowledged mode, but they all contain a sequence number as well as a few extra header fields. *Status PDUs* carry the receiver's acknowledgements.

Figure 12.8 shows the architecture. The transmitter sends data PDUs in a similar way to unacknowledged mode, but it also stores them in a re-transmission buffer until it knows that they have arrived correctly. Whenever the number of transmitted bytes or PDUs exceeds a threshold, the transmitter sets a *polling bit* in one of the data PDU headers. In response, the receiver returns a status PDU, which identifies the data PDUs that have and have not arrived by means of their sequence numbers and segment offsets. The transmitter can then discard the data PDUs that have arrived and can re-transmit the ones that have not.

When carrying out a re-transmission, the transmitter can apply further segmentation to any PDUs that it has already delivered, in order to satisfy the MAC protocol's request for the total amount of data to send. That process uses the same segmentation function that was introduced for unacknowledged mode, so it does not add much further complexity.

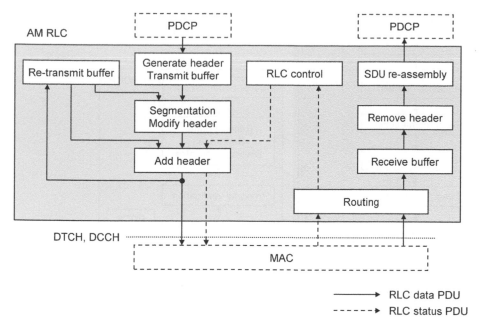

Figure 12.8 Architecture of the RLC protocol in acknowledged mode. Source: Adapted from 3GPP TS 38.322.

12.3 Packet Data Convergence Protocol

12.3.1 Protocol Architecture

The packet data convergence protocol routes radio bearers through either or both of the master and secondary cell groups, and processes them using higher-level functions that are outside the scope of the MAC and RLC [9]. It is only used by the DTCH and DCCH, for which the underlying RLC protocol is operating in unacknowledged or acknowledged mode. There is one pair of PDCP transmit and receive entities in the master node for each MN-terminated bearer, and one pair in the secondary node for each SN-terminated bearer.

Figure 12.9 shows the internal architecture of the PDCP. We will discuss the functions of packet routing and header compression in this chapter, as well as the protocol's role in preventing packet loss during a change of node. We will leave the security functions of ciphering and integrity protection until Chapter 14.

12.3.2 Transmission and Reception

The PDCP transmitter labels each outgoing packet with a PDCP sequence number, which forms part of the PDCP header. It then stores the packet in a re-transmission buffer, before processing it and delivering it to the RLC. The packet is discarded once a timer (10–1500 ms) expires.

The PDCP receiver stores the incoming packets in a receive buffer. If the bearer is configured for *in-sequence delivery*, then the receiver uses the sequence numbers to return

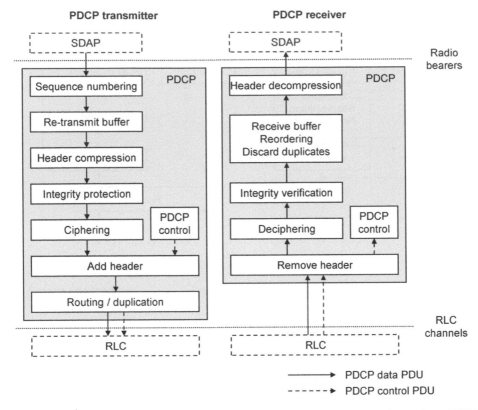

Figure 12.9 Architecture of the packet data convergence protocol. Source: Adapted from 3GPP TS 38.323.

the corresponding packets to their correct order. That procedure completes the process of hybrid ARQ re-ordering, and also corrects any ordering problems that arose due to re-transmissions in the RLC protocol.

12.3.3 PDCP Duplication

As we noted in Section 12.2.1, a radio bearer can optionally be associated with two RLC entities in Release 15, or with a maximum of four in Release 16. 5G can exploit these in a technique known as *PDCP duplication*, which is configured in two stages. Using RRC signalling, the base station first associates a bearer with multiple RLC entities and indicates whether the bearer is allowed to use PDCP duplication. In the case of a data radio bearer, the base station can then activate and deactivate the use of PDCP duplication dynamically by means of a MAC control element.

If a bearer is using PDCP duplication, then the transmitter delivers each PDCP PDU through all of the associated RLC entities. That increases the reliability of the air interface, notably for URLLC bearers that are operating in RLC unacknowledged mode. If a bearer has multiple RLC entities but is not using PDCP duplication (a split bearer, for example), then the transmitter delivers each PDCP PDU through only one. There are rules that influence

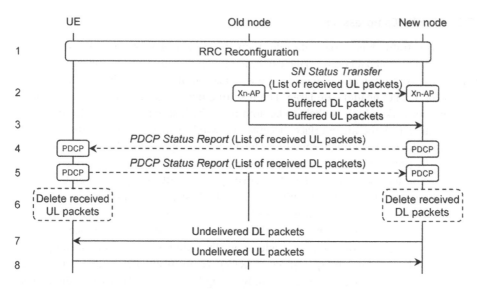

Figure 12.10 PDCP status reporting and re-transmission procedures.

the mobile's choice of RLC entity on the uplink, while the choice on the downlink is a matter for the base station.

12.3.4 Prevention of Packet Loss during a Change of Node

If the master or secondary node changes, for example during a handover, then the mobile and network delete the RLC entities that they were using in the old node, and re-establish them in the new one. As a result, any partially delivered RLC SDUs are lost. However, the mobile and network both maintain their PDCP configurations, and can use that protocol to prevent packet loss in the manner shown in Figure 12.10. The procedure begins when the network initiates a change of node using a signalling message known as *RRC Reconfiguration* that we will discuss in Chapter 16 (Figure 12.10, step 1).

Subsequently, the messages in solid lines are available for any data radio bearer. In step 3, the old node forwards the downlink packets in its PDCP re-transmission buffer over to the new one, and also forwards any uplink packets that arrived out of sequence and are in its PDCP receive buffer. The new node can then re-transmit the downlink packets so as to prevent packet loss (7). Similarly, the mobile re-transmits the uplink packets that are in its own re-transmission buffer (8).

By themselves, these measures can lead to the unnecessary duplication of packets which were still in the re-transmission buffers but had already reached the receiver. That problem can be avoided using the messages in dashed lines, which are only available for data radio bearers in RLC acknowledged mode. At the start of the procedure, the old node sends an X2-AP or Xn-AP *SN Status Transfer* message to the new one, which lists the uplink packets that have already arrived (2) [10]. The new node forwards that information to the mobile by means of a *PDCP status report* (4), and the mobile sends a similar status report in reply (5). The two PDCP transmitters can now delete the unwanted packets from their re-transmission buffers (6) before re-transmitting the remainder.

12.3.5 Header Compression

If a stream of traffic has a low data rate, then the packet header can make up a large proportion of the total packet size. That can be a problem for applications such as voice and machine-type communications, particularly when communicating with IP data networks in which the packet headers are large. The resulting overhead does not matter so much in the core network, which is likely to be dominated by fast data services. However, it is inappropriate across the air interface, because the wireless link acts as a bottleneck, and because an individual cell might sometimes be dominated by voice calls.

To solve the problem, the PDCP can be configured to use an Internet Engineering Task Force (IETF) protocol known as *robust header compression* (ROHC) [11]. When using that protocol, the transmitter sends the full header in the first packet, but subsequently only sends any differences from expectations. Most of the header either stays the same from one packet to the next or changes in a predictable way, so the resulting difference fields are small. As an example, a voice over IP header typically contains 40 to 60 bytes, comprising 12 bytes from the *real-time transport protocol* (RTP), 8 bytes from UDP, and either 20 bytes from IP version 4 or 40 bytes from version 6. By using ROHC, the PDCP can reduce those figures to as little as 1 and 3 bytes respectively.

12.4 Service Data Adaptation Protocol

The service data adaptation protocol maps the 5G core network's QoS flows onto the data radio bearers in the next-generation radio access network [12]. There is one pair of SDAP transmit and receive entities in the master node for each PDU session containing MN-terminated bearers, and one pair in the secondary node for each PDU session containing SN-terminated bearers.

Figure 12.11 shows the protocol architecture. In the transmitter, the protocol receives an SDU that belongs to a particular QoS flow. The protocol adds a header to identify the QoS flow if it has been configured to do so, and then delivers the resulting PDU on the corresponding data radio bearer. The process is reversed in the receiver. On the downlink,

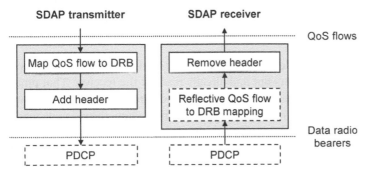

Figure 12.11 Architecture of the service data adaptation protocol. Source: Adapted from 3GPP TS 37.324.

the mobile receiver can also implement a function known as reflective QoS, which we will discuss in Chapter 15.

Usually, there is a one-to-one mapping between data radio bearers and QoS flows, so that the former can implement the QoS targets that are associated with the latter. However, a data radio bearer can also handle packets from multiple QoS flows. That might be appropriate if, for example, two QoS flows only differ in a field known as the allocation and retention priority (ARP), which describes how important it is to retain the QoS flow in cases of network congestion.

References

1 3GPP TS 38.321 (2019) NR; Medium access control (MAC) protocol specification (Release 15), December 2019.
2 Dahlman, E., Parkvall, S., and Sköld, J. (2018). *5G NR: The Next Generation Wireless Access Technology*. Academic Press, Section 6.4.4.
3 3GPP TS 38.321 (2019) NR; Medium access control (MAC) protocol specification (Release 15), December 2019, Sections 5.18, 6.1.3.
4 3GPP TS 38.213 (2019) NR; Physical layer procedures for control (Release 15), December 2019, Section 7.7.
5 3GPP TS 26.114 (2019) IP multimedia subsystem (IMS); Multimedia telephony; Media handling and interaction (Release 15), September 2019, Section 10.7.
6 3GPP TS 38.322 (2019) NR; Radio link control (RLC) protocol specification (Release 15), March 2019.
7 3GPP TR 38.825 (2019) Study on NR industrial internet of things (IoT); (Release 16), March 2019.
8 Dahlman, E., Parkvall, S., and Sköld, J. (2018). *5G NR: The Next Generation Wireless Access Technology*. Academic Press, Section 6.4.3.
9 3GPP TS 38.323 (2019) NR; Packet data convergence protocol (PDCP) specification (Release 15), June 2019.
10 3GPP TS 38.423 (2019) NG-RAN; Xn application protocol (XnAP) (Release 15), December 2019, Sections 8.2.2, 9.1.1.4.
11 IETF RFC 4995 (2007) The RObust Header Compression (ROHC) framework, July 2007.
12 3GPP TS 37.324 (2018) E-UTRA and NR; Service data adaptation protocol (SDAP) specification (Release 15), September 2018.

13

Registration Procedures

We will now move on to a set of seven chapters that cover the end-to-end operation of 5G. Throughout these chapters, the main focus will be on architectural options that involve the 5G core network, although we will restrict some aspects to architectures that require a master gNB. To keep things simple, we will mainly assume the use of an integrated gNB, but we will discuss what happens if the gNB is split into central and distributed units as part of Chapter 18. We will also address inter-operation with the evolved packet core in Chapter 19.

The main specification for these discussions is TS 23.502 [1], which covers the signalling procedures within the 5G core. The equivalent procedures in the radio access network (RAN) form part of TS 38.300 [2], while full details of the resulting signalling are in the corresponding stage 3 specifications, which were laid out in Chapters 2 and 3. The use of service-based interfaces and HTTP/2 makes the core network's signalling more intricate than in previous generations, and readers may wish to review the discussion of HTTP/2 in Chapter 2 before proceeding.

In this chapter, we will address the procedures by which the mobile registers with the 5G core network when it switches on. In doing so, we will cover the procedures for network and cell selection, for the establishment of a radio resource control (RRC) signalling connection with the RAN and for registration itself. As part of the discussion, we will address how a mobile chooses between a 4G cell controlled by an eNB and a 5G cell controlled by a gNB, and how it chooses between the evolved packet core and the 5G core. As well as the specifications listed in the previous paragraph, two others are particularly relevant for this chapter, namely the specifications for idle mode procedures in the non-access stratum (NAS), TS 23.122 [3], and the access stratum (AS), TS 38.304 [4].

13.1 Power-on Sequence

Figure 13.1 summarizes the procedure by which a mobile establishes communications with a master gNB and registers with the 5G core, under architectural option 2. To begin the procedure, the mobile selects a public land mobile network (PLMN) that it will register with, and a 5G cell that belongs to the PLMN. In doing so, the mobile is said to *camp* on the cell.

Using the selected cell, the mobile contacts the corresponding gNB using the random access procedure from Chapter 9. It then initiates a procedure known as RRC connection

An Introduction to 5G: The New Radio, 5G Network and Beyond, First Edition. Christopher Cox.
© 2021 John Wiley & Sons Ltd. Published 2021 by John Wiley & Sons Ltd.

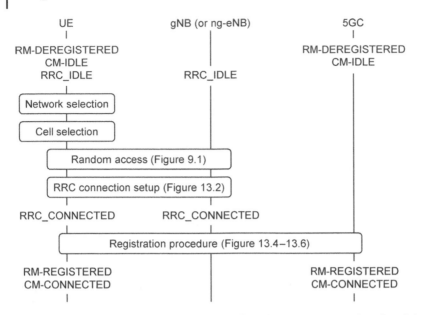

Figure 13.1 Overview of the power-on procedures in architectural options 2 and 4.

establishment, in which it establishes a signalling connection with the selected gNB and moves to the state of RRC_CONNECTED. As part of that procedure, the gNB configures the mobile with a set of parameters for communication over the air interface, including a set of uplink and downlink bandwidth parts, and configuration parameters for signalling radio bearer 1 (SRB 1).

Using the selected gNB, the mobile then initiates the registration procedure. During that procedure, the mobile registers with an access and mobility management function (AMF) in the 5G core network, and moves to the states of RM-REGISTERED and CM-CONNECTED. The mobile also receives a set of configuration parameters for SRB 2.

Subsequently, the mobile can establish connectivity with an external data network, using signalling procedures that we will cover in Chapter 15. If it does not do so, or after any later lapse in data communications, then the network can transfer the mobile to the states of CM-IDLE and RRC_IDLE in the manner described in Chapter 16.

As an alternative to this procedure, the mobile might select a 4G cell and establish a signalling connection with the corresponding eNB. Depending on how that cell is configured, the mobile can then register either with the 5G core network using the procedure introduced above (architectural option 5) or with the evolved packet core using the attach procedure from LTE (option 1).

13.2 Network and Cell Selection

13.2.1 Network Selection

The network selection procedure is much the same as in previous generations [5]. After switching on, the mobile reads several parameters from the Universal Subscriber Identity Module (USIM) [6]. These include the tracking area in which the mobile was last registered, from which it extracts the identity of the corresponding network, known as the *registered PLMN*.

The mobile now searches for cells that belong to the registered PLMN, using all of the radio access technologies that it supports. The search begins on any radio frequencies that the mobile stored when it was last switched on. If the mobile does not find the registered PLMN on any of those radio frequencies, then it scans all the frequency bands that it supports, using all of its access technologies.

In the case of 5G cells, the mobile runs the acquisition procedure from Chapter 8 and identifies the strongest cell in each frequency band, ignoring cells which are barred or which are not transmitting system information block 1 (SIB 1) [7]. It then reads the PLMN identity list from SIB 1. There is a similar procedure for 4G cells, except that they can advertise two independent lists of network identities: one list of 5G core networks that can be reached over the NG reference point, and one list of evolved packet core networks that can be reached over S1.

If the mobile cannot find the registered PLMN, then it moves to one of two modes for network selection, with the choice configured by the user. In *automatic mode*, the mobile first looks through a prioritized list of network identities that it should treat as home networks, and then tries prioritized lists of network identities that have been defined by the user and by the operator. Each of those networks can be associated with a prioritized list of radio access network technologies, which sets up a priority order between the evolved UMTS terrestrial radio access network (E-UTRAN) and the next-generation radio access network (NG-RAN), but does not distinguish between 4G and 5G cells within the NG-RAN. If the mobile cannot find any of those networks, then it selects any network whose received signal power is sufficiently high or, failing that, any network at all.

In the last two cases, the selected network might not have a roaming agreement with the home network operator. In that situation, the registration procedure will fail. If that happens for every network, then the mobile enters a *limited service* state, in which it can only make emergency calls, or receive warnings from the earthquake and tsunami warning system and the commercial mobile alert service.

In *manual mode*, the mobile presents the user with the networks that it has found, with the networks listed in the same priority order as in automatic mode. The network is then selected by the user.

13.2.2 Cell Selection

During the *cell selection* procedure, the mobile chooses a *suitable cell* that belongs to the selected network and to the selected radio access network technology. A suitable cell is one that satisfies several criteria. In the case of a 5G cell, the most important are two criteria on the strength of the received signal [8, 9]. Firstly, the received signal power has to satisfy the following condition:

$$S_{rxlev} = Q_{rxlevmeas} - (Q_{rxlevmin} + Q_{rxlevminoffset}) - Q_{offset,temp} - P_{compensation} > 0 \quad (13.1)$$

where $Q_{rxlevmeas}$ is the *synchronization signal reference signal received power* (SS-RSRP), which is the average power per resource element that the mobile receives on the cell's secondary synchronization signal (SSS); and $Q_{rxlevmin}$ is the minimum value of SS-RSRP, which the base station advertises in SIB 1. Thus, to a first approximation, the equation simply states that the received signal power should be sufficiently high.

There are several refinements. Firstly, if the cell is using multiple beams, then the mobile will measure a different value of SS-RSRP on each beam. During cell selection, the mobile can convert those into a cell-specific value in any way that it pleases, for example by choosing the largest. Secondly, if the cell has a supplementary uplink (SUL), then it can advertise a second, lower value of $Q_{rxlevmin}$, which the mobile chooses if it supports the SUL. That allows the mobile to select a weaker cell than it usually would, safe in the knowledge that it can still reach the cell by means of the SUL. Thirdly, $Q_{rxlevminoffset}$ and $Q_{offset, temp}$ are both zero during initial cell selection: they are respectively used in the procedure for network reselection (Section 17.3.2), and after repeated failures to establish communications with the cell.

In basic deployments, the final parameter is calculated as follows:

$$P_{compensation} = \max(P_{EMAX1} - P_{PowerClass}, \ 0) \tag{13.2}$$

In this equation, $P_{PowerClass}$ is the mobile's maximum transmit power. The 5G specifications define different power classes for different types of mobile, the default being power class 3 [10, 11]. For those mobiles, the transmit power is limited to 23 dB relative to 1 mW (dBm), in other words about 200 mW, while the equivalent isotropic radiated power is limited to 43 dBm in frequency range 2 to account for the transmit array gain. P_{EMAX1} is the maximum uplink power that is allowed within the cell, which the base station advertises as part of SIB 1. By combining these quantities, $P_{compensation}$ makes the mobile less likely to select the cell if its transmit power is lower than the maximum value that the base station is expecting, and reduces the risk of uplink coverage problems. In more complex deployments, the base station can also enforce an *additional maximum power reduction*, which limits the mobiles' emissions into adjacent frequency bands.

In addition, the received signal quality has to satisfy the following condition:

$$S_{qual} = Q_{qualmeas} - (Q_{qualmin} + Q_{qualminoffset}) - Q_{offset,temp} > 0 \tag{13.3}$$

where $Q_{qualmeas}$ is the *synchronization signal reference signal received quality* (SS-RSRQ), which equals the value of SS-RSRP divided by the average power per resource element that the mobile receives from all sources, including signal, interference and noise; $Q_{qualmin}$ is a minimum value that the base station advertises in SIB 1, with the same value used for the normal and supplementary uplink; and $Q_{qualminoffset}$ and $Q_{offset, temp}$ are both zero during initial cell selection, as before. This criterion prevents a mobile from selecting a cell with a strong received signal, if the cell is subject to high levels of interference.

Furthermore, a suitable cell must not be barred or reserved for operator use, and must not lie in any forbidden tracking areas that are listed in SIB 1. The criteria for selection of a suitable 4G cell are similar [12].

13.3 RRC Connection Establishment

13.3.1 RRC Connection Establishment with a gNB

If the mobile selected a 5G cell, then it contacts the corresponding gNB using the random access procedure from Chapter 9. During initial registration, message 3 of that procedure is

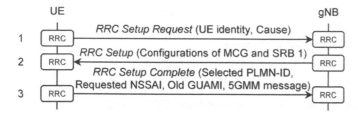

Figure 13.2 RRC connection establishment procedure. Source: Adapted from 3GPP TS 38.331.

an *RRC Setup Request*. That initiates the procedure of *RRC connection establishment*, which is illustrated in Figure 13.2 [13].

As part of its RRC Setup Request (step 1), the mobile identifies itself using the unique identity that we introduced as part of the random access procedure. It also states the establishment cause, in other words its reason for sending the request. In this example, the cause is mobile-originated signalling, while other causes include mobile-originated data, an emergency call and a response to mobile-terminated paging.

The base station receives the message, takes on the role of the mobile's master gNB and replies with the message *RRC Setup* (2). That message configures the mobile's master cell group, perhaps taking the mobile into an immediate state of carrier aggregation. It also provides each cell with a set of downlink bandwidth parts, a set of uplink bandwidth parts and a set of parameters for each one. Other parameters apply to each individual cell, for example the TDD configuration, and to the master cell group as a whole, for example the configuration of the medium access control (MAC) protocol. The message also configures SRB 1 and its associated radio link control (RLC) and packet data convergence protocol (PDCP) entities. The mobile applies the requested configuration, and moves into the state of RRC_CONNECTED.

Having done so, the mobile acknowledges its configuration by sending the message *RRC Setup Complete* (3). That message includes a 5G mobility management (5GMM) message that the base station should deliver to the core network, in this case a Registration Request. It also states the selected network identity, the requested network slices, and the globally unique identity of the AMF with which the mobile was previously registered. The base station uses these information elements for the procedure of AMF selection, which is described further in this chapter.

13.3.2 Initial UE Message

The gNB now chooses a suitable AMF for the mobile [14]. If it is connected to the mobile's previous AMF, then it chooses that one. Otherwise, the base station chooses an AMF from the ones to which it is connected, using information such as the capacity of each one, the network and slices that the mobile requested, and the mobile's previous AMF set. If the base station cannot identify a suitable AMF, then it chooses any AMF that is itself configured for AMF selection.

The gNB then forwards the mobile's registration request to the AMF by embedding it into an NG application protocol (NG-AP) *Initial UE Message*, as shown in Figure 13.3 [15]. That message requests the establishment of an NG signalling connection for the mobile

Figure 13.3 Initial UE message procedure. Source: Adapted from 3GPP TS 38.413.

and supplies a 32-bit field that the base station will use to identify the mobile over NG. The message also includes the selected network identity, the tracking area code of the selected cell and the establishment cause from earlier. The AMF extracts the registration request and continues with the registration procedure in Section 13.4.

13.3.3 RRC Connection Establishment with an eNB

If the mobile selected a 4G cell, then it contacts the corresponding eNB using the random access and RRC connection establishment procedures from LTE [16]. The base station then takes on the role of the mobile's master eNB.

In the LTE message *RRC Connection Setup Complete*, the mobile states whether the selected network is one of the 5G core networks that the base station advertised in SIB 1, or one of the evolved packet cores. In the first case (option 5), the eNB forwards the message to an AMF in the manner described in Section 13.4, and the mobile continues with the 5G registration procedure below. In the second case (option 1), the eNB forwards the message to a mobility management entity in the evolved packet core, and the mobile continues with the attach procedure from LTE [17].

13.4 Registration Procedure

13.4.1 Registration Without AMF Change

During the registration procedure, the mobile registers itself with a serving AMF in the 5G core network, through a set of network slices that is known as the allowed network slice selection assistance information (NSSAI). Figure 13.4 shows the simplest version of the procedure, in which the mobile is not roaming, and the serving AMF is unchanged from the mobile's previous registration [18, 19]. Solid lines are for mandatory messages, while dashed lines are for ones that are optional or conditional. The numbering scheme is the same as in the stage 2 specification, a convention that we will apply throughout the chapters that follow. (Several messages are omitted: some of them are unusual, but others will appear in more complex versions of the procedure.)

To begin the procedure, the mobile composes a 5GMM *Registration Request*. In the message, the mobile requests an initial registration, and identifies itself using the 5G globally unique temporary identity (5G-GUTI) from its previous registration, if available, or its subscription concealed identifier (SUCI) otherwise. It also states its requested NSSAI, in other words the network slices with which it would like to register.

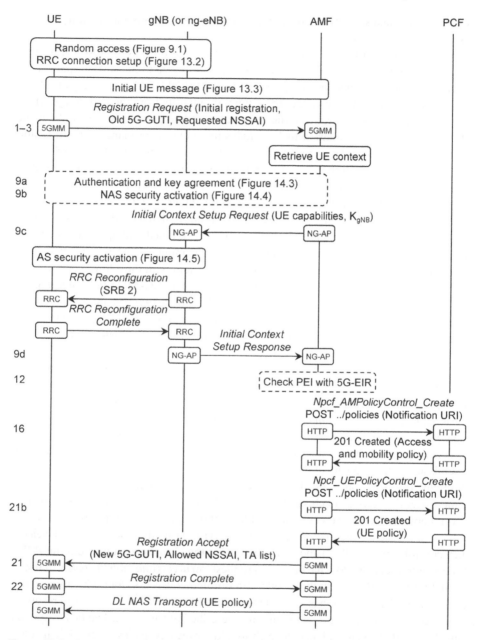

Figure 13.4 Registration procedure if the AMF is unchanged. Source: Adapted from 3GPP TS 23.502.

As part of the message, the mobile includes a *NAS key set identifier*, which identifies the security keys that it was using when it was last switched on. The mobile uses those keys to secure the message by means of integrity protection, and to cipher most of its information elements [20]. However, it leaves in plain text any information elements that are required for AMF selection and security establishment, for example its temporary identity and the NAS key set identifier itself.

The mobile is unable to send the message right away, so it triggers the RRC connection establishment procedure from earlier. The message is delivered to the AMF as part of the earlier messages RRC Connection Setup Complete and Initial UE Message (steps 1–3).

In this example, the AMF recognizes the temporary identity from the mobile's previous registration. That allows it to retrieve the information it had previously stored about the mobile, known as the *UE context*, which includes the mobile's permanent identities and radio access capabilities, and the previous values of its security keys and allowed NSSAI. (If the UE context is missing, perhaps due to a database failure in the network, then the AMF retrieves the subscription concealed identifier from the mobile itself.)

Using the retrieved security keys, the AMF checks the integrity of the mobile's message. It can now authenticate the mobile, using a security procedure that we will cover in Chapter 14 (9a). The procedure is mandatory if the integrity check failed and optional otherwise, but is generally called because it triggers an update of the mobile's NAS security keys (9b). The AMF also checks that the requested NSSAI is consistent with the allowed NSSAI.

The AMF contacts the mobile's master node by means of an NG-AP *Initial Context Setup Request* to establish an NG signalling connection for the mobile (9c). The message includes the mobile's radio access capabilities and a security key that is denoted K_{gNB}. The message also includes *core network assistance information*, which helps the master node during a later transition to RRC_INACTIVE or RRC_IDLE.

Using the security key, the master node initializes access stratum security for the mobile. It then contacts the mobile using a general-purpose message that reconfigures the mobile's air interface, which in this case is used to establish SRB 2. In the case of a master gNB the message is a 5G *RRC Reconfiguration*, while a master ng-eNB uses the equivalent 4G message *RRC Connection Reconfiguration*. (The master node can also retrieve the mobile's radio access capabilities if it did not receive them from the AMF.) Having done so, the master node sends an acknowledgement to the AMF (9d).

Optionally, the AMF can send the mobile's permanent equipment identity to the 5G equipment identity register, to check that the mobile equipment has not been stolen (12). However, network operators do not often implement the EIR, so this step is often omitted.

We have now reached the first procedure on a service-based interface. The AMF selects a policy control function (PCF) and contacts it using the *Create* operation from the PCF's *AMPolicyControl* service (16). That operation requests an *access and mobility policy* that will constrain how the AMF controls the mobile, and simultaneously asks for notifications if the policy is updated. The operation is implemented as an HTTP POST to the uniform resource identifier (URI) `../policies`, in the manner described in Chapter 2. The PCF responds with an HTTP 201 Created, and includes a child URI for the new policy as an HTTP header. It also supplies two fields in the form of JavaScript Object Notation (JSON) data, namely a *service area restriction* containing a list of allowed tracking areas for the mobile, and a *RAT/frequency selection priority* (RFSP) that influences the mobile's choice of radio access technology in RRC_IDLE.

Using a similar HTTP request, the AMF also retrieves a *UE policy* that will constrain the mobile's internal behaviour (21b). The PCF's reply includes a *UE route selection policy*, which contains application-specific rules that tell the mobile how to manage its connections with external data networks. It also includes an *access network discovery and selection policy*, which helps the mobile use non-3GPP access networks such as WiFi [21].

The AMF can now accept the mobile's registration request (21). The message includes a new temporary identity, the mobile's allowed NSSAI and tracking area list, a timer for periodic registration updates, and a list of networks that the mobile can assume are equivalent to the current one. On receiving the mobile's acknowledgement of its new temporary identity (22), the AMF can forward any UE policy transparently to the mobile [22].

13.4.2 Registration with a New AMF

If the mobile has moved since its previous registration, then the AMF may have to change. The resulting procedure is more complex, and is shown in Figure 13.5.

The mobile sends a Registration Request in the same way as before, but this time the master node forwards the message to a new AMF, which does not recognize the mobile's temporary identity (steps 1–3). If the network has deployed an unstructured data storage function (UDSF), then the AMF may be able to retrieve the UE context directly from there. Otherwise, the AMF uses the temporary identity to extract the globally unique identity of the mobile's previous AMF. It then looks up that identity in the list of AMFs that it had previously discovered from the network repository function (NRF), and identifies the corresponding domain name.

The new AMF can now contact the old one using the *UEContextTransfer* operation from the AMF's *Communication* service (4). The service is implemented as a custom operation, using an HTTP POST to the URI `../ue-contexts/{ueContextId}/transfer`, where `{ueContextID}` is constructed from the mobile's temporary identity. Along with its HTTP request, the new AMF includes the mobile's registration request as a binary message body. The old AMF checks the integrity of the mobile's message, retrieves the UE context and returns it (5).

The new AMF can now authenticate the mobile and configure the RAN in the manner described earlier (9). Having done so, the new AMF notifies the old one that the mobile registered successfully (10). Optionally, the AMF can check that the device has not been stolen, as before (12).

The AMF now selects a network function for unified data management (UDM). It tells the UDM that it is now handling the mobile's registration, by requesting the *Registration* operation from the UDM's *UEContextManagement* service (14a). That operation is implemented as an HTTP PUT to `../{supi}/registrations/amf-3gpp-access`, where `{supi}` is the mobile's subscription permanent identifier, so as to create a new URI for the mobile's registration upon the UDM. As part of the request, the AMF asks to be notified if the UDM deregisters the mobile, and supplies a notification URI.

The UE context does not include all of the information about a mobile, so the AMF also contacts the UDM to request its *access and mobility subscription data* (14b). That information includes the slices in the mobile's subscribed NSSAI, and identifies one of those slices as the default. It also includes a periodic registration update timer, as well as a service area restriction and a RAT/frequency selection priority. (Although the last two fields appear to duplicate the information retrieved earlier from the PCF, they have different origins: the PCF's fields are determined by network policy, including that of the visited network

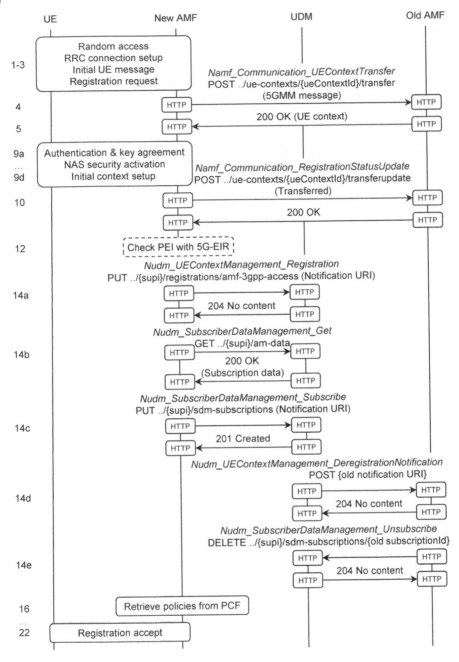

Figure 13.5 Registration procedure if the new AMF is different from the old one. Source: Adapted from 3GPP TS 23.502.

if the mobile is roaming, while the UDM's are from the mobile's subscription.) In the same message, the AMF requests a dataset for use when selecting a session management function (SMF), which is known as the *SMF selection subscription data*. That dataset lists the allowed data network names (DNNs) for each network slice, and identifies one of those DNNs as the default. The AMF also asks the UDM for notifications if the subscription data change (14c). In response, the UDM creates a new subscription, and returns a subscription URI that identifies it.

The UDM cancels the mobile's registration with the old AMF, using the notification URI that the old AMF had previously supplied (14d). In turn, the old AMF cancels its own request for notifications, using the subscription URI that it had previously received from the UDM (14e). At the same time, the new AMF selects a PCF, retrieves the appropriate policies (16) and accepts the mobile's registration request (21, 22).

13.4.3 Registration with AMF Re-allocation

Sometimes, an AMF receives a registration request but is unable to serve the mobile. Instead, it re-routes the request to another AMF in the manner shown in Figure 13.6 [23].

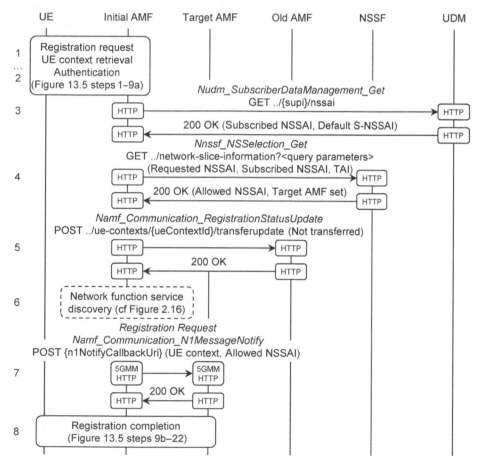

Figure 13.6 Registration procedure, including re-allocation of the AMF. Source: Adapted from 3GPP TS 23.502.

As in the previous example, the initial AMF receives a registration request, retrieves the UE context from the old AMF and optionally authenticates the mobile (steps 1 and 2). This time, however, the requested NSSAI is inconsistent either with the allowed NSSAI from the UE context, or with the AMF's capabilities, or with the mobile's new tracking area.

The UE context does not necessarily contain the subscribed NSSAI. If it does not, then the AMF contacts the UDM to retrieve the mobile's *slice selection subscription data* (3). The UDM's reply includes part of the access and mobility subscription data from earlier, by listing the slices in the subscribed NSSAI and identifying one of those slices as the default. The AMF then contacts the network slice selection function (NSSF), supplies the subscribed and requested NSSAIs, and identifies the mobile's tracking area (4). In response, the NSSF returns an allowed NSSAI for the mobile, and a suitable AMF set.

The initial AMF now notifies the old one that the registration has not yet completed (5). It may also need to contact the network repository function to discover the domain names of the AMFs that are within the target AMF set (6).

The initial AMF selects a target AMF, and it now has two possibilities. If network policy permits, then it can forward the mobile's registration request directly to the target AMF, using an HTTP POST to a notification URI that it retrieved during network function service discovery (7). The initial AMF includes the UE context and the allowed NSSAI, and the procedure continues from the step of NAS security activation. Alternatively, the initial AMF can ask the NG-RAN to re-route the mobile's message, and can include the identity of the target AMF.

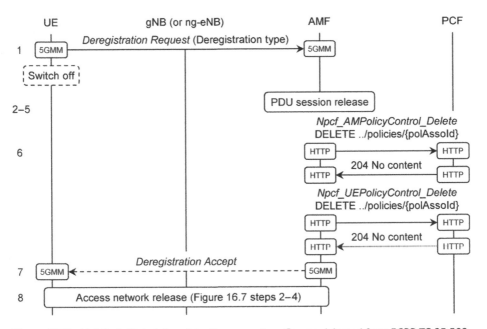

Figure 13.7 Mobile-initiated deregistration procedure. Source: Adapted from 3GPP TS 23.502.

13.5 Deregistration Procedure

The deregistration procedure takes down a mobile's access to the 5G core network. It can be triggered either by the mobile (for example, if the user selects aeroplane mode or switches the mobile off) or by the network. Figure 13.7 shows an example for a mobile-initiated deregistration from the state of RRC_CONNECTED [24].

To begin the procedure, the mobile sends a 5GMM *Deregistration Request* to the AMF and states its reason for doing so, either a normal deregistration or switching off (step 1). If the mobile is switching off, then it can do so now without waiting for the network's response.

If the mobile has any PDU sessions, then the AMF instructs the relevant SMFs to tear them down (2–5). The AMF also deletes the relevant policy associations with the PCF (6). However, it does not delete its record of the mobile's registration in the UDM. Instead, the AMF and mobile both retain information such as the mobile's identities and security keys, for use during the next registration procedure. If the mobile is not switching off, then the AMF accepts the deregistration request (7). Finally, the AMF instructs the NG-RAN to release its own resources for the mobile (8).

References

1 3GPP TS 23.502 (2019) Procedures for the 5G system (5GS); Stage 2 (Release 15), December 2019.

2 3GPP TS 38.300 (2019) NR; NR and NG-RAN overall description; Stage 2 (Release 15), December 2019.

3 3GPP TS 23.122 (2019) Non-access-stratum (NAS) functions related to mobile station (MS) in idle mode (Release 15), March 2019.

4 3GPP TS 38.304 (2019) NR; User equipment (UE) procedures in idle mode and RRC inactive state (Release 15), December 2019.

5 3GPP TS 23.122 (2019) Non-access-stratum (NAS) functions related to mobile station (MS) in idle mode (Release 15), March 2019, Sections 3.1, 3.5, 4.4, 5.

6 3GPP TS 31.102 (2019) Characteristics of the Universal Subscriber Identity Module (USIM) application (Release 15), December 2019, Sections 4.2.5, 4.2.53, 4.2.54, 4.2.84, 4.2.91, 4.4.11.2.

7 3GPP TS 38.304 (2019) NR; User equipment (UE) procedures in idle mode and RRC inactive state (Release 15), December 2019, Section 5.1.1.

8 3GPP TS 38.304 (2019) NR; User equipment (UE) procedures in idle mode and RRC inactive state (Release 15), December 2019, Sections 5.2.1, 5.2.3, 5.3.

9 3GPP TS 38.215 (2019) NR; Physical layer measurements (Release 15), December 2019, Sections 5.1.1, 5.1.3.

10 3GPP TS 38.101-1 (2019) NR; User equipment (UE) radio transmission and reception; Part 1: Range 1 standalone (Release 15), December 2019, Section 6.2.

11 3GPP TS 38.101-2 (2019) NR; User equipment (UE) radio transmission and reception; Part 2: Range 2 standalone (Release 15), December 2019, Section 6.2.

12 3GPP TS 36.304 (2019) Evolved Universal Terrestrial Radio Access (E-UTRA); User equipment (UE) procedures in idle mode (Release 15), December 2019, Sections 5.2.3, 5.3.

13 3GPP TS 38.331 (2019) NR; Radio resource control (RRC) protocol specification (Release 15), December 2019, Section 5.3.3.

14 3GPP TS 23.501 (2019) System architecture for the 5G system (5GS); Stage 2 (Release 15), December 2019, Section 6.3.5.

15 3GPP TS 38.413 (2019) NG-RAN; NG Application Protocol (NGAP) (Release 15), December 2019, Section 8.6.1.

16 3GPP TS 36.331 (2019) Evolved Universal Terrestrial Radio Access (E-UTRA); Radio resource control (RRC); Protocol specification (Release 15), December 2019, Section 5.3.3.

17 3GPP TS 23.401 (2019) General Packet Radio Service (GPRS) enhancements for Evolved Universal Terrestrial Radio Access Network (E-UTRAN) access (Release 15), December 2019, Section 5.3.2.1.

18 3GPP TS 23.502 (2019) Procedures for the 5G system (5GS); Stage 2 (Release 15), December 2019, Section 4.2.2.2.2.

19 3GPP TS 38.300 (2019) NR; NR and NG-RAN overall description; Stage 2 (Release 15), December 2019, Section 9.2.1.3.

20 3GPP TS 33.501 (2019) Security architecture and procedures for 5G system (Release 15), December 2019, Section 6.4.6.

21 3GPP TS 24.526 (2019) User equipment (UE) policies for 5G system (5GS); Stage 3 (Release 15), June 2019.

22 3GPP TS 24.501 (2019) Non-access-stratum (NAS) protocol for 5G system (5GS); Stage 3 (Release 15), December 2019, Sections 5.4.5, 8.2.11, Annex D.

23 3GPP TS 23.502 (2019) Procedures for the 5G system (5GS); Stage 2 (Release 15), December 2019, Section 4.2.2.2.3.

24 3GPP TS 23.502 (2019) Procedures for the 5G system (5GS); Stage 2 (Release 15), December 2019, Section 4.2.2.3.2.

14

Security

This chapter covers the security procedures in the 5G system. To help define those procedures, the 3GPP specifications break down the 5G system into a number of different security domains. In previous generations, two of those have been particularly important: network access security, which protects the mobile's communications with the network over the air interface; and network domain security, which protects the network itself. 5G also introduces the concept of service-based architecture domain security, which protects the network's service-based interfaces using procedures that are different from the ones used elsewhere.

In this chapter, we will start by reviewing the underlying security techniques that 5G uses, and then work our way through each of its security domains in turn. The most important specification is the stage 2 description, TS 33.501 [1].

14.1   Security Principles

The 5G system is secured using several techniques whose underlying principles are common to all of its security domains. During *authentication*, two devices confirm that each other is a trusted device, not an intruder, and set up security keys for use by the procedures that follow. *Ciphering*, also known as *encryption*, ensures that intruders cannot read the data and signalling messages that two devices exchange.

Some national regulations restrict the use of encryption, which makes two other techniques important. *Integrity protection* detects any attempt by an intruder to modify the data or signalling messages that two devices exchange. It protects the system against problems such as man-in-the-middle attacks, in which an intruder intercepts a sequence of signalling messages and modifies them in an attempt to take control of the target device. As part of that process, *replay protection* detects any attempt to play back a signalling message that an intruder has previously recorded, so as to prevent the intruder from spoofing the identity of the device that transmitted the message.

A final technique, particularly relevant for network access security, is *confidentiality*. The subscription permanent identifier (SUPI) is one of the quantities that an intruder needs in order to clone a mobile, so 5G avoids broadcasting it over the air interface. Instead, the network identifies the user by means of temporary identities. If the 5G core network knows the AMF region that the mobile is in (for example, during paging), then it uses the 40-bit

An Introduction to 5G: The New Radio, 5G Network and Beyond, First Edition. Christopher Cox.

5G-S-TMSI (temporary mobile subscriber identity). Otherwise (for example, during registration), it uses the longer 5G globally unique temporary identity (5G-GUTI). In a feature that is new to 5G, a mobile can also identify itself using an encrypted version of the SUPI, which is known as the subscription concealed identifier (SUCI).

The 5G system also supports *lawful interception* [2–4]. This allows the police and security services to intercept traffic and signalling messages for individual users, subject to the appropriate national regulations.

14.2 Network Access Security

14.2.1 Network Access Security Architecture

Network access security refers to all the security procedures that involve the mobile, and involves a number of additional network functions. The most important is the authentication server function (AUSF), which manages the authentication procedures in the home network. The AUSF has three main roles, which are most apparent when the mobile is roaming. Firstly, the AUSF verifies that the visited network is a genuine network that has been authorized to serve the mobile, and is not a spoof network. Secondly, the AUSF authenticates the mobile on behalf of the home network, alongside the access and mobility management function's (AMF's) authentication on behalf of the visited network. Thirdly, the AUSF allows 5G to use the same authentication architecture for the cases of 3GPP and non-3GPP access. That situation is different from LTE, in which a roaming mobile was authenticated by the visited network's mobility management entity when using 3GPP access, but by the home network's *authentication, authorization and accounting* (AAA) *server* when using non-3GPP access technologies such as WiFi.

The other security functions are currently co-located with network functions that we have already discussed in this book, although they might eventually be implemented separately. The *security anchor function* (SEAF) is co-located with the AMF and manages the authentication procedures in the visited network. The *authentication credential repository and processing function* (ARPF) is co-located with the function for unified data management (UDM). It stores the most important security keys and carries out security-related calculations, so it has a similar role to the *authentication centre* (AuC) from 2G and 3G. Finally, the *subscription identifier de-concealing function* (SIDF) is also co-located with the UDM. Its role is to decrypt the SUCI.

Figure 14.1 shows the high-level architecture. Authentication takes place between the mobile and the home network, specifically between the universal integrated circuit card (UICC) and the ARPF. At a minimum, the mobile has to support two authentication procedures: an enhanced version of the legacy 3GPP procedure that is known as *5G authentication and key agreement* (5G AKA), and an Internet Engineering Task Force (IETF) protocol known as the *improved extensible authentication protocol for 3G authentication and key agreement* (EAP-AKA′) [5]. If the mobile is roaming, then the SEAF has to support those procedures as well. On the other hand, a private 5G network can implement other authentication schemes, using algorithms that are downloaded onto the UICC.

Whenever regulations permit, non-access stratum signalling messages between the mobile and the AMF are protected by encryption. They are also protected by integrity and

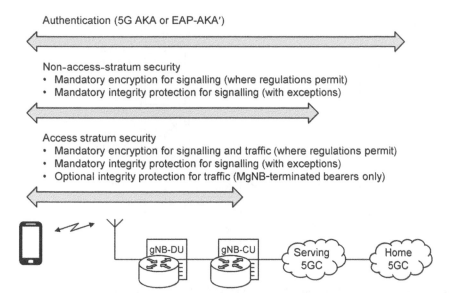

Figure 14.1 Network access security architecture.

replay protection, except for a few special cases such as emergency calls from mobiles that are in a limited service state. Access stratum signalling messages between the mobile and the master node are secured in the same way, with the protection extending to the gNB central unit on the grounds that the distributed unit may be at an insecure location.

Access stratum traffic is encrypted in the same way as signalling. A new feature in 5G is that access stratum traffic can also be protected by integrity and replay protection, although that is only supported in architectural options 2 and 4 for bearers that terminate in a master gNB.

14.2.2 Key Hierarchy

Network access security relies on shared knowledge of a user-specific key, K, which is securely stored in the ARPF and securely distributed within the UICC. During the authentication procedure, the mobile and network confirm that each other has the correct value of K. They then compute a hierarchy of lower-level keys, which are illustrated in Figure 14.2 and which are used by the lower-level procedures that follow [6].

From K, the ARPF and UICC derive two further keys, denoted CK and IK. 3G systems used those keys directly for ciphering and integrity protection. In 5G, they are used to derive a sequence of lower-level keys, which are denoted K_{AUSF}, K_{SEAF}, K_{AMF} and K_{gNB}, and which are passed to the AUSF, SEAF, AMF and master node respectively. From K_{AMF}, the mobile equipment and the AMF derive two further keys, denoted K_{NASenc} and K_{NASint}, which are used for encryption and integrity protection of non-access stratum signalling messages. Similarly, the mobile equipment and the master node derive four further keys, denoted K_{RRCenc}, K_{RRCint}, K_{UPenc} and K_{UPint}, which are used for encryption and integrity protection of radio resource control (RRC) signalling messages and of user plane traffic. Each set of keys is identified by means of a *key set identifier*, denoted ngKSI.

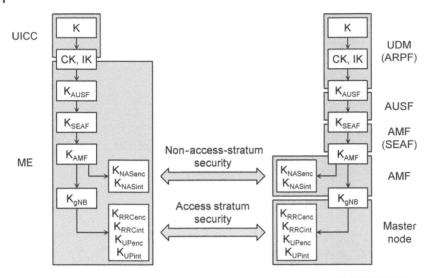

Figure 14.2 Network access security keys. Source: Adapted from 3GPP TS 33.501.

K has either 128 or 256 bits, while CK and IK have 128 bits each. The other keys all have 256 bits, but the current algorithms for encryption and integrity protection only use the least significant 128. Future advances in computing technology may make 128-bit keys insecure, but may also make the use of 256-bit keys more feasible. If that happens, then 5G will be able to upgrade its algorithms to support 256-bit keys with ease.

14.3 Network Access Security Procedures

14.3.1 Subscription Concealed Identifier

Usually, the mobile identifies itself to the 5G core network using a 5G globally unique temporary identity, from which its serving AMF can retrieve the corresponding subscription permanent identifier. However, there are a few situations in which that procedure fails. Examples include the very first use of the Universal Subscriber Identity Module (USIM), when the temporary identity does not yet exist, or after a database failure within the network.

In previous generations, the mobile reacted by quoting its permanent identity. That left a security weakness, which 5G solves by introducing the SUCI. Within the SUCI, the mobile's identity is concealed by means of public key encryption, using a public key that the home network operator has provisioned within the USIM. The mobile network and mobile country codes are left in plain text, so the information can be routed towards the correct home network, which recovers the mobile's identity using a private key that the UDM stores within the ARPF. The algorithm used is known as the *elliptic curve integrated encryption scheme* (ECIES) [9–11], which has the advantage over other public key techniques of achieving the same level of security using significantly shorter keys.

14.3.2 Authentication and Key Agreement

Depending on its internal policies, the AMF can initiate authentication during any procedure that takes the mobile from CM-IDLE into CM-CONNECTED, for example during the registration procedure from Chapter 13. Figure 14.3 shows one of the authentication procedures that can result, namely 5G authentication and key agreement [12].

Triggered by a signalling message such as a registration request, the AMF selects an AUSF in the mobile's home network, and asks it to authenticate the mobile (step 1). The information elements include a serving network name that is constructed from the corresponding mobile network and mobile country codes, and the subscriber's permanent or concealed identity. On receiving the request, the AUSF first checks that the AMF is entitled to use the serving network name. It then forwards the information to the UDM, which creates a new resource that includes the subscriber's identity.

The next few steps take place inside the UDM. If the AUSF supplied the subscription concealed identifier, then the ARPF asks the SIDF to return the corresponding permanent identifier. The ARPF then looks up the network's copy of the user-specific key K, and generates an *authentication vector* that contains four elements. The first of these is a random number, which is used as an authentication challenge to the mobile and is denoted RAND. The other three are all computed using RAND, K and the serving network name. XRES* is the expected response to the authentication challenge, which can only be correctly computed by a mobile that has the same value of K. AUTN is an *authentication token*, which demonstrates to the mobile that the network has the same value of K, and which includes a sequence number for use in replay protection. Finally, K_{AUSF} is the home network's anchor key from Figure 14.2. The ARPF returns the authentication vector to the AUSF, along with any SUPI that it retrieved (2). Unlike in previous generations, 5G only supports the return of one authentication vector at a time.

The AUSF stores XRES* for later use by the home network's authentication procedure, and computes a derived version, denoted HXRES*, which is used for authentication in the visited network (3, 4). (Here, the prefix H has nothing to do with the home network: instead, it indicates that HXRES* is computed using a hashing algorithm.) The AUSF also stores any permanent identity that it received from the UDM, and computes the visited network's anchor key K_{SEAF}. It then forwards the relevant authentication material to the AMF, and supplies a uniform resource identifier (URI) that the AMF should use for the subsequent confirmation (5). In turn, the AMF sends the random number and authentication token to the mobile (6).

Inside the mobile, the mobile equipment sends the random number and authentication token to the UICC. In the UICC, the USIM application examines the authentication token to check that the network has the same value of K and that the enclosed sequence number has not been used before. If it is happy, then it calculates the authentication response that was used in earlier 3GPP generations, denoted RES, by combining RAND with its own copy of K. It passes that response back to the mobile equipment, along with the values of CK and IK from Figure 14.2. The mobile equipment combines RES with its own understanding of the serving network name, so as to compute the 5G authentication response RES* (7). It also uses CK and IK to compute the mobile's values of K_{AUSF}, K_{SEAF} and K_{AMF}. Although a little complicated, this process allows a mobile to access the 5G core network using a

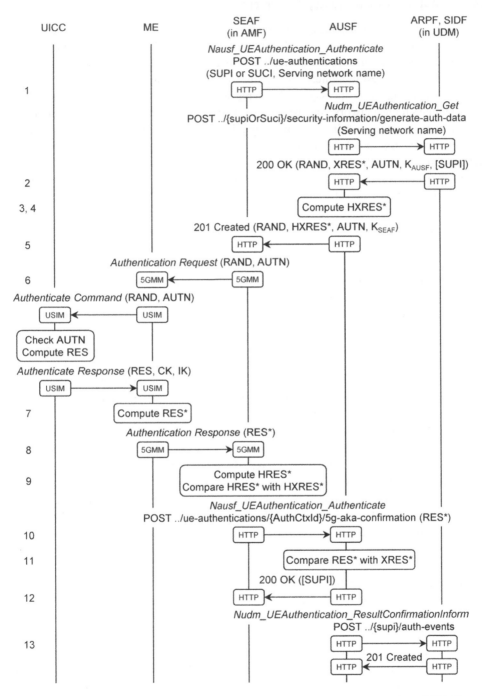

Figure 14.3 5G authentication and key agreement procedure. Source: Adapted from 3GPP TS 33.501.

legacy USIM, and ensures that 5G network operators do not necessarily have to replace their subscribers' SIM cards.

The mobile then returns the authentication response to the AMF (8). That computes a hashed version of the mobile's response, denoted HRES*, and compares it with the value of HXRES* from earlier (9). If the two are the same, then it concludes that the mobile has the correct value of K and has therefore been authenticated successfully. The AMF confirms the authentication using the URI that it received from the AUSF and includes the mobile's original response RES* (10).

The AUSF compares the mobile's response with the expected response XRES* (11). If the two are the same, then the AUSF concludes not only that the mobile is authenticated, but also that the serving network quoted the same network and country codes to the home network and to the mobile. By combining this with its previous conclusion that the AMF was entitled to use those codes, the AUSF concludes that the serving network is genuine. The AUSF returns an acknowledgement to the AMF, and includes any permanent identity that it retrieved earlier (12). On receiving that acknowledgement, the AMF computes the key K_{AMF}.

To conclude the procedure, the AUSF sends a confirmation to the UDM that the mobile has been authenticated successfully (13). The UDM can then link that confirmation to subsequent procedures. For example, it might only accept a serving network's request to register a mobile (Figure 13.5, step 14a), if the mobile has recently been authenticated by the same serving network.

14.3.3 Activation of Non-access Stratum Security

At the end of the authentication procedure, the mobile and AMF both have new values of K_{AMF}. During the security activation procedures, the two devices derive the keys for ciphering and integrity protection, and bring those keys into action [13].

The AMF activates non-access stratum security immediately after authentication and key agreement, as shown in Figure 14.4. From K_{AMF}, the AMF calculates the ciphering and

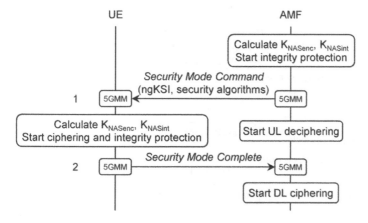

Figure 14.4 Non-access stratum security mode command procedure. Source: Adapted from 3GPP TS 33.501.

integrity keys K_{NASenc} and K_{NASint}. It then sends the mobile a 5G mobility management (5GMM) message known as a *Security Mode Command* (step 1), which identifies the new key set and selects the security algorithms to use. The message is secured by integrity protection using the new value of K_{NASint}, but is not ciphered, even if the mobile already has a valid set of security keys.

The mobile computes its own copies of K_{NASenc} and K_{NASint}, and checks the integrity of the message in the manner described later in this chapter. If the message passes the integrity check, then the mobile starts uplink ciphering, and acknowledges the AMF's command using a 5GMM *Security Mode Complete* (2). On receiving the message, the AMF starts downlink ciphering.

14.3.4 Activation of Access Stratum Security

Access stratum security is activated in much the same way (Figure 14.5), but the steps are in a slightly different order. To trigger the process, the AMF computes the value of K_{gNB}, and passes it to the master node as part of an NG application protocol (NG-AP) Initial Context Setup Request (Figure 13.4, step 9c). The master node computes the access stratum's ciphering and integrity keys, and sends the mobile an RRC *Security Mode Command* (1) that selects the security algorithms to use. The mobile computes its own copies of the security keys, checks the integrity of the message and replies with an RRC *Security Mode Complete* (2). Unlike in the non-access stratum procedure, neither of these messages is ciphered.

In dual connectivity, the master node uses its copy of K_{gNB} to generate a new key, denoted K_{SN}. Using Xn application protocol (Xn-AP) signalling, the master node passes K_{SN} to the secondary, which takes the key into action as its own value of K_{gNB}. The secondary node might support different security algorithms from the master, so it can select different algorithms for communications with the mobile.

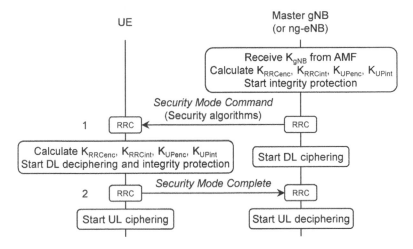

Figure 14.5 Access stratum security mode command procedure. Source: Adapted from 3GPP TS 33.501.

14.3.5 Key Handling During Mobility

Other procedures deal with key handling during mobility and state transitions [14]. If the master node's central unit remains unchanged as part of a handover, then it can optionally retain the old value of K_{gNB}, depending on network policy. Otherwise, there are two ways to compute a new value. The first is a *horizontal key derivation*, in which the old master node uses the old value of K_{gNB} to compute a new quantity, which is denoted K_{gNB}^*. It passes that quantity to the new master node, which brings it into action as the new value of K_{gNB}. The second is a *vertical key derivation*, in which the AMF uses the value of K_{AMF} to compute two intermediate quantities, known as *next hop* (NH) and *next hop chaining counter* (NCC). It passes those quantities to the new master node, which uses them to compute the new value of K_{gNB}. The Xn-based handover procedure in Chapter 16 uses both types of derivation, while an NG-based handover uses vertical key derivation alone.

If the AMF changes, then the old AMF can optionally use another type of horizontal key derivation to compute a new value of K_{AMF} from the old one, and can then pass the new key to the new AMF. If that happens, then the new AMF brings the new key into action using the same security mode command procedure as before, but indicates in step 1 that the mobile should perform its own horizontal key derivation, so as to update its own value of K_{AMF}.

14.3.6 Key Handling During State Transitions

If the network releases a mobile's RRC connection and moves it to RRC_IDLE, then the mobile's access stratum security keys are all deleted. If, however, the network merely suspends a mobile's RRC connection and moves it to RRC_INACTIVE, then the mobile, master node and any secondary all retain their copies of the integrity key K_{RRCint}, and of the access stratum security key K_{gNB}. Later, they can use the first of those keys to check their subsequent access stratum signalling messages, and can use the second to resume the RRC connection without the need to communicate with the core.

During the deregistration procedure, the mobile and AMF delete their copies of K_{NASenc} and K_{NASint}, but retain their copies of K_{AMF}. During a subsequent registration request, the mobile re-calculates its previous copies of K_{NASenc} and K_{NASint}. The mobile uses the first key to encrypt most of the information elements in the registration request, leaving in plain text the ones that are needed for message routing and security establishment. It uses the second key to secure the message by integrity protection.

14.3.7 Ciphering

Ciphering ensures that intruders cannot read the information that is exchanged between the mobile and the network [15]. The packet data convergence protocol ciphers data and signalling messages in the air interface's access stratum, while the 5G mobility management protocol ciphers signalling messages in the non-access stratum.

Figure 14.6 shows the ciphering process. The transmitter uses its ciphering key and other information fields to generate a pseudo-random key stream, and mixes that with the outgoing data using an exclusive-OR operation. The receiver generates its own copy of the key stream and repeats the mixing process so as to recover the original data.

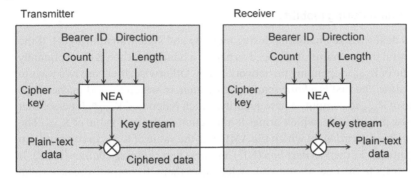

Figure 14.6 Ciphering. Source: Adapted from 3GPP TS 33.501.

5G currently supports four encryption algorithms, denoted NEA, which are the same ones used by LTE. It is mandatory for the mobile to support two algorithms, namely SNOW 3G, which was originally used in the Release 7 standards for UMTS, and the *Advanced Encryption Standard* (AES). It is optional for the mobile to support a third algorithm, known as ZUC, which is mainly intended for use in China. The fourth is a null encryption algorithm in which the air interface does not implement ciphering at all.

14.3.8 Integrity Protection

Integrity protection allows a receiver to detect modifications to data packets or signalling messages, as a protection against problems such as man-in-the-middle attacks. Integrity protection is implemented in the same protocols as ciphering, and is illustrated in Figure 14.7 [16].

The transmitter uses several information fields, notably the outgoing information, the integrity key, and a counter that is incremented for each transmission. Using these fields, the transmitter computes a 32-bit *message authentication code* (MAC) and appends it to the outgoing data. The receiver separates the integrity field from the rest of the incoming data, and computes the expected authentication code XMAC. If the observed and expected

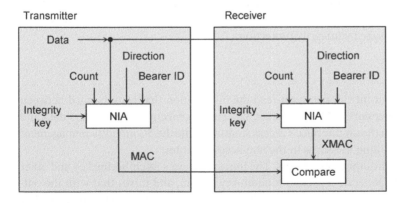

Figure 14.7 Integrity protection. Source: Adapted from 3GPP TS 33.501.

codes differ, then the receiver concludes either that the data have been modified or that they have been replayed using an old value of the counter. In both cases, the incoming data are discarded.

Integrity protection is mandatory for almost all of the signalling messages that the mobile exchanges with the network. There are four algorithms, denoted NIA, which are the same ones used for encryption. The null integrity algorithm is only allowed in a few special cases, for example emergency calls from mobiles that are in a limited service state with no access to the home network. Integrity protection is optional for traffic, and is only supported for bearers that terminate in a master gNB.

14.4 Network Domain Security

14.4.1 Network Domain Security Architecture

The term *network domain security* (NDS) applies to reference points in the 5G fixed network that do not have a corresponding service-based interface. Some of those reference points may already be physically secure, for example if they are confined within a physical site that the network operator controls. The others are vulnerable to intrusion and have to be secured. Examples include the interfaces between different sites in the radio access network, which are often managed by a third party, and the ones between different sites in the network operator's core.

Figure 14.8 shows the architecture used [17]. In most cases, authentication, encryption, integrity protection and replay protection are mandatory for any reference point that is not physically secure. The only exceptions are the legacy signalling reference points to the evolved packet core and an external application function, namely N26 and Rx, for which it is mandatory to support those procedures but optional to use them.

14.4.2 Network Domain Security Protocols

Network domain security can be implemented using two main protocols, namely *IP network layer security* and *transport layer security* (TLS) [18, 19]. Support of IP network layer security is mandatory for all the reference points in Figure 14.8. To use it, two devices first authenticate each other using a certificate-based protocol known as *Internet Key Exchange version 2* (IKEv2) [20], and set up a *security association* that includes lower-level keys for encryption and integrity protection.

The devices then implement those procedures using the *internet protocol security* (IPSec) *encapsulating security payload* (ESP) [21]. It is optional for a network operator to support ESP *transport mode*, which encrypts the payload of an IP packet but leaves the headers in plain text for routing purposes. It is mandatory for the operator to support ESP *tunnel mode*, in which the original IP headers are also encrypted, and new headers are added for routing.

In the core network, an IPSec tunnel can be implemented using a separate device known as a *security gateway* (SEG). The originating gateway receives an outgoing IP packet from a network function, encrypts it and adds a new IP header that routes the packet to the destination gateway. That device then decrypts the original packet and sends it on. In the radio access network, the gateway's functions usually form part of the base station itself.

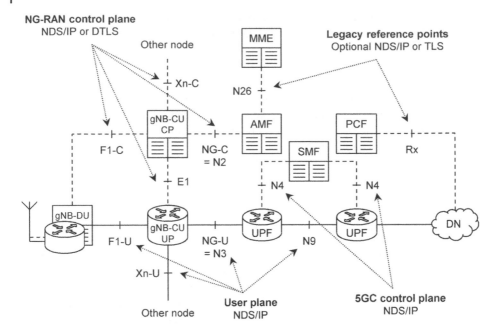

Figure 14.8 Network domain security architecture.

Transport layer security is an alternative protocol which implements authentication, encryption and integrity protection immediately above a reliable TCP transport layer. *Datagram transport layer security* (DTLS) is a variant which was originally designed for use over UDP [22], and which was adapted further for use over the stream control transmission protocol (SCTP) [23]. Support for DTLS over SCTP is mandatory for the control plane reference points in the radio access network. The original form of TLS is only used by the N26 and Rx reference points, and in some cases by the service-based interfaces that we will discuss in Section 14.5. If the protocol is implemented, then support for versions 1.3 and 1.2 is mandatory [24, 25], while version 1.1 is optional.

14.5   Service-based Architecture Domain Security

14.5.1   Security Architecture

The term *service-based architecture domain security* applics to the service-based interfaces in the 5G core. From the viewpoint of security, there are three types of service-based interface, which are illustrated in Figure 14.9. The first covers the service-based interfaces inside a single operator's network, which are secured using IP network layer security or TLS in the manner described in Section 14.4.2 [26].

Service-based interfaces between two networks are implemented over the N32 reference point, which connects the security edge protection proxies (SEPPs) that we introduced in Chapter 2. If the N32 reference point is a direct interconnection that does not traverse the IP packet exchange (IPX), then it is secured using TLS alone.

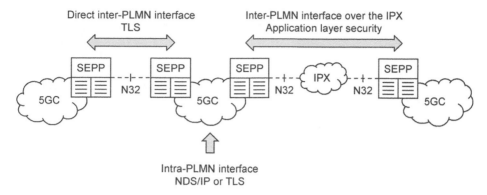

Figure 14.9 Service-based interface security architecture.

However, a different scenario arises if the N32 reference point traverses the IPX [27]. Each network operator communicates with its own trusted IPX service provider, and has a business relationship through which the service provider may wish to read, and even modify, individual information elements in the HTTP/2 and JavaScript Object Notation (JSON) signalling messages. (One example is the modification of HTTP/2 headers for routing purposes, while other examples might cover the provision of value-added services by the IPX.) Other service providers could also be involved, but they just forward the messages transparently.

For those interfaces, security at the network or transport layer would be inappropriate. Instead, the SEPPs secure most of the signalling messages at the application layer, by applying encryption and integrity protection to individual HTTP/2 and JSON information elements, and by allowing the IPX service providers to make authorized modifications to them.

Figure 14.10 shows the architectural details. In the diagram, a consumer network function (cNF) requests a service from a producer network function (pNF) in another operator's network. The networks communicate using SEPPs – denoted cSEPP and pSEPP respectively – and have business relationships with their own IPX service providers, which are denoted cIPX and pIPX.

The N32 reference point has two components. N32-c carries HTTP/2 signalling messages that manage the relationship between the two SEPPs. Those messages are secured by TLS, so they cannot be read or modified by the intervening IPXs. N32-f forwards subsequent HTTP/2 signalling messages between the two network functions. Those messages are secured at the application layer by *JSON object signing and encryption* (JOSE), a security framework that supports integrity and replay protection by means of *JSON web signature* (JWS) [28] and supports encryption by means of *JSON web encryption* (JWE) [29]. We will discuss the details of both components next.

14.5.2 Initial Handshake Procedures over N32-c

After establishing communications, the two SEPPs authenticate each other and set up a secure TLS connection over N32-c. They then configure their subsequent communications

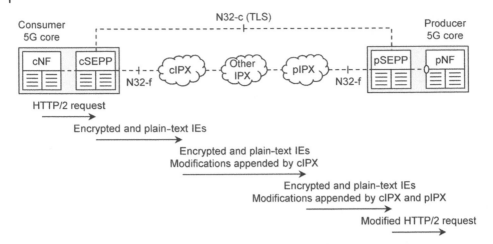

Figure 14.10 Principles of secure message delivery over N32-f.

over N32-f in an HTTP/2 signalling procedure that has three steps [30]. During the first, the SEPPs agree the mechanism that they will use, either transport layer or application layer security. In the case of application layer security, they proceed to the second step of the procedure, in which they agree the security algorithms to use.

The richest step is the third one, in which the two SEPPs agree a policy for protecting the information that they will exchange. That policy has two parts. The first part lists the information elements that the SEPPs will protect by encryption and integrity protection. These include authentication vectors, cryptographic material, any location data such as a mobile's cell identity, and usually the subscription permanent identifier. The second part lists the information elements that the two IPXs can modify and is derived from the modification policies that the network operators already have with their IPXs.

14.5.3 Forwarding of JOSE Protected Messages over N32-f

Once the handshake has completed, the two SEPPs protect subsequent HTTP/2 signalling messages using the procedure shown in Figure 14.11 [31, 32].

To start the procedure, the consumer network function sends an HTTP/2 request to the producer, which the consumer's network routes through the corresponding SEPP (step 1). The SEPP reformats the request by copying individual HTTP/2 and JSON information elements into two JSON objects, which are denoted `dataToIntegrityProtect` and `dataToIntegrityProtectAndCipher`. As the names imply, the SEPP secures the first of those objects by integrity protection, and the second by integrity protection and ciphering (2). The SEPP then forwards the information to the consumer's IPX as the payload of an HTTP/2 POST (3).

The IPX modifies the `dataToIntegrityProtect` object, in accordance with the rules defined by its business relationship with the consumer's network. It then appends the modifications to the payload as a new JSON object, denoted `modifications` (4). That object is signed by means of JSON web signature so as to certify its origin. The IPX then forwards the message towards the producer's IPX (5), which evaluates the first IPX's modifications

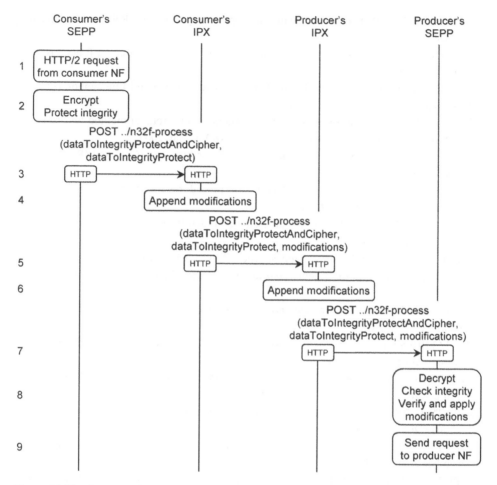

Figure 14.11 Procedure for secure message delivery over N32-f. Source: Adapted from 3GPP TS 33.501.

and applies its own in a similar way (6). In turn, the producer's IPX forwards the message to the producer's SEPP (7).

The producer's SEPP checks the integrity of the incoming message, decrypts the `data-ToIntegrityProtectAndCipher` object and reconstructs the original HTTP/2 request (8). It then checks the integrity of the two IPXs' modifications, verifies that they are compliant with the modification policy established earlier and applies them. Finally, the SEPP delivers the HTTP/2 request to the producer network function (9). The HTTP/2 response is handled in a similar way.

References

1 3GPP TS 33.501 (2019) Security architecture and procedures for 5G system (Release 15), December 2019.

2 3GPP TS 33.126 (2019) Security; Lawful interception requirements (Release 15), December 2018.

3 3GPP TS 33.127 (2019) Security; Lawful interception (LI) architecture and functions (Release 15), September 2019.

4 3GPP TS 33.128 (2019) Security; Protocol and procedures for lawful interception (LI); Stage 3 (Release 15), December 2019.

5 IETF RFC 5448 (2009) Improved extensible authentication protocol method for 3rd generation authentication and key agreement (EAP-AKA'), May 2009.

6 3GPP TS 33.501 (2019) Security architecture and procedures for 5G system (Release 15), December 2019, Section 6.2.

7 3GPP TS 33.501 (2019) Security architecture and procedures for 5G system (Release 15), December 2019, Section 6.12.2, Annex C.

8 3GPP TS 23.003 (2019) Numbering, addressing and identification (Release 15), September 2019, Section 2.2B.

9 Gayoso, M.,.V., Hernández, E.,.L., and Sánchez, Á.,.C. (2010). A survey of the elliptic curve integrated encryption scheme. *Journal of Computer Science and Engineering* 2 (2): 7–13.

10 SECG (2009) Standards for efficient cryptography 1 (SEC 1): Elliptic curve cryptography, Version 2.0, May 2009.

11 SECG (2010) Standards for efficient cryptography 2 (SEC 2): Recommended elliptic curve domain parameters, Version 2.0, January 2010.

12 3GPP TS 33.501 (2019) Security architecture and procedures for 5G system (Release 15), December 2019, Sections 6.1.2, 6.1.3.2, 6.1.4.

13 3GPP TS 33.501 (2019) Security architecture and procedures for 5G system (Release 15), December 2019, Section 6.7.

14 3GPP TS 33.501 (2019) Security architecture and procedures for 5G system (Release 15), December 2019, Sections 6.8, 6.9, 6.10.

15 3GPP TS 33.501 (2019) Security architecture and procedures for 5G system (Release 15), December 2019, Annex D.2.

16 3GPP TS 33.501 (2019) Security architecture and procedures for 5G system (Release 15), December 2019, Annex D.3.

17 3GPP TS 33.501 (2019) Security architecture and procedures for 5G system (Release 15), December 2019, Section 9.

18 3GPP TS 33.210 (2019) 3G security; Network domain security (NDS); IP network layer security (Release 15), September 2019.

19 3GPP TS 33.310 (2018) Network domain security (NDS); Authentication framework (AF) (Release 15), December 2018.

20 IETF RFC 7296 (2014) Internet Key Exchange Protocol Version 2 (IKEv2), October 2014.

21 IETF RFC 4303 (2005) IP encapsulating security payload (ESP), December 2005.

22 IETF RFC 6347 (2012) Datagram Transport Layer Security Version 1.2, January 2012.

23 IETF RFC 6083 (2011) Datagram transport layer security (DTLS) for stream control transmission protocol (SCTP), January 2011.

24 IETF RFC 8446 (2018) The Transport Layer Security (TLS) Protocol Version 1.3, August 2018.

25 IETF RFC 5246 (2008) The Transport Layer Security (TLS) Protocol Version 1.2, August 2008.

26 3GPP TS 33.501 (2019) Security architecture and procedures for 5G system (Release 15), December 2019, Section 13.1.

27 3GPP TS 33.501 (2019) Security architecture and procedures for 5G system (Release 15), December 2019, Section 13.2.

28 IETF RFC 7515 (2015) JSON web signature (JWS), May 2015.

29 IETF RFC 7516 (2015) JSON web encryption (JWE), May 2015.

30 3GPP TS 29.573 (2019) 5G System; Public land mobile network (PLMN) interconnection; Stage 3 (Release 15), October 2019, Sections 5.2, 6.1.

31 3GPP TS 33.501 (2019) Security architecture and procedures for 5G system (Release 15), December 2019, Section 13.2.4.

32 3GPP TS 29.573 (2019) 5G system; Public land mobile network (PLMN) interconnection; Stage 3 (Release 15), October 2019, Sections 5.3, 6.2.

15

Session Management, Policy and Charging

In this chapter, we will investigate how the 5G user plane delivers data packets between the mobile and an external data network. We will begin by elaborating on the concepts of protocol data unit (PDU) sessions and quality of service (QoS) flows, and by discussing how those concepts are actually implemented by means of data radio bearers and tunnels. We will then introduce some additional network functions into the 5G core, which mediate its interactions with external application servers, and which determine the quality of service that the user will receive and the amount that the user will be charged. The chapter closes with some examples of the relevant signalling procedures, including the establishment of a PDU session, and traffic steering under the influence of an application server.

The main sources of information for this chapter are the stage 2 specifications for the architecture of the 5G core network and the resulting signalling procedures [1, 2]. Those are supplemented by the corresponding specifications for policy and charging control (PCC), namely TS 23.503 [3] and TS 29.513 [4].

15.1   Types of PDU Session

15.1.1   IP PDU Sessions

A PDU session is a bidirectional connection between the mobile and an external data net-work, in which the packets travel through one or more user plane functions (UPFs) that are themselves controlled by a session management function (SMF). Every PDU session has a *PDU session type*, which depends on the external network's routing protocols, and therefore on the type of address that is used to identify the mobile [5].

The most common type is an IP PDU session, in which the mobile is connected to an external IP network such as the internet or the IP multimedia subsystem (IMS). That network identifies the mobile using an *IP version 4* (IPv4) address and/or an *IP version 6* (IPv6) address, so there are three sub-types of IP PDU session, namely IPv4, IPv6 and IPv4v6.

IPv4 addresses are 32 bits long. Usually, the SMF assigns a dynamic IPv4 address to the mobile during the establishment of a PDU session, either by allocating the address by itself or by acquiring a suitable address from a *dynamic host configuration protocol*

An Introduction to 5G: The New Radio, 5G Network and Beyond, First Edition. Christopher Cox.
© 2021 John Wiley & Sons Ltd. Published 2021 by John Wiley & Sons Ltd.

version 4 (DHCPv4) server. Alternatively, the mobile can itself use DHCP after PDU session establishment is complete, so as to acquire a dynamic IPv4 address over the 5G user plane. To achieve this, the SMF acts as a DHCP server towards the mobile, and the UPF forwards DHCP packets between the two. The IP address is also sent to the UPF as part of these procedures, for later use in identifying the packets that the mobile sends and receives.

Because of the shortage of IPv4 addresses, the allocated address is usually a private IP address that is invisible to the outside world. Using *network address translation* (NAT) [6–8], the system can map that address to a public IP address that is shared amongst several mobiles, plus a mobile-specific TCP or UDP port number.

IPv6 addresses are 128 bits long and have two parts, namely a 64-bit network prefix and a 64-bit interface identifier. They are allocated using a procedure known as *IPv6 stateless address autoconfiguration* [9]. In 5G's implementation of that procedure, the SMF assigns the mobile a globally unique IPv6 prefix during PDU session establishment, as well as a temporary interface identifier. After PDU session establishment is complete, the mobile uses the temporary address to contact the SMF in a procedure known as router solicitation, with the packets once again forwarded by the UPF. The SMF returns the IPv6 prefix, from which the mobile constructs its full IP address. Because the prefix is globally unique, the mobile can do so using any interface identifier that it likes.

A mobile can also use a static IPv4 address and/or IPv6 prefix, which can be specific to an individual data network name and/or network slice. However, the mobile does not store them permanently: instead, the SMF retrieves them during PDU session establishment, either from the mobile's subscription data, or from configuration data in an external authentication, authorization and accounting (AAA) or DHCP server. Static IP addresses are unusual in the case of IPv4, due to the shortage of IPv4 addresses.

In the case of roaming with home-routed traffic, the mobile has two SMFs, one each in the home and visited networks. In that scenario, IP address allocation is handled by the home SMF, while the visited SMF forwards the relevant session management signalling messages between the home SMF and the mobile.

15.1.2 Ethernet PDU Sessions

5G also supports Ethernet PDU sessions, in which the mobile exchanges Ethernet frames with an external Ethernet network over the 5G core. In an Ethernet PDU session, the mobile is identified by a 48-bit Ethernet medium access control (MAC) address, or possibly by multiple addresses. (Those addresses have nothing to do with the MAC protocol on the 5G air interface: instead, they are named after an Ethernet protocol that has the same role but a different implementation.) This time, however, the network does not allocate a MAC address to the mobile. Instead, it discovers the mobile's permanent MAC addresses by recording the addresses that the mobile uses for its uplink data traffic.

Ethernet frames are smaller than IP packets because they do not contain any IP, UDP or TCP headers. That makes Ethernet PDU sessions especially valuable for machine-type and low-latency communications, in which the data packets are usually small, and the overhead of the IP header can be troublesome.

15.1.3 Unstructured PDU Sessions

The final type of PDU session is *unstructured*. In this case, the 5G core network knows nothing about the routing protocols that the data network is using, but instead communicates with the data network using a point-to-point tunnel. In the most common solution, the tunnel is implemented over an interconnecting IPv6 network. During PDU session establishment, the SMF assigns a globally unique IPv6 prefix that will identify the mobile over the interconnecting network, in much the same way as before. This time, however, it only sends that prefix to the UPF, not to the mobile, as the mobile might not understand IP at all.

15.2 Quality of Service

15.2.1 Packet Flows, Service Data Flows, and QoS Flows

A PDU session handles three types of information stream, namely packet flows, service data flows (SDFs) and QoS flows. Figure 15.1 illustrates the relationship between them.

A *packet flow* is a unidirectional flow of packets within a PDU session, which is known to the application. Its packets are identified by a *packet filter*. Amongst other fields, an IP packet filter includes the packet flow's direction (uplink or downlink), layer 4 protocol identity (for example, UDP or TCP), source and destination IP address or range, and source and destination port number. An Ethernet packet filter includes the packet flow's direction, layer 3 protocol (the *Ethertype*), and source and destination MAC address or range.

A *service data flow* is a bidirectional set of packet flows which forms the basis for quality of service and charging in 5G. A service data flow is created by the policy control function

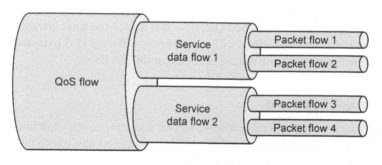

	QoS flow	Service data flow	Packet flow
Role	Transports traffic over the 5G core	Basis for policy and charging control	Underlying data flow
Composition	Service data flows with the same QCI and ARP	Packet flows from one service with the same QCI and ARP	Packets with the same address range
Scope	UE ⇔ SMF	UE ⇔ PCF	UE ⇔ AF
Identification	QoS rule	SDF template	Packet filter

Figure 15.1 Packet flows, service data flows and QoS flows.

(PCF), by grouping packet flows from the same PDU session that share the same parameters for QoS and charging. Its packets are identified by a *service data flow template*, which contains the packet filters from all of its constituent packet flows.

These parameters are the most important part of a dataset known as a *policy and charging control* (PCC) *rule* [10, 11], which is a complete description of the service data flow. There are two types of PCC rule. *Dynamic PCC rules* are composed on the fly by the PCF, and are essential when interworking with external application functions. *Predefined PCC rules* are permanently stored within the SMF.

A *QoS flow* is a bidirectional set of service data flows which transports data over the 5G system [12]. A QoS flow is created by the SMF by grouping service data flows from the same PDU session that share the same parameters for quality of service, but not necessarily for charging. A QoS flow is identified by a 6-bit *QoS flow identifier* (QFI). Its packets are identified by a *QoS rule*, which contains the packet filters from all its constituent service data flows.

When a PDU session is established, the network sets up an initial QoS flow which is identified by a *default QoS rule*, and which stays in place for the lifetime of the PDU session. This is equivalent to a default bearer from LTE, and to a primary *packet data protocol* (PDP) context from 2G or 3G. Subsequently, the network can establish other QoS flows within the same PDU session, which are identified by different QoS rules and have different QoS parameters. These are equivalent to LTE dedicated bearers and to 2G/3G secondary PDP contexts.

Some useful examples come from the field of IMS voice and video. An IMS voice call is usually implemented using two packet flows, one each for the uplink and downlink, which together make up a single service data flow and a single QoS flow. If the user puts that call on hold and starts another, then the second call might have different charging characteristics, so it might be implemented using a second service data flow. However, the SMF could still map that onto the first QoS flow, if it shared the same parameters for QoS. On the other hand, an audio-visual call might require different QoS parameters for voice and video, which would require the use of two separate QoS flows.

15.2.2 QoS Parameters

A PCC rule contains several quality-of-service parameters, which define how the system should deliver the corresponding packets. The parameters apply both to service data flows and to QoS flows: our description will refer to the latter.

At the highest level, there are two types of QoS flow, which are summarized in Table 15.1. A *guaranteed bit rate* (GBR) QoS flow is suitable for real-time services such as voice over IP. It is associated with a pair of *guaranteed flow bit rates* (GFBRs), one each for the uplink and downlink, which define the minimum data rate that the network will deliver. It is also associated with a pair of *maximum flow bit rates* (MFBRs), which define the maximum data rate that the mobile is allowed to send or receive. A *delay-critical* GBR QoS flow is a type of GBR QoS flow with more stringent requirements for packet delay, which is suitable for ultra-reliable low-latency communication (URLLC).

A *non-guaranteed bit rate* (non-GBR) QoS flow is suitable for non-real-time services such as web browsing. Each PDU session receives a pair of *session aggregate maximum bit rates*

Table 15.1 Quality-of-service (QoS) parameters.

Parameter	Description	Use by GBR QoS flows	Use by non-GBR QoS flows
ARP	Allocation and retention priority	✓	✓
5QI	5G QoS identifier	✓	✓
QNC	QoS notification control	✓	✗
RQA	Reflective QoS attribute	✗	✓
GFBR	Guaranteed flow bit rate	✓	✗
MFBR	Maximum flow bit rate	✓	✗
Session AMBR	Session aggregate maximum bit rate	✗	One per PDU session
UE AMBR	User equipment aggregate maximum bit rate	✗	One per UE

(AMBRs), which limit the total uplink and downlink data rates on the mobile's non-GBR QoS flows within that session. The mobile also receives a pair of *UE aggregate maximum bit rates*, which limit the total data rates on all of its non-GBR QoS flows.

Every QoS flow is associated with an *allocation and retention priority* (ARP), which has three fields. The *ARP priority level* is a number from 1 to 15, where 1 has the highest priority, and levels 1 to 8 are intended for priority services such as emergency calls. It indicates the importance of accepting or rejecting a request to establish a new QoS flow, in cases of network congestion. The *pre-emption capability* states whether a QoS flow can grab resources from another with a lower priority, while the *pre-emption vulnerability* states whether a QoS flow can lose resources to another with a higher priority. The last two fields are valuable in emergency situations. To handle these, the 5G system can first set the pre-emption capability for QoS flows that are carrying real-time emergency voice calls, while setting the pre-emption vulnerability for all other QoS flows. If the cells become congested, then the 5G system can free up resources by tearing down the non-emergency QoS flows, to ensure that the emergency QoS flows are maintained.

Every QoS flow is also associated with a *5G QoS identifier* (5QI). This is an integer from 0 to 255, which is based on the *QoS class identifier* (QCI) from LTE. It serves as an entry into a look-up table, which defines how the network should deliver packets within the QoS flow. Some values have been standardized, and they are listed in Table 15.2. A network operator can define other values, but those might not be understood by a visited network if the mobile is roaming.

Most of the 5QI fields are similar to the ones in LTE. The *packet error rate* (PER) limits the proportion of packets that are lost between the mobile and the PDU session anchor due to errors in transmission and reception. The figure applies to GBR QoS flows as it is, but non-GBR QoS flows might suffer additional packet loss if the network becomes congested. The *packet delay budget* (PDB) limits the time taken to deliver a packet between the mobile and the PDU session anchor. For non-GBR and normal GBR QoS flows, an uncongested network should deliver 98% of packets within the delay budget, but a few packets can be delayed for longer. In the case of delay-critical GBR QoS flows, however, any packet that exceeds the delay budget is counted as lost. Finally, the *5QI priority level* is an integer from

Table 15.2 Standardized values of the 5G QoS identifier.

5QI	Type	Default priority level	Packet delay budget (ms)	Packet error rate	Default maximum data burst	Default averaging window (ms)	Examples
1	GBR	20	100	10^{-2}	—	2000	Conversational voice
2		40	150	10^{-3}	—	2000	Conversational video
3		30	50	10^{-3}	—	2000	Real-time gaming
4		50	300	10^{-6}	—	2000	Buffered video
65		7	75	10^{-2}	—	2000	MCPTT voice
66		20	100	10^{-2}	—	2000	Other PTT voice
67		15	100	10^{-3}	—	2000	Mission-critical video
75		25	50	10^{-2}	—	2000	V2X messages
5	Non-GBR	10	100	10^{-6}	—	—	IMS signalling
6		60	300	10^{-6}	—	—	High-priority internet
7		70	100	10^{-3}	—	—	Voice, video, gaming
8		80	300	10^{-6}	—	—	Mid-priority internet
9		90	300	10^{-6}	—	—	Low-priority internet
69		5	60	10^{-6}	—	—	MCPTT signalling
70		55	200	10^{-2}	—	—	Mission-critical data
79		65	50	10^{-6}	—	—	V2X messages
80		68	10	10^{-4}	—	—	Low-latency eMBB
82	Delay-critical GBR	19	10	10^{-4}	255 bytes	2000	Discrete automation
83		22	10	10^{-4}	1358 bytes	2000	Discrete automation
84		24	30	10^{-5}	1354 bytes	2000	Intelligent transport
85		21	5	10^{-5}	255 bytes	2000	Electricity distribution

Source: Adapted from 3GPP TS 23.501.

1 to 127, where 1 has the highest priority. It helps the network share its resources amongst different QoS flows, for example by ensuring that high-priority QoS flows are still delivered within the delay budget if the network becomes congested, while allowing longer delays for lower-priority flows.

Two subsidiary fields are new to 5G. The *averaging window* is the time over which the GFBR and MFBR are calculated, and is only used for GBR QoS flows. The *maximum data burst volume* defines the maximum amount of data that the access network has to deliver within the delay budget, so as to help its implementation of delay-critical GBR QoS flows. (The values of 1354 and 1358 bytes avoid fragmentation in a transport network that can handle a packet size of 1500 bytes, after allowing for the expected packet headers [13].) For these fields and for the 5QI priority level, the 5G QoS identifier supplies a default which the network can over-write with a value that is specific to the QoS flow.

Most of the standardized 5QI values were also supported by LTE. The most common are values from 1 to 9, which are inherited from LTE Release 8. Typically, for example, a network operator uses values of 8 and 9 for the internet, 5 to carry signalling messages between the mobile and the IMS, and 1 for IMS voice calls. Values from 65 to 70 were added from LTE Release 12 to carry *mission-critical* (MC) communications such as *push-to-talk* (PTT) voice in a public safety network. Values of 75 and 79 appeared in Release 14 to carry basic vehicle-to-everything (V2X) communications, while the value of 80 appeared in Release 15 to carry low-latency enhanced mobile broadband (eMBB). The remaining 5QI values (82–85) carry URLLC services that exceed the capabilities of LTE and are only supported by 5G.

Referring back to Table 15.1, two final QoS parameters complete the set. For a GBR QoS flow, the *QoS notification control* (QNC) indicates whether the access network should notify the 5G core if it can no longer deliver the guaranteed flow bit rate, for example if the mobile loses coverage. The core network might respond by tearing down the QoS flow, which releases resources and ensures that the user is no longer charged. For a non-GBR QoS flow, the *reflective QoS attribute* (RQA) indicates whether some of the traffic might eventually be extracted into a new QoS flow with a new set of QoS parameters, in a process that is dynamically controlled within the user plane.

The same description applies to service data flows, subject to one clarification. Within a GBR QoS flow, the constituent service data flows can have different values for the guaranteed and maximum flow bit rates. These can then be added in order to calculate the corresponding figures for the QoS flow. Otherwise, the QoS parameters of the constituent service data flows should be the same.

15.2.3 Charging Parameters

A PCC rule also contains a number of charging parameters, which define how the user should be charged. The *charging method* can be *offline charging*, which is suitable for simple monthly billing, or *online charging*, which supports more complex scenarios such as pre-paid services. The *measurement method* determines whether the network should monitor the volume or duration of the service data flow, or both. Finally, the *charging key*, also known as the *rating group*, indicates the tariff that the charging system will eventually use.

15.3 Implementation of PDU Sessions

15.3.1 Bearers and Tunnels

Within a PDU session, the 5G system has to deliver packets across several different interfaces. As a minimum these include the air interface and the N3 backhaul, and they can also include the N9 reference point between two UPFs, the Xn reference point between two nodes, and the F1 reference point between the central and distributed units of a single node. To help the system manage this process, PDU sessions and QoS flows are implemented using the data radio bearers that we introduced in Chapter 3, and using *tunnels* [14–16].

A tunnel is a short section of data pipe in the fixed network which transports packets in a particular PDU session from one network function to another. Each tunnel is identified using a pair of 32-bit *tunnel endpoint identifiers* (TEIDs), one for the uplink and one for the downlink. The TEID is allocated by the tunnel's terminating network function, and tells the originating network function how to label its outgoing packets so that the terminating network function will recognize them. In the 5G core network, TEIDs can be allocated by either the UPF or the corresponding SMF. In the discussions that follow in this chapter, we will assume the use of the UPF.

Figure 15.2 illustrates the relationships that are involved. On any instance of the N9 reference point, there is one N9 tunnel for each PDU session. On the backhaul, there are one or two N3 tunnels per PDU session, with one tunnel carrying any bearers that terminate in the master node, and one carrying any bearers that terminate in the secondary. Within the radio access network, each data radio bearer handles one or more QoS flows from a single PDU session. If any of those bearers traverses the Xn reference point, then there is one Xn tunnel per bearer, and there is also one F1 tunnel per bearer in the case of a split gNB.

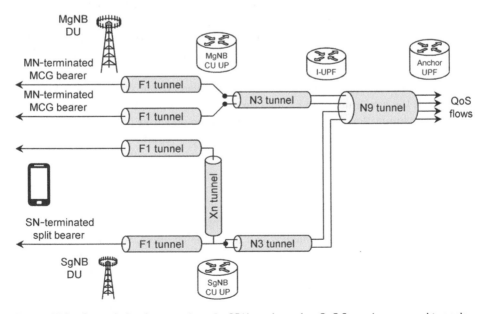

Figure 15.2 Example implementation of a PDU session using QoS flows, bearers and tunnels.

Release 16 adds two new features that improve the reliability of data delivery for the case of URLLC [17]. If the underlying transport network is unreliable, then the 5G system can duplicate each of its N3 and N9 tunnels, and can deliver a copy of every packet through each of those duplicated tunnels. If the network functions are unreliable as well, then the 5G system can duplicate the entire PDU session, with the duplicate sessions implemented by means of different UPFs.

15.3.2 User Plane Protocols

The fixed network forwards packets from one network function to another with the help of three user plane protocols, which are illustrated in Figure 15.3. There is one instance of each protocol per tunnel.

The *GPRS tunnelling protocol user part* (GTP-U) is inherited from LTE and is used on all the reference points in Figure 15.3 [18]. The protocol adds a GTP-U header to each data packet, whose most important role is to identify the associated tunnel by means of its TEID.

The *PDU session user plane protocol* is new to 5G [19]. The protocol is used on the N9 and N3 reference points, and also on Xn to support the temporary forwarding tunnels that are used during changes to the master and secondary nodes. It adds an additional header to each data packet, whose most important role is to identify the QoS flow by means of its QFI. The new header is delivered as an extension field in the GTP-U header.

The *NR user plane protocol* [20] is used on the F1 reference point, and also on Xn to support the long-term tunnels that are required for dual connectivity. It has two roles. The first is downlink flow control, which regulates a bearer's downlink data rate and prevents the distributed unit's receive buffers from overflowing. (There is no need for uplink flow control, because the central unit's receive buffers are much larger.) The second is the provision of information to the central unit about the conditions on the air interface, for example the number of hybrid ARQ failures, to help with tasks such as routing for dual connectivity. As before, the new header is delivered as a GTP-U header extension. Each tunnel only carries one bearer, so there is no need for the bearer to be identified.

The fixed network then transports the packets by means of an IP-based protocol stack. In that stack, the internet protocol is enhanced by means of *differentiated services*

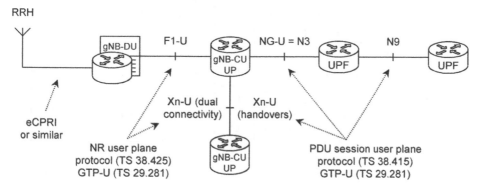

Figure 15.3 User plane protocols.

(DiffServ) [21–24]. That protocol classifies packets into a small number of classes, depending on the quality of service that they require, and identifies each class using a *differentiated services code point* (DSCP) in the IP header. DiffServ routers can then use the code point to support their algorithms for queueing, packet forwarding and packet dropping. Layers 1 and 2 can use any protocol at all, but a common choice is Ethernet combined with *multi-protocol label switching* (MPLS) [25, 26].

15.3.3 End-to-end Protocol Stack

Figure 15.4 shows an example end-to-end protocol stack for the 5G user plane [27]. In this example, the mobile is communicating with a server in an IP data network. The core network is using an intermediate UPF (I-UPF), and the gNB is split into central and distributed units. There is a single data radio bearer, which is implemented as a master cell group (MCG) bearer terminating in the master node.

If the server sends a data packet towards the mobile, then the data network routes that packet towards the mobile's PDU session anchor. That carries out a process of *deep packet inspection*, in which it compares the packet's addressing information with the packet filters that it stored during PDU session establishment. By doing so, it can identify the destination mobile, and the appropriate QoS flow and tunnel. The anchor adds a GTP-U header containing the tunnel endpoint identifier, which itself includes a PDU session user plane protocol header containing the QFI. It then forwards the packet to the next network function, in this case an intermediate UPF.

The intermediate UPF extracts the packet's TEID and QFI, and compares them with its own internal packet filters. By doing so, it can itself identify the destination mobile and the appropriate QoS flow and tunnel, and can forward the packet to the next network function. The packet eventually reaches the distributed unit, which delivers it using the air interface protocols from Chapter 7.

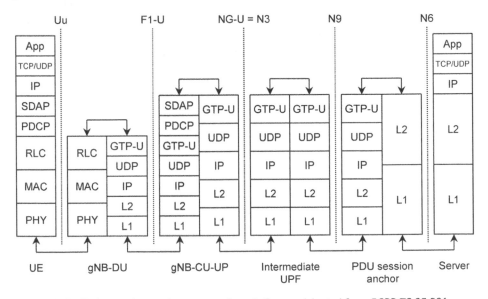

Figure 15.4 End-to-end user plane protocol stack. Source: Adapted from 3GPP TS 23.501.

In the case of an Ethernet PDU session, the fixed network still carries out its internal packet forwarding by means of IP. Although that process adds IP and UDP headers to the Ethernet frames, those extra headers are absent on the air interface, so they do not cause trouble in the part of the system that is usually the bottleneck. Another subtlety is that Ethernet frames begin with a 56-bit preamble and an 8-bit start-of-frame delimiter, which are standardized bit sequences that are used for synchronization. They have no value in the 5G system, so they are stripped out on the path between the mobile and the anchor UPF.

In the case of an unstructured PDU session, the anchor forwards mobile-originated packets to the external data network by means of IPv6, and identifies the mobile using the IPv6 prefix that was assigned during PDU session establishment. The data network can then use the same process to send mobile-terminated packets towards the anchor, which examines the destination IPv6 address, identifies the target mobile and forwards the packet towards it as before.

15.3.4 Multiple PDU Session Anchors

In 5G, a PDU session can have more than one PDU session anchor [28]. This feature is useful for low-latency communications, in which the mobile might exchange some of its packets with the data network using an anchor that is centrally located, and others using a second anchor that is closer to the base station.

Figure 15.5 shows the architecture used. The mobile has one SMF, or two for the case of routing with home-routed traffic. It also has an intermediate UPF which forwards uplink packets to the appropriate anchors and collects downlink packets from them. The intermediate UPF can be co-located with one of the PDU session anchors, and, in the low-latency communication example, it usually would be.

The 5G core network can implement multiple PDU session anchors in two ways. In the first, which is supported by IP and Ethernet PDU sessions, the intermediate UPF acquires a role known as an *uplink classifier* (UL CL). The mobile is unaware that the uplink classifier is there, and communicates with the 5G network in the same way as before. However, the SMF provides the uplink classifier with a set of uplink packet filters which define the correct choice of anchor for each individual packet within the PDU session. The original

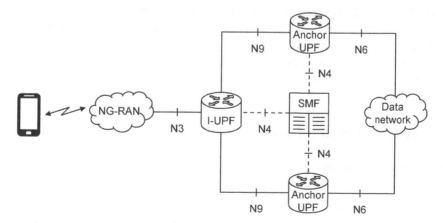

Figure 15.5 Multiple PDU session anchors.

anchor remains in place throughout the lifetime of the PDU session, and is the only one that can quote the mobile's IP or Ethernet address to the data network. Any subsequent anchors communicate using mechanisms which are outside the 3GPP specifications, but which might include network address translation or tunnelling.

The second way is known as *IPv6 multi-homing* [29], in which the intermediate UPF acquires a role known as a *branching point*. In this technique, the mobile has multiple IPv6 prefixes, one for each anchor. During PDU session establishment, the SMF configures the mobile with a set of routing preferences, which determine the originating IPv6 prefix that the mobile should use when sending uplink packets to the outside world. The branching point uses that prefix to forward uplink packets to the appropriate anchor, which can quote the prefix to the data network without making any changes.

15.3.5 PDU Session Anchor Relocation

Another new feature in 5G is the ability to relocate the PDU session anchor [30]. This is valuable if the mobile establishes a PDU session and then moves, triggering a wish to shorten the user plane path through the core, and is also valuable for the case of low-latency communications.

To help support this technique, each PDU session is associated with a *session and service continuity* (SSC) mode, which indicates whether the anchor can be relocated and, if so, how. The mobile requests an SSC mode during PDU session establishment based on the requirements of the application, while the actual choice is made by the SMF. The SSC mode then stays unchanged throughout the life of the PDU session.

SSC mode 1 is the legacy behaviour from LTE. The anchor cannot be changed but, in the case of IP PDU sessions, the SMF promises to preserve the mobile's IP address. Mode 1 can be used by any type of PDU session.

SSC mode 2 is a break-before-make capability, which can also be used by any type of PDU session. The SMF relocates the anchor first by tearing down the old PDU session, and then by asking the mobile to establish a new PDU session to the same data network. If this happens, then the mobile's IP address will change, and there will be a break in connectivity.

SSC mode 3 is a make-before-break capability that has two implementations. In the first, which can only be used by IP PDU sessions, the SMF instructs the mobile to establish a new PDU session to the same data network, and then tears down the old PDU session. In the second, which can only be used by IPv6 PDU sessions, the SMF uses IPv6 multi-homing to set up a new anchor within the original PDU session, and then tears down the old anchor. The mobile's IP address will still change, but there is no break in connectivity.

15.4 Policy and Charging Control Architecture

15.4.1 High-level Architecture

The 5G system includes a number of network functions which make decisions about quality of service, charging and other aspects of network policy, and which handle the interactions with external application servers that can trigger those decisions. Figures 15.6 and 15.7

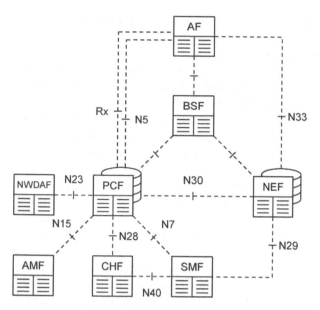

Figure 15.6 Representation of the policy and charging control architecture using reference points. Source: Adapted from 3GPP TS 23.503.

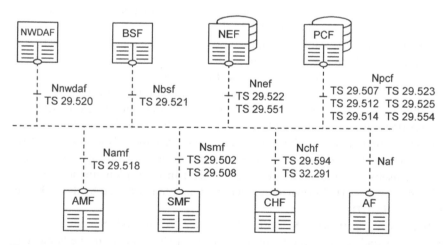

Figure 15.7 Representation of the policy and charging control architecture using service-based interfaces. Source: Adapted from 3GPP TS 23.503.

show the resulting architecture, in terms of reference points and service-based interfaces respectively [31, 32].

An *application function* (AF) is an external server that can interact with the 5G core. There are two types. If the application function is fully trusted by the network operator, then it can interact directly with the operator's PCF. The best example is a *proxy call session control function* (P-CSCF) in the operator's IP multimedia subsystem. Otherwise (for example, a server controlled by a third-party industrial client), the network operator might allow a more limited set of interactions with the operator's *network exposure function* (NEF).

The PCF decides how the mobiles in the network should be operating, and sends the resulting policy information to the other network functions. That includes the access and mobility policy [33] and UE policy [34] that it supplied to the access and mobility management function (AMF) in Chapter 13, and a session management policy [35] that it supplies to the SMF. To help it achieve this, the PCF interacts directly with trusted application functions, using either the HTTP/2 protocol over the N5 reference point [36], or the *Diameter* protocol over the legacy Rx reference point from LTE [37]. It also communicates with the NEF [36, 38, 39], to support the NEF's own interactions.

The NEF interacts with external application functions over the northbound application programming interface (API) described in Section 15.4.3, so as to offer them some influence over the mobiles that they are serving [40]. It also interacts with the SMF [41]. It is mainly intended for machine-type communications, and is based on the *service capability exposure function* (SCEF) from later releases of LTE.

These tasks are supported by two other network functions. The *binding support function* (BSF) maps a mobile's identity (for example, its IP address) onto the relevant policy control function to ensure that an external request for a specific mobile reaches the correct PCF [42]. The *network data analytics function* (NWDAF) supplies information about the current state of the network [43]. In Release 15, that information is limited to the load on individual network slices, which helps the network decide whether to admit individual mobiles and service data flows. Additional information in Release 16 includes the locations and communication patterns of individual mobiles [44].

The *charging function* (CHF) supplies information about a user's spending limits to the PCF, collects charging information from the other network functions and charges the user [45, 46]. It is described in more detail further in Section 15.4.4.

Some of the other network functions from Chapter 2 are also relevant here. The SMF implements the PCF's decisions about session management policy, and maps service data flows onto QoS flows. It also notifies the PCF about events that take place within the network, such as a failure to maintain the guaranteed flow bit rate for a GBR flow. Similarly, the AMF implements the PCF's decisions about access and mobility management policy, which might themselves be influenced by interactions with an application function. Finally, the unified data repository (UDR) can store structured data that are used by the PCF and NEF, as well as by the UDM.

If the mobile is roaming, then the AF, NEF, BSF and NWDAF are all in the same network as the PDU session anchor. However, there is a PCF and a CHF in both of the home and visited networks, so that both networks have some control over network policy and can collect charging information about the user.

15.4.2 Support for 3GPP Services

The 5G core network supports two other services by means of additional network functions. Firstly, a mobile can send and receive SMS messages using the architecture in Figure 15.8. Within the usual architecture for SMS, the *SMS service centre* (SMS-SC) receives a mobile-originated message, stores it and forwards it as a mobile-terminated message to the destination. It communicates with the 5G core network through two

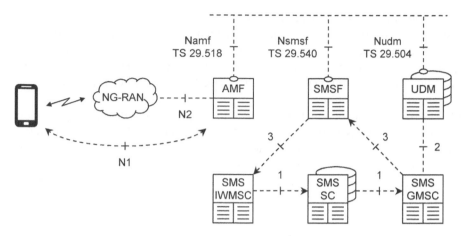

Figure 15.8 Architecture for SMS over the non-access stratum. Source: Adapted from 3GPP TS 23.501.

devices that respectively handle the mobile-originated and mobile-terminated cases, namely the *SMS interworking mobile switching centre* (SMS-IWMSC) and the *SMS gateway mobile switching centre* (SMS-GMSC) [47]. Reference point 1 is not defined by the 3GPP specifications, while reference points 2 and 3 use either the *mobile application part* (MAP) [48] or a *Diameter* application [49].

Within the 5G core network, the point of contact with those devices is the *short message service function* (SMSF) [50]. The SMSF exchanges SMS messages with the AMF, which in turn exchanges them with the mobile by embedding them into mobility management signalling. A new 5G messaging service is also being developed as part of Release 17 to focus on the needs of machine-type communications [51, 52].

Secondly, the 5G core network can discover the physical location of a user and can provide that location to authorized external clients. The interactions are handled by the *location management function* (LMF), which we will discuss as part of Chapter 20.

15.4.3 Northbound API

The network exposure function allows third-party application functions to monitor and influence the operation of the 5G core. In doing so, it acts as the point of contact to a 5G software-defined network over a northbound API [40]. It is mainly intended for machine-type communications, for which the ability to monitor and influence the network is especially relevant.

Some capabilities of the northbound API are inherited from LTE [41, 53, 54]. By way of examples, an application function can supply the expected communication patterns for one or more devices, for example whether they transmit periodically or on demand, so as to help the network manage its resources. It can negotiate a future background data transfer for a time when the network is uncongested and the charging rates are low, such as for a night-time software update. It can ask the NEF to trigger a device's application by means of a mobile-terminated SMS, which is valuable if the device does not have an IP or Ethernet address. It can ask to be notified about any relevant events in the network, such as the geographical location, reachability or any loss of connectivity for one or more devices.

An application function can also send a *packet flow description* (PFD) to the NEF, which asks it to set up a new packet flow. Using a pre-defined service-level agreement, the NEF translates the description into parameters such as media type and bandwidth, and forwards the request to the PCF. The resulting capability is more limited than it is for a trusted application function, which can interact with the PCF directly, and which can state the precise media parameters that it requires.

The NEF's traffic influence functions are new to 5G [55]. Using these, an application function can send requests to the NEF that influence the network's decisions on traffic routing. In turn, that can influence the network's choice of PDU session anchor, and can trigger the selection of a nearby anchor to handle low-latency communications between the mobile and the server. The resulting signalling is covered in more detail towards the end of this chapter.

15.4.4 Charging and Billing System

The 5G system introduces a converged architecture for charging and billing, illustrated in Figure 15.9, which unifies the previous architectures for offline and online charging [56–58].

When the SMF sets up or modifies a service data flow, it tells the UPF to return *usage reports* that describe the amount of data transferred, its duration and any events such as the beginning or end of a traffic flow. Inside the SMF, a *charging trigger function* (CTF) collates that information and delivers it to the CHF by means of HTTP/2 signalling. The details depend on the type of charging method being used.

Offline charging involves a one-way interaction, in which the trigger function collates the usage reports and sends them to the CHF. That combines the usage reports into *charging data records* (CDRs) and sends them to a *charging gateway function* (CGF), which

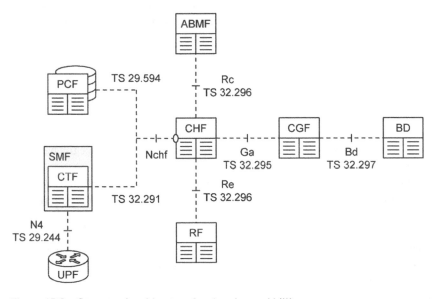

Figure 15.9 Converged architecture for charging and billing.

post-processes the charging data records and combines them into *CDR files*. Finally, the *billing domain* (BD) determines how much the resources have cost and sends an invoice to the user. The gateway function can be a separate device, or it can be integrated with the CHF or the billing domain.

Online charging involves a two-way interaction, in which the trigger function asks the CHF to authorize a certain resource allocation, which is expressed using the requested duration or data volume of the service data flow. The CHF retrieves the cost of the requested resource from the *rating function* (RF), retrieves the balance of the subscriber's account from the *account balance management function* (ABMF) and returns the requested information. If the subscriber approaches the end of the original allocation, then the trigger function asks for additional resources. At the end of the service data flow, the trigger function notifies the CHF about any resources that remain, and the CHF returns the credit to the ABMF.

15.5 PDU Session Establishment Procedures

15.5.1 PDU Session Establishment

Let us now examine some of the resulting signalling procedures. The most fundamental is PDU session establishment, which sets up communications between an application in the mobile and an external data network. Figure 15.10 shows how the signalling begins for the case of an IP data network, a mobile that is not roaming and no intermediate UPF [59].

To start the procedure, the mobile inspects the UE route selection policy that it received during registration, and looks up the policy rules that are associated with the application. Using the 5G session management (5GSM) protocol, the mobile composes a *PDU session establishment request* that includes the associated PDU session type and SSC mode. Using the 5G mobility management (5GMM) protocol, it also composes an *UL NAS transport* message that includes the associated network slice and data network name. It then embeds the first message into the second, and delivers them to the AMF (step 1). (If no suitable policy rule is available, then the mobile can fall back to a default policy rule or to its own configuration. If any of the information elements are omitted, then the network chooses default values from the mobile's subscription data.)

Using the mobile's 5GMM message and its SMF selection subscription data, the AMF selects a network slice and a data network name for the mobile, and also selects an SMF that supports them [60]. It then asks the SMF to establish the PDU session, includes the selected network slice and data network name as a JavaScript Object Notation (JSON) message body, and includes the mobile's session management message as binary data (3).

The SMF contacts the UDM, which creates a new registration for the mobile's PDU session (4a). It then retrieves the *session management subscription data* for the selected network slice and data network name (4b). These include the default and allowed PDU session types and SSC modes; default values for the ARP, 5QI and session AMBR; any static IP address; and a *mapped EPS bearer context* for any PDU session that can be transferred to the evolved packet system using signalling messages over a reference point known as N26. The SMF also asks to be notified if the subscription data change (4c).

Once that procedure has completed, the SMF accepts the AMF's request from earlier (5), and selects an anchor UPF for the mobile. If configured to do so, the SMF can also trigger a procedure of secondary authentication and authorization between the mobile and

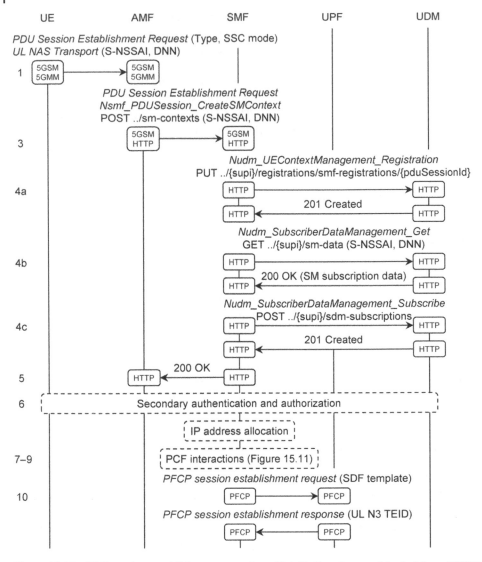

Figure 15.10 PDU session establishment procedure. (1) Initiation. Source: Adapted from 3GPP TS 23.502.

the external data network (6). During that procedure, the data network authenticates the mobile using messages that are conveyed by way of the UPF and SMF [61].

The SMF can now allocate an IPv4 address and/or IPv6 prefix for the mobile that lies in the address range of the UPF. Having done so, it contacts the PCF, using messages that we will cover in Section 15.5.2, to tell it about the mobile's IP address and to retrieve a PCC rule for the first service data flow (7–9). It then maps the service data flow onto a corresponding QoS flow, and initializes the UPF (10). As part of that procedure, the network functions allocate an uplink TEID for the PDU session's N3 tunnel.

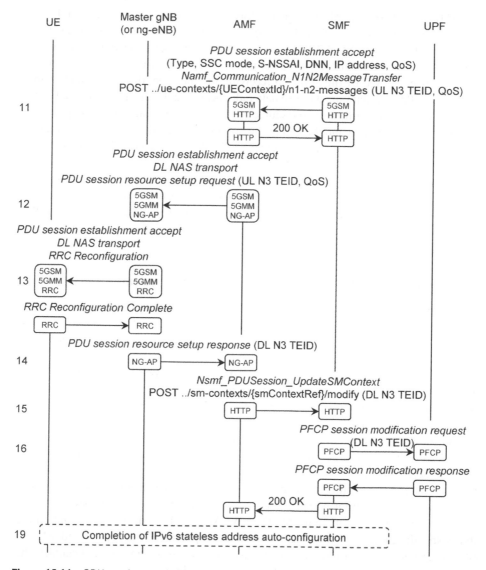

Figure 15.11 PDU session establishment procedure. (2) Completion. Source: Adapted from 3GPP TS 23.502.

The procedure concludes in Figure 15.11. Using the 5GSM protocol, the SMF composes a *PDU session establishment accept* for delivery to the mobile, which includes several parameters: the selected PDU session type, SSC mode, network slice and data network name; the mobile's IP address; a description of the first QoS flow; and any mapped EPS bearer context. It also composes an NG application protocol (NG-AP) information element for delivery to the master node, which includes the UPF's IP address, the uplink tunnel endpoint identifier and the QoS parameters. It sends both sets of information to the AMF in the form of binary data (11).

The AMF extracts the 5GSM message and forwards it towards the mobile. To do so, it embeds the message first into a *DL NAS transport* message that is written using the 5GMM protocol, and then into an NG-AP message (12). That last message tells the master node to establish a PDU session for the mobile, and includes the NG-AP information element that was originally composed by the SMF. On receiving the message, the master node implements the QoS flow by mapping it onto a data radio bearer, and delivers the original 5GSM message by embedding it into a radio resource control (RRC) message that sets up the new bearer (13). On receiving the mobile's reply, the master node sends an acknowledgement to the AMF, and includes the radio access network's IP address and TEID for downlink data forwarding over N3 (14).

To complete the procedure, the AMF tells the SMF that the access network has accepted the PDU session, and includes the access network's IP address and TEID (15). The SMF can then forward that information to the UPF, for use in downlink data forwarding (16). Once the procedure has ended, the mobile can complete any allocation of an IPv6 address (19).

15.5.2 Interactions with the Policy and Charging Control System

Earlier in this chapter, we skipped over the interactions with the policy control function. They are only required when the system is using dynamic policy and charging control, and are illustrated in Figure 15.12 [62, 63]. As before, the figure assumes that the mobile is not roaming.

To start the procedure, the SMF selects a PCF that supports the requested network slice and data network name (7a), and asks for a set of network policies that will govern the PDU session (7b). Its request includes several parameters: the mobile's subscription permanent identifier and IP address; the selected network slice, data network name and PDU session type; and the QoS and charging parameters from the subscription data. The request also includes a uniform resource identifier (URI), on which the PCF will notify the SMF if the network policies change.

There are now a few conditional steps. If the PCF's decisions depend on network load, but it has no up-to-date information about the load on the requested network slice, then the PCF retrieves that information from the NWDAF (7c). If its decisions depend on the mobile's spending limits, then it asks the CHF for a spending limit report, and asks for notifications when the information changes (7d). If the BSF is separate, then it contacts the BSF so as to register its association with the mobile's IP address (7e).

Once those procedures have completed, the PCF creates a PCC rule, which contains the QoS and charging characteristics of the first service data flow in the PDU session. It then returns the information to the SMF. If the PCC rule indicates the need for online charging, then the SMF asks the CHF to reserve charging resources for the user, in terms of the allowed duration and/or data volume of the service data flow (7f). The CHF replies with the resources that it has allocated.

15.5.3 PDU Session Release

Various network functions can trigger the release of a PDU session [64]. For example, the mobile can do so if it is itself triggered by the user or by an application, the AMF does so

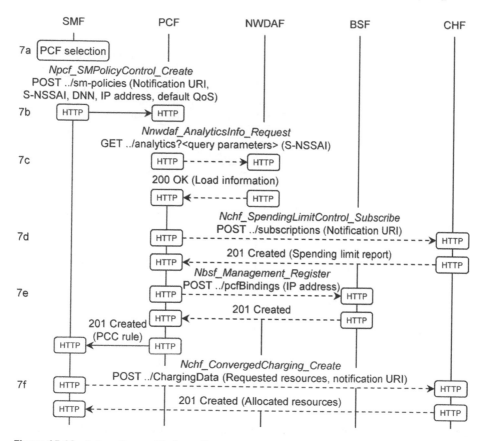

Figure 15.12 Interactions with the policy control function during PDU session establishment. Source: Adapted from 3GPP TS 29.513.

during the deregistration procedure from Chapter 13, and the SMF does so during some of the procedures for changing the PDU session anchor in Section 15.6. As part of the procedure, the radio access network releases the data radio bearers that were associated with the PDU session, and the fixed network releases the tunnels. The SMF releases any dynamic IP address that it assigned to the mobile, and releases the UPF. It also deregisters from the UDM and terminates its policy association with the PCF.

15.6 Traffic Steering

15.6.1 Traffic Steering Request

The procedures for traffic steering are a useful way to illustrate the network exposure function's interactions with an external application function. Using these procedures, an application function can send requests to the 5G core network, which influence the traffic

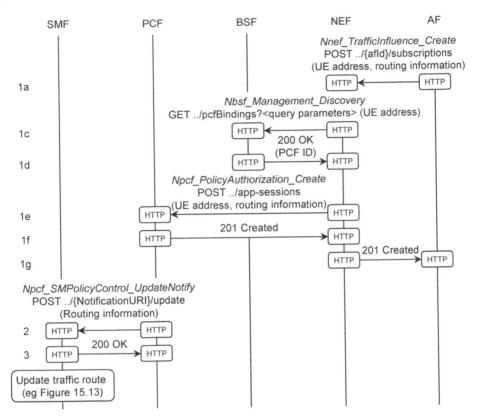

Figure 15.13 Delivery of a traffic steering request to the 5G core network. Source: Adapted from 3GPP TS 29.513.

routing decisions in the SMF. Figure 15.13 shows the resulting signalling for a case in which the application function is sending a request for a specific mobile [65].

To start the procedure, the application function sends a traffic routing request to the NEF, in which it identifies the mobile and supplies suitable routing information (step 1a). That information includes a *data network access identifier* (DNAI), which identifies the desired point of contact between the 5G system and the external data network. The information also includes either the IP address and port number of the mobile's application server, or a predefined routing profile that will be recognized by the 5G core.

The NEF passes the mobile's address to the BSF, retrieves the domain name and/or IP address of the mobile's policy control function (1c, 1d) and forwards the application function's request there (1e, 1f). It can then acknowledge the application function's original request (1g).

Meanwhile, the PCF sends a notification to the URI that the SMF supplied during PDU session establishment, and includes the application function's routing information (2, 3). The SMF reacts by updating the mobile's traffic path, for example by adding a PDU session anchor in the manner described in Section 15.6.2, or by relocating the mobile's original anchor.

15.6.2 Addition of a PDU Session Anchor

Figure 15.14 shows the procedure for the addition of a second PDU session anchor [66]. The figure assumes the use of IPv6 multi-homing with a branching point that is co-located with the second anchor, that being a likely architecture if the second anchor is close to the mobile. The procedure is triggered by a traffic steering request such as the one we described in Section 15.6.1, and is initiated by the SMF.

The SMF starts by selecting a new UPF for the mobile, which will double up as the second PDU session anchor and as the branching point. In the case of IPv6 multi-homing, the SMF also allocates a new IPv6 prefix and informs the PCF (step 1). The SMF then provides the new UPF with various sets of information (2, 3). These include a service data flow template

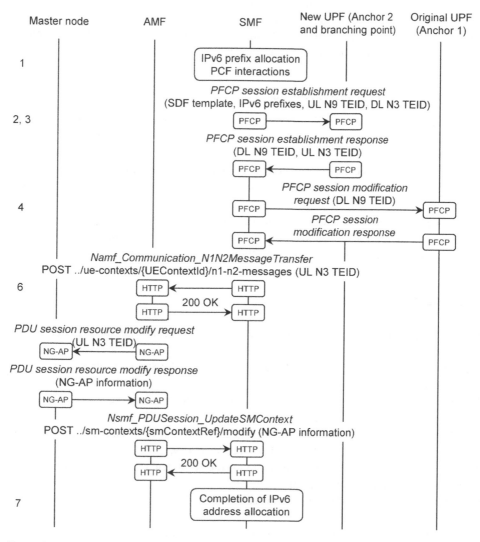

Figure 15.14 Addition of a PDU session anchor. Source: Adapted from 3GPP TS 29.502.

that states how to recognize downlink traffic in its role as the new PDU session anchor, and the mobile's originating IPv6 prefixes for use in its role as the uplink branching point. The information also includes the IP address and N9 TEID for the delivery of uplink traffic to the original anchor, and the IP address and N3 TEID for the delivery of downlink traffic to the radio access network. Both of those are known from the PDU session establishment procedure.

In response, the UPF returns two new TEIDs, one for downlink data delivery over N9, and the other for uplink data delivery over N3. The SMF forwards the first of these to the original PDU session anchor (4), and forwards the second to the radio access network by way of the AMF (6). The AMF forwards the radio access network's response to the SMF, but there is no need to inform the UPF if nothing in the access network has changed.

However, there is one remaining task: the SMF notifies the mobile about the new IPv6 prefix, and provides it with routing rules that describe how to use its prefixes when sending uplink data (7). The information reaches the mobile by means of IPv6 Router Advertisement messages, which are sent over the 5G user plane by way of the UPF.

15.6.3 Change of PDU Session Anchor

The SMF can also react to a traffic steering request by initiating a change of PDU session anchor [67]. If the PDU session is in SSC mode 2, then the SMF releases the original PDU session, but tells the mobile to establish a new PDU session that replicates the old one. When the mobile does so, the SMF can select a different anchor from before.

If the mobile is in SSC mode 3, then there are two possible implementations. In the first, the SMF tells the mobile to establish a duplicate of the original PDU session. Once it has done so, the original session is released. The second implementation exploits the use of IPv6 multi-homing: the SMF first establishes a new anchor within the original PDU session in the manner described in Section 15.6.2, and then releases the old one.

References

1 3GPP TS 23.501 (2019) System architecture for the 5G system (5GS); Stage 2 (Release 15), December 2019.

2 3GPP TS 23.502 (2019) Procedures for the 5G system (5GS); Stage 2 (Release 15), December 2019.

3 3GPP TS 23.503 (2019) Policy and charging control framework for the 5G system (5GS); Stage 2 (Release 15), December 2019.

4 3GPP TS 29.513 (2019) 5G system; Policy and charging control signalling flows and QoS parameter mapping; Stage 3 (Release 15), December 2019.

5 3GPP TS 23.501 (2019) System architecture for the 5G system (5GS); Stage 2 (Release 15), December 2019, Section 5.6.10.

6 IETF RFC 2663 (1999) IP network address translator (NAT) terminology and considerations, August 1999.

7 IETF RFC 3022 (2001) Traditional IP network address translator (traditional NAT), January 2001.

8 IETF RFC 3242 (2011) Mobile networks considerations for IPv6 deployment, August 2011.

9 IETF RFC 4862 (2007) IPv6 stateless address autoconfiguration, September 2007.

10 3GPP TS 23.503 (2019) Policy and charging control framework for the 5G system (5GS); Stage 2 (Release 15), December 2019, Section 6.3.

11 3GPP TS 29.512 (2019) 5G system; Session management policy control service; Stage 3 (Release 15), December 2019, Section 4.1.4.

12 3GPP TS 23.501 (2019) System architecture for the 5G system (5GS); Stage 2 (Release 15), December 2019, Section 5.7.

13 3GPP TS 23.060 (2018) General packet radio service (GPRS); Service description; Stage 2 (Release 15), December 2018, Annex C.

14 3GPP TS 23.501 (2019) System architecture for the 5G system (5GS); Stage 2 (Release 15), December 2019, Sections 5.8.2.3, 5.8.2.10, 5.11.

15 3GPP TS 38.300 (2019) NR; NR and NG-RAN overall description; Stage 2 (Release 15), December 2019, Section 12.

16 3GPP TS 37.340 (2019) Evolved Universal Terrestrial Radio Access (E-UTRA) and NR; Multi-connectivity; Stage 2 (Release 15), December 2019, Sections 4.2.2, 4.3.2, 8.

17 3GPP TR 23.725 (2019) Study on enhancement of ultra-reliable low-latency communication (URLLC) support in the 5G core network (5GC) (Release 16), June 2019.

18 3GPP TS 29.281 (2019) General packet radio system (GPRS) tunnelling protocol user plane (GTPv1-U) (Release 15), December 2019.

19 3GPP TS 38.415 (2018) NG-RAN; PDU session user plane protocol (Release 15), December 2018.

20 3GPP TS 38.425 (2019) NG-RAN; NR user plane protocol (Release 15), July 2019.

21 IETF RFC 2474 (1998) Definition of the differentiated services field (DS field) in the IPv4 and IPv6 headers, December 1998.

22 IETF RFC 2475 (1998) An architecture for differentiated services, December 1998.

23 IETF RFC 2597 (1999) Assured forwarding PHB group, June 1999.

24 IETF RFC 3246 (2002) An expedited forwarding PHB (per-hop behavior), March 2002.

25 IETF RFC 3031 (2001) Multiprotocol label switching architecture, January 2001.

26 IETF RFC 3032 (2001) MPLS label stack encoding, January 2001.

27 3GPP TS 23.501 (2019) System architecture for the 5G system (5GS); Stage 2 (Release 15), December 2019, Section 8.3.1.

28 3GPP TS 23.501 (2019) System architecture for the 5G system (5GS); Stage 2 (Release 15), December 2019, Section 5.6.4.

29 IETF RFC 7157 (2014) IPv6 multihoming without network address translation, March 2014.

30 3GPP TS 23.501 (2019) System architecture for the 5G system (5GS); Stage 2 (Release 15), December 2019, Section 5.6.9.

31 3GPP TS 23.503 (2019) Policy and charging control framework for the 5G system (5GS); Stage 2 (Release 15), December 2019, Section 5.

32 3GPP TS 29.513 (2019) 5G system; Policy and charging control signalling flows and QoS parameter mapping; Stage 3 (Release 15), December 2019, Section 4.

33 3GPP TS 29.507 (2019) 5G system; Access and mobility policy control service; Stage 3 (Release 15), December 2019.

34 3GPP TS 29.525 (2019) 5G system; UE policy control service; Stage 3 (Release 15), December 2019.

35 3GPP TS 29.512 (2019) 5G system; Session management policy control service; Stage 3 (Release 15), December 2019.

36 3GPP TS 29.514 (2019) 5G system; Policy authorization service; Stage 3 (Release 15), December 2019.

37 3GPP TS 29.214 (2019) Policy and charging control over Rx reference point (Release 15), September 2019.

38 3GPP TS 29.523 (2019) 5G system; Policy control event exposure service; Stage 3 (Release 15), June 2019.

39 3GPP TS 29.554 (2019) 5G system; Background data transfer policy control service; Stage 3 (Release 15), June 2019.

40 3GPP TS 29.522 (2019) 5G system; Network exposure function northbound APIs; Stage 3 (Release 15), December 2019.

41 3GPP TS 29.551 (2019) 5G system; Packet flow description management service; Stage 3 (Release 15), September 2019.

42 3GPP TS 29.521 (2019) 5G system; Binding support management service; Stage 3 (Release 15), September 2019.

43 3GPP TS 29.520 (2019) 5G system; Network data analytics services; Stage 3 (Release 15), September 2019.

44 3GPP TR 23.791 (2019) Study of enablers for network automation for 5G (Release 16), June 2019.

45 3GPP TS 29.594 (2019) 5G system; Spending limit control service; Stage 3 (Release 15), September 2019.

46 3GPP TS 32.291 (2019) Telecommunication management; Charging management; 5G system, charging service; Stage 3 (Release 15), December 2019.

47 3GPP TS 23.040 (2019) Technical realization of the short message service (SMS) (Release 15), March 2019, Annex K.

48 3GPP TS 29.002 (2019) Mobile application part (MAP) specification (Release 15), June 2019.

49 3GPP TS 29.338 (2018) Diameter based protocols to support short message service (SMS) capable mobile management entities (MMEs) (Release 15), December 2018.

50 3GPP TS 29.540 (2019) 5G system; SMS services; Stage 3 (Release 15), December 2019.

51 3GPP TR 22.824 (2018) Feasibility study on 5G message service for MIoT; Stage 1 (Release 16), September 2018.

52 3GPP TS 22.262 (2018) Message service within the 5G system; Stage 1 (Release 16), December 2018.

53 3GPP TS 23.682 (2019) Architecture enhancements to facilitate communications with packet data networks and applications (Release 15), December 2019.

54 3GPP TS 29.122 (2019) T8 reference point for northbound APIs (Release 15), December 2019.

55 3GPP TS 23.501 (2019) System architecture for the 5G system (5GS); Stage 2 (Release 15), December 2019, Section 5.6.7.

56 3GPP TS 32.240 (2019) Telecommunication management; Charging management; Charging architecture and principles (Release 15), December 2019, Section 4.3.3.

57 3GPP TS 32.255 (2019) Telecommunication management; Charging management; 5G data connectivity domain charging; stage 2 (Release 15), December 2019, Section 4.2.

58 3GPP TS 32.290 (2019) Telecommunication management; Charging management; 5G system; Services, operations and procedures of charging using service based interface (SBI) (Release 15), December 2019, Section 4.

59 3GPP TS 23.502 (2019) Procedures for the 5G system (5GS); Stage 2 (Release 15), December 2019, Section 4.3.2.2.

60 3GPP TS 23.501 (2019) System architecture for the 5G System (5GS); Stage 2 (Release 15), December 2019, Section 6.3.2.

61 3GPP TS 33.501 (2019) Security architecture and procedures for 5G system (Release 15), December 2019, Section 11.

62 3GPP TS 23.502 (2019) Procedures for the 5G system (5GS); Stage 2 (Release 15), December 2019, Section 4.16.4.

63 3GPP TS 29.513 (2019) 5G system; Policy and charging control signalling flows and QoS parameter mapping; Stage 3 (Release 15), December 2019, Sections 5.2.1, 5.3.2, 5.4.3, 8.5.2.

64 3GPP TS 23.502 (2019) Procedures for the 5G system (5GS); Stage 2 (Release 15), December 2019, Section 4.3.4.2.

65 3GPP TS 29.513 (2019) 5G system; Policy and charging control signalling flows and QoS parameter mapping; Stage 3 (Release 15), December 2019, Section 5.5.3.2.

66 3GPP TS 23.502 (2019) Procedures for the 5G system (5GS); Stage 2 (Release 15), December 2019, Section 4.3.5.4.

67 3GPP TS 23.502 (2019) Procedures for the 5G system (5GS); Stage 2 (Release 15), December 2019, Sections 4.3.5.1, 4.3.5.2, 4.3.5.3.

16

Mobility Management in RRC_CONNECTED

Mobility management is the process in which the network keeps track of the mobile's location as it moves around, and controls the cells with which it is communicating. There are different procedures for mobility management in each of the mobile's radio resource control (RRC) states, so we will devote a chapter to each one. This chapter covers the state of RRC_CONNECTED, in which the mobile is transmitting and receiving, and the network is in full control of the mobile.

Throughout this chapter, we will assume that the mobile is controlled by the next-generation radio access network (NG-RAN) and by the 5G core. When discussing the measurement procedures in Section 16.2, we will also assume that the mobile is communicating with a gNB, which might be acting as either a master node or a secondary node. Despite those assumptions, the discussion of dual connectivity in Section 16.4 will also be relevant if the mobile is controlled by the evolved UMTS terrestrial radio access network (E-UTRAN) and the evolved packet core in architectural option 3, with only minor changes.

As part of this chapter, we will discuss mobility between a gNB and an ng-eNB within the NG-RAN. The mobile might also move between the NG-RAN and the E-UTRAN, but we will leave that issue until Chapter 19. The main specifications are the usual ones for signalling in the NG-RAN and the 5G core [1, 2], while dual connectivity is covered in TS 37.340 [3].

16.1   Introduction to RRC_CONNECTED

16.1.1   Principles

In the state of RRC_CONNECTED, the mobile is transmitting and receiving, possibly at a very high data rate. The network has to control those transmissions as accurately as it can, to maximize their spectral and energy efficiency, and minimize the interference that is being delivered into nearby cells.

To achieve those objectives, the radio access network (RAN) is in full control of the mobile. In the absence of dual connectivity, the serving node tells the mobile to measure the signals that it receives from neighbouring cells and to report back the results. Based on those results, it can make changes to the mobile's serving cells, and can hand the mobile over to a neighbouring node.

An Introduction to 5G: The New Radio, 5G Network and Beyond, First Edition. Christopher Cox.
© 2021 John Wiley & Sons Ltd. Published 2021 by John Wiley & Sons Ltd.

16.1.2 Dual Connectivity

If the mobile is in dual connectivity, then the radio access network's responsibilities are shared between the master and secondary nodes. Both nodes can tell the mobile to make measurements of neighbouring cells and to report the results. Using those results, the master node (MN) can make changes to the master cell group (MCG) and can relocate the master node in the equivalent of a traditional handover procedure. In addition, the master node can add, change or remove the mobile's secondary node (SN), subject in the first two cases to the agreement of any new secondary. The secondary node can make changes to the secondary cell group (SCG) without any need for the master to be involved.

Bearers are handled in a similar way. The master node decides whether each individual data radio bearer should terminate in the master or the secondary, with the latter requiring the agreement of the secondary node. In addition, the master node decides whether to implement an MN-terminated bearer as an MCG, split or SCG bearer, with the last two requiring the agreement of the secondary. Similarly, the secondary node decides whether to implement an SN-terminated bearer as an MCG, split or SCG bearer, with the first two requiring the agreement of the master.

In the case of the signalling radio bearers (SRBs), the master node decides whether to implement SRB 1 and/or SRB 2 as split bearers, subject to the agreement of the secondary. A secondary gNB can decide by itself whether to set up SRB 3 without the need to involve the master node.

16.1.3 PDU Sessions

In RRC_CONNECTED, the behaviour of the mobile's PDU sessions depends on the architectural option that we are using. If the mobile is controlled by the evolved packet core and the E-UTRAN, then it follows the legacy behaviour from LTE, in which all of its PDU sessions are *active*. If a PDU session is active, then it is implemented in the manner shown in Figure 15.2, in which its tunnels and data radio bearers are all in place, and traffic can flow.

If the mobile is controlled by the 5G core network and the NG-RAN, then the behaviour is more flexible, as each individual PDU session can be either active or *inactive*. If a PDU session is inactive, then its data radio bearers are torn down; its N3, Xn and F1 tunnels are torn down as well; and no traffic can flow. We will cover inactive PDU sessions in more detail during our discussion of RRC_IDLE in Chapter 17.

16.2 Measurement Configuration and Reporting

16.2.1 Measurement Configuration and Reporting Procedure

Figure 16.1 shows the first of the signalling procedures, in which a master gNB configures the measurements that a mobile is making and tells it to report the results [4].

The gNB instructs the mobile to measure the signals received from neighbouring cells as part of the general-purpose signalling message *RRC Reconfiguration* (step 1). In this

Figure 16.1 Measurement configuration and reporting procedure. Source: Adapted from 3GPP TS 38.331.

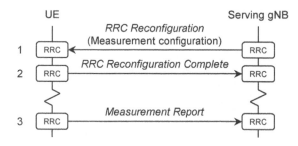

procedure, the message contains an information element known as a *measurement configuration*, which has three main components. Firstly, there is a list of *measurement objects*, each of which identifies a possible set of neighbouring cells for the mobile to measure. Secondly, there is a list of *reporting configurations*, each of which identifies a possible trigger for the delivery of measurement reports. Thirdly, there is a list of *measurement identities*, each of which pairs up a measurement object with a reporting configuration, so as to define exactly what the mobile should measure and report. There may also be one or two sets of *measurement gaps*, which define time intervals in which the mobile should carry out the measurements.

The mobile acknowledges the message, and starts its measurements (2). When one of the reporting configurations is triggered, the mobile sends an RRC *Measurement Report* to the serving node that initiated the procedure (3). Based on that report, the node can initiate a change to the master and/or secondary cell groups.

In cases of multi-RAT dual connectivity, there are a few more issues to consider [5]. Firstly, a master gNB can instruct the mobile to measure the signals received from neighbouring eNBs, and vice versa. Secondly, the high-level procedure remains the same if the master gNB is replaced by a master ng-eNB, except that the 5G signalling messages are replaced by a 4G *RRC Connection Reconfiguration* and a 4G *Measurement Report* respectively.

Thirdly, a secondary node can communicate with the mobile in two different ways. If the secondary node is a gNB, and SRB 3 is configured, then it can communicate directly with the mobile in the same way as the master. Otherwise, the messages are forwarded transparently by way of the master node, by embedding them into RRC signalling messages between the mobile and the master that are transported using SRB 1, and into Xn application protocol (Xn-AP) signalling messages between the master and the secondary. Figure 16.2 shows the resulting signalling for the non-standalone example of a master ng-eNB and a secondary gNB [6, 7]. The master node also forwards the secondary's messages if it has any involvement in the signalling procedure, for example in several of the handover and dual connectivity procedures that follow.

Finally, the signalling procedures in the E-UTRAN are exactly the same, except that the Xn-AP messages in Figure 16.2 are replaced by their X2-AP equivalents [8].

16.2.2 Measurement Objects

A measurement object identifies a set of neighbouring cells that the mobile should measure. The definition of 4G neighbours is straightforward, the main parameters being the centre

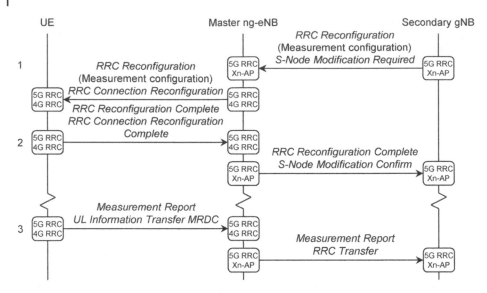

Figure 16.2 Forwarding of measurement configuration and reporting messages between the mobile and a secondary node.

frequency of the neighbouring cells and the bandwidth to measure. The mobile can identify individual 4G neighbours by itself, so it does not require any neighbour list.

In the case of 5G neighbours, there are two possible reference signals to measure. As the first of these, the serving node can tell the mobile to measure the neighbours' synchronization signal/physical broadcast channel (SS/PBCH) blocks, by supplying not only their centre frequency but also their subcarrier spacing, transmission period (5–160 ms) and timing offset within that period. Optionally, the serving node can also supply a bitmap that identifies the precise SS/PBCH blocks that the neighbours are transmitting. The serving node can discover the required information using RRC information elements that its neighbours forward over F1 and Xn [9]. As before, the mobile can identify the individual neighbours by itself.

Alternatively, the serving node can tell the mobile to measure any periodic channel state information reference signal (CSI-RS) resources that the neighbouring cells are transmitting. To do so, it identifies the frequency of point A in the neighbours' common resource grids, and the subcarrier spacing of their CSI reference signals. It then lists the individual cells by means of their physical cell identities and measurement bandwidths, and defines the CS-RS resources by means of their resource mapping, transmission period (4–40 ms) and timing offset within that period. This time, however, the specifications do not define how the serving node discovers the required information, so it might be unavailable.

In both cases, the serving node can tell the mobile to measure the reference signal received power (RSRP), reference signal received quality (RSRQ) or signal-to-interference plus noise ratio (SINR) of the incoming reference signal, the last of these being the RSRP divided by the average power per resource element due to noise and interference alone. The mobile first measures these quantities on each individual beam, and then converts the results to cell-specific measurements as follows. If the strongest beam is weaker than a threshold supplied by the serving node, then the mobile simply applies its measurement of that beam to the cell as a whole. Otherwise, the mobile averages its measurements of the beams that are stronger than the threshold, subject to an upper limit on the number of beams to consider.

Optionally, the serving node can supply a list of neighbours that the mobile should measure, or a list of neighbours to ignore. It can also supply optional cell-specific offsets which encourage or discourage the delivery of measurement reports about individual neighbours.

16.2.3 Reporting Configurations

Measurement reports in 5G use event-triggered periodic reporting, in which the mobile starts sending periodic measurement reports if a signal crosses over a threshold, and stops if the signal returns. By simplifying that procedure, the mobile can also send one-off event-triggered reports to the network, as well as unconditional periodic and one-off reports.

The 5G specifications define eight types of measurement event. Table 16.1 summarizes the first six, which are denoted A1–A6, and are triggered by measurements of 5G cells alone. Table 16.2 summarizes the remaining two, which are denoted B1 and B2, and are triggered

Table 16.1 Measurement events involving 5G cells alone.

Event	Description	Possible applications
A1	Serving cell > Threshold	Stop measuring lower-priority carriers
A2	Serving cell < Threshold	Start measuring lower-priority carriers
		RRC release and redirection to another carrier
		Release PSCell or SCell
A3	Neighbour > SpCell + Offset	Replace SpCell by equal-priority neighbour
A4	Neighbour > Threshold	Replace SpCell by higher-priority neighbour
		Add PSCell or SCell
A5	SpCell < Threshold 1, and SCell or Neighbour > Threshold 2	Replace SpCell by lower-priority neighbour
		Replace SpCell by SCell
A6	Neighbour > SCell + Offset	Replace SCell by neighbour

PSCell: Primary SCG cell; SCell: secondary cell; SpCell: special cell.

Table 16.2 Measurement events involving other radio access technologies.

Event	Description	Possible applications
B1	Inter-RAT neighbour > Threshold	Replace 5G PCell by higher-priority 4G neighbour
		Add 4G PSCell alongside 5G PCell
B2	PCell < Threshold 1, and Inter-RAT neighbour > Threshold 2	Replace 5G PCell by lower-priority 4G neighbour

PCell: Primary cell; PSCell: primary SCG cell.

by neighbours that are using other radio access technologies such as LTE. (The LTE specifications define two equivalent measurement events which are relevant for the measurement of 5G neighbours.)

As an example, measurement event A3 might trigger a change of special cell, that being the collective term for a primary cell or a primary SCG cell, to a neighbour that is on the same carrier frequency. The mobile enters a measurement reporting state if the following condition is met, for a time of at least *timeToTrigger* (0–5120 ms):

$$M_n + Of_n + Oc_n > M_p + Of_p + Oc_p + Off + Hys \qquad (16.1)$$

Whilst in this state, the mobile sends measurement reports to the serving node that configured the measurement, with a period of *reportInterval* (120 ms to 30 minutes), up to a maximum of *reportAmount* reports (1–64 or unlimited). Usually, those measurement reports trigger a change of cell. Failing that, the mobile leaves the reporting state if the following condition is met, for a time of at least TimeToTrigger:

$$M_n + Of_n + Oc_n > M_p + Of_p + Oc_p + Off - Hys \qquad (16.2)$$

In these equations, M_p is a measurement of the serving node's primary or primary SCG cell, while M_n is a measurement of the neighbour. Both of these quantities are cell-specific values of RSRP, RSRQ or SINR, which are obtained by measurements of the cells' SS/PBCH blocks or CSI reference signals in the manner defined in Section 16.2.2 [10]. As for the other parameters, Hys (0–15 dB) is a hysteresis parameter, which discourages a mobile from bouncing in and out of the measurement reporting state, while Off (also 0–15 dB) acts as a hysteresis parameter for any resulting handover. Of_p and Of_n are optional frequency-specific offsets, while Oc_p and Oc_n are optional cell-specific offsets.

16.2.4 Measurement Gaps

A *measurement gap* is a mobile-specific time interval, with a duration of 1.5–6 ms and a period of 20–160 ms, during which the radio access network promises not to schedule the mobile. During the measurement gap, the mobile can break communications with its serving cells and measure their neighbours, confident that it will not miss any downlink data or uplink transmission opportunities. Depending on its capabilities, a mobile might support a single pattern of measurement gaps that suppresses communications with all of its serving cells, or two independent patterns that apply to serving cells in frequency ranges 1 and 2. The radio access network provides a mobile with measurement gaps if it has instructed the mobile to measure its neighbouring cells, but is not confident that the mobile can communicate with its serving cells at the same time.

The need for measurement gaps depends on the type of measurement that the mobile is making [11–13]. At the most basic level, an *intra-frequency measurement* is on the same radio frequency as the serving cell, and uses the same radio access technology (5G or LTE). Conversely, an *inter-frequency measurement* is on a different radio frequency and may also use a different RAT. Adding some more detail, a measurement of a neighbour's SS/PBCH blocks is intra-frequency if the SS/PBCH blocks of the serving and neighbouring cells have the same centre frequency and subcarrier spacing. Similarly, a measurement of a neighbour's CSI reference signal is intra-frequency if the neighbour's CSI-RS resource lies entirely within the bandwidth of the serving cell's CSI-RS resource and has the same subcarrier spacing. All other measurements are inter-frequency.

If a mobile is measuring its neighbours' SS/PBCH blocks, then the radio access network provides measurement gaps in three situations. The first is for intra-frequency measurements, if the neighbour's SS/PBCH blocks stray outside one or more of the mobile's downlink bandwidth parts. The second is for inter-frequency measurements if the mobile only supports one measurement gap pattern. The third is for inter-frequency measurements if the mobile supports two measurement gap patterns, but any of the serving cells lies in the same frequency range as the neighbours. If none of these conditions applies, then the measurement gaps can be skipped. Similar considerations apply to measurements of a neighbour's CSI reference signals and to measurements of 4G neighbours.

Even if measurement gaps are provided, a mobile may still be able to ignore them. If, for example, the mobile only has one serving cell, and supports downlink carrier aggregation between the serving and neighbouring frequency bands, then it may be able to measure the neighbour and communicate with the serving cell at the same time. Such decisions are a matter for the mobile's implementation.

16.2.5   Measurement Reporting

Once a report is triggered, the mobile sends its results to the serving node by means of an RRC *Measurement Report*. The information elements include the measurement identity that triggered the report, and the mobile's measurements of its serving cells and its strongest neighbours.

The exact quantities reported depend on the mobile's reporting configuration. In the case of a 5G cell, the mobile reports the physical cell identity, and any or all of the cell-specific measurements of RSRP, RSRQ and SINR, from either the cell's SS/PBCH blocks or its CSI reference signals. If it is configured to do so, then the mobile also reports the SS/PBCH block indexes or CSI-RS resource indexes of the best beams that it has discovered, along with the corresponding beam-specific measurements of RSRP, RSRQ and/or SINR. For measurements of a 4G neighbour, the mobile reports the neighbour's 4G physical cell identity, and the RSRP, RSRQ and/or SINR of the neighbour's downlink cell-specific reference signals.

If the serving node does not recognize a physical cell identity that the mobile reported, then it asks the mobile to read the corresponding global cell identity from the neighbour's system information, using the procedure for *automatic neighbour relations* that we will discuss in Chapter 20. Otherwise, the serving node looks up the global cell identity from its internal records. It can then initiate a change of serving cell.

16.3   Handover Procedures

16.3.1   Xn-based Handover Procedure

A measurement report can trigger several changes to the mobile's serving cells. Perhaps the most important is a change of primary cell to one that is controlled by another node within

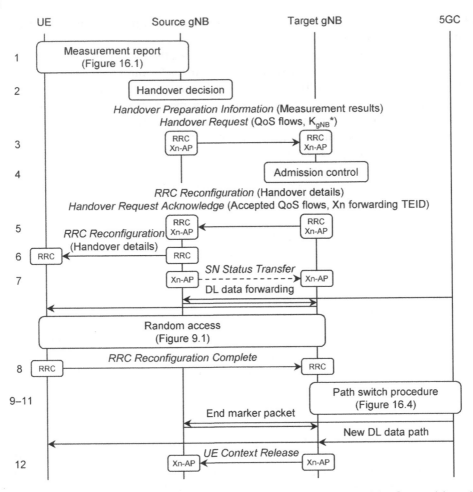

Figure 16.3 Xn-based handover procedure in the absence of dual connectivity. Source: Adapted from 3GPP TS 38.300.

the NG-RAN. If the mobile is not in dual connectivity, then this is a traditional handover procedure. Figure 16.3 shows the most common implementation, in which the base stations communicate over the Xn reference point [14]. The figure assumes that the source and target nodes are both gNBs, but it also applies to the case where either node is an ng-eNB with just a few changes.

The procedure begins when the mobile sends a measurement report to its serving node about a neighbouring cell, for example using measurement event A3, A4 or A5 (step 1). Based on that report, the original node looks up the node that is controlling the neighbouring cell, and decides to initiate a handover (2).

The old node asks the new one to take control of the mobile by sending it an Xn-AP *Handover Request* (3). The message identifies the target cell and the mobile's access and mobility management function (AMF), and includes details of the mobile's PDU sessions and QoS flows. The message also includes a new access stratum security key, K_{gNB}^{*}, which the old base station has computed by means of a horizontal key derivation. Finally, the message includes an embedded RRC message known as *Handover Preparation Information*. That is only ever delivered by embedding it into Xn-AP signalling, and includes the cell- and beam-specific information from the earlier measurement report, as well as the mobile's radio access capabilities and air interface configuration.

The new node decides whether to admit the mobile and, if so, whether to admit each of the mobile's PDU sessions and QoS flows (4). It might, for example, reject a PDU session that is associated with a network slice that it does not support. If the target cell is congested, then it might also reject an individual QoS flow with a high guaranteed bit rate and a low allocation and retention priority, such as the video component of an IMS audio-visual call.

If it does admit the mobile, then the new node replies using an Xn-AP *Handover Request Acknowledge* (5). That message states the PDU sessions and QoS flows that the base station has accepted, and includes a tunnel endpoint identifier (TEID) for data forwarding over Xn. It also includes an embedded RRC Reconfiguration message that configures the mobile's radio communications with the target cell. As part of the RRC message, the node can provide the mobile with a set of dedicated resources for the physical random access channel (PRACH), including an association between the cell's PRACH resources and the SS/PBCH block indexes and/or CSI-RS resource indexes that the mobile will eventually report. It can also provide details of any secondary cells, so as to place the mobile into an immediate state of carrier aggregation.

The source node extracts the RRC message that it received from the target and sends it to the mobile as a handover command (6). It also starts forwarding undelivered data packets on all of the mobile's data radio bearers towards the target. If any of those bearers are configured to avoid packet duplication during a handover, then the source node lists the uplink packets that have already arrived by means of an Xn-AP *SN Status Transfer* (7). Meanwhile, the mobile establishes communications with the target cell by means of the random access procedure, and acquires an initial value for the uplink timing advance. It then sends an RRC acknowledgement to the target node (8).

On receiving the mobile's acknowledgement, the target node informs the 5G core that it is now serving the mobile, using the path switch procedure that we will discuss in Section 16.3.2 (steps 9–11). As part of that procedure, the core network indicates the end of the old downlink data stream by sending one or more *end marker* packets towards the source node, and then switches its downlink data path towards the target. The source node forwards the end marker packet to the target, which continues to transmit forwarded packets to the mobile until the end marker packet arrives. At that point, the target node can start transmitting data that it has received directly from the core.

On receiving the core network's signalling response, the target node informs the source node that the handover has completed, by means of an Xn-AP *UE Context Release* (12). Once the source node has received that message, it can release the resources that it had assigned to the mobile.

If the handover is between two cells that are controlled by the same node, then the signalling procedure is much simpler. The only steps required are the air interface's measurement report and RRC reconfiguration: the Xn-AP messages are all redundant, as is the path switch request to the 5G core.

16.3.2 Path Switch Procedure

Figure 16.4 shows the details of the path switch procedure, by which the target node redirects the data and signalling messages that the mobile is exchanging over the NG backhaul [15]. The figure assumes that there are no intermediate user plane functions (UPFs), either before the handover or after.

To begin the procedure, the target node sends an NG application protocol (NG-AP) *Path switch request* to the mobile's AMF (step 1). The message identifies the QoS flows that the node has accepted, and includes an IP address and downlink TEID for each of the mobile's N3 tunnels. On receiving the request, the AMF redirects its downlink signalling messages towards the target node.

For each of the mobile's PDU sessions, the AMF identifies the corresponding session management function (SMF), and sends it the IP addresses and TEIDs that it received from the target node (2). The SMF forwards the relevant information to the UPF (3), which delivers an end marker packet to the source node on each of the mobile's N3 tunnels, and then redirects its downlink data path towards the target (4, 5).

On receiving the SMF's response (6), the AMF sends an acknowledgement to the target node (7). That includes intermediate values for a new access stratum security key, which the AMF has computed by means of a vertical key derivation, and a new set of core network assistance information. The target node can either compute the new security key right away and take it into use by means of an RRC reconfiguration procedure, or store the information until the next Xn-based handover.

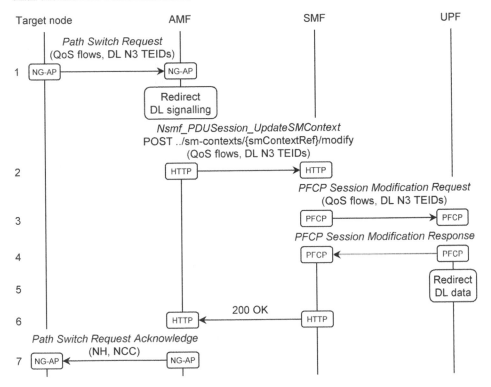

Figure 16.4 Path switch procedure. Source: Adapted from 3GPP TS 23.502.

16.3.3 NG-based Handover Procedure

In the Xn-based handover procedure described earlier in this chapter, the two nodes communicated with each other using signalling messages over the Xn reference point. As an alternative, the nodes can communicate indirectly, using signalling messages that are delivered by way of the 5G core network using two instances of NG [16]. The resulting NG-based procedure is essential if the two nodes have no connectivity over Xn, or if the target node has no connectivity with the mobile's serving AMF.

At least at a high level, the procedure is very like the Xn-based procedure from earlier. The mobile's access stratum security keys are updated by means of a vertical key derivation alone, and the procedure can change the mobile's serving AMF to one that is connected with the target. As part of the procedure, the source node can forward undelivered data packets to the target either directly or by way of the 5G core, depending on whether the two nodes have IP connectivity.

16.3.4 Handovers Between a gNB and an ng-eNB

The NG-RAN can control the mobile using a master gNB or a master ng-eNB, and can hand the mobile over between the two. That causes a few complications.

A handover from a gNB to an ng-eNB is triggered by 5G measurement event B1 or B2. On receiving the measurement report, the source gNB looks up the target cell and contacts the corresponding ng-eNB in the same way as before. As part of its Handover Request Acknowledge (Figure 16.3, step 5), the target ng-eNB sends a 4G RRC Connection Reconfiguration message to the source. In place of step 6, the source gNB delivers that message to the mobile by embedding it into a 5G RRC *Mobility From NR Command*, which tells the mobile to hand over to another radio access technology. The mobile extracts the 4G RRC message and establishes communications with the target ng-eNB as before.

A handover from an ng-eNB to a gNB proceeds in a similar way. In the case of a handover between two ng-eNBs, we can simply replace the 5G RRC signalling messages in Figure 16.3 with their 4G equivalents.

16.4 Dual Connectivity Procedures

16.4.1 Secondary Node Addition

The most important procedure for dual connectivity is the addition of a secondary node. During that procedure, the radio access network redirects one or more of a mobile's data radio bearers so that they travel through the secondary. If any of those bearers are reconfigured as SCG or split bearers, then the secondary node sets up radio communications with the mobile through a primary SCG cell, and optionally through one or more secondary cells. Similarly, if any data radio bearers are reconfigured as SN-terminated bearers, then the secondary node sets up a new N3 tunnel for each of the corresponding PDU sessions. Figure 16.5 shows a non-standalone example, in which the master node is an ng-eNB and the secondary is a gNB [17].

The procedure is usually triggered when the mobile sends a measurement report to its master node about a neighbouring cell controlled by a different node, for example using

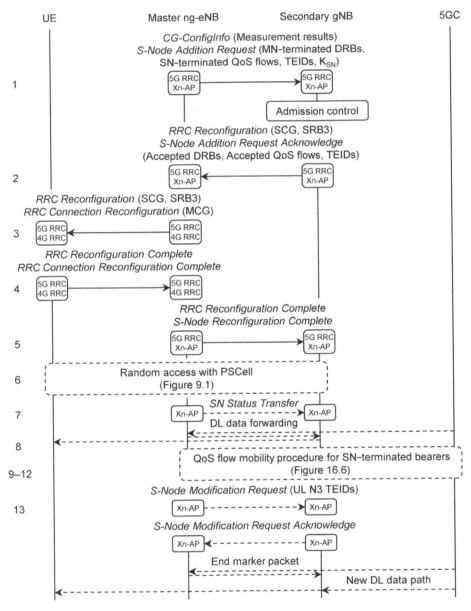

Figure 16.5 Addition of a secondary node. Source: Adapted from 3GPP TS 37.340.

measurement event A4 or B1. It could also be triggered in the absence of a measurement report, if the only change is the addition of an MCG bearer that terminates in the secondary node.

To begin the procedure, the master node asks its neighbour to act as the mobile's secondary node, by means of an Xn-AP *S-Node Addition Request* (step 1). The message includes the secondary node's access stratum security key K$_{SN}$, and optionally a request to implement SRBs 1 and/or 2 as split SRBs. It also includes an embedded RRC message known as *CG-ConfigInfo*, which states the mobile's radio capabilities, measurement configuration and measurement results. If the secondary node admits the mobile, then it replies with an Xn-AP *S-Node Addition Request Acknowledge* (2).

Within these two messages, the nodes redirect QoS flows and data radio bearers by collecting them into two groups. To handle the first group, the master node lists any data radio bearers that it would like to implement using MN-terminated SCG or split bearers as part of message 1. For each one, it lists the parameters of their constituent QoS flows and includes a TEID for uplink data delivery over Xn. In message 2, the secondary lists the bearers that it has accepted and includes the corresponding TEIDs for downlink data delivery over Xn.

To handle the second group, the master node also lists any QoS flows that it would like to map onto SN-terminated bearers as part of message 1. For each one, it lists the QoS parameters, states whether it is willing to implement the bearer as an MCG or split bearer and includes a TEID for the delivery of downlink data over Xn. In message 2, the secondary lists the QoS flows that it has accepted, states how it has mapped them onto bearers and includes TEIDs for the delivery of uplink data over Xn and downlink data over N3.

As part of its reply, the secondary node also supplies an embedded RRC Reconfiguration message for the mobile. That message reconfigures the service data adaptation protocol (SDAP) and packet data convergence protocol (PDCP) of any SN-terminated bearers, and reconfigures the mobile's air interface to handle any split or SCG bearers. A secondary gNB can also supply the mobile with a configuration for SRB 3.

The master node forwards the secondary's RRC message to the mobile by embedding it into an RRC message of its own that reconfigures the mobile's communications with the master (3). The mobile reconfigures itself as instructed, acknowledges the master and includes an embedded acknowledgement for the secondary (4). The master node extracts the latter acknowledgement and forwards it to the secondary as part of an Xn-AP *S-Node Reconfiguration Complete* (5), which itself acknowledges the bearer configuration that the secondary supplied earlier. If the secondary node reconfigured the mobile's air interface, then the mobile runs the random access procedure to establish communications with the primary SCG cell and to acquire an uplink timing advance (6).

If there are any SN-terminated bearers, then the master node forwards undelivered data packets to the secondary, and lists the uplink packets that have already arrived on bearers configured for duplication avoidance (7, 8). Using the QoS flow mobility procedure that we discuss in Section 16.4.2, the master node also tells the core network to deliver future downlink data on the corresponding QoS flows towards the secondary, using the N3 TEIDs that the secondary supplied earlier (9–12). The core network returns a set of uplink N3 TEIDs as part of that procedure, which the master node forwards to the secondary (13).

16.4.2 QoS Flow Mobility Procedure

The QoS flow mobility procedure is similar to the path switch request procedure from earlier in this chapter, but it has a narrower scope: it simply redirects the downlink data path

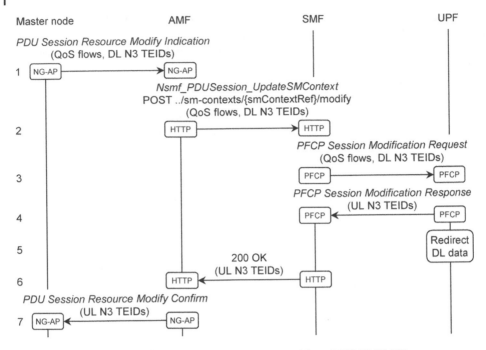

Figure 16.6 QoS flow mobility procedure. Source: Adapted from 3GPP TS 23.502.

for selected QoS flows, making no change either to the signalling pathway or to the access stratum security keys [18]. Figure 16.6 shows the procedure.

Using an NG-AP *PDU Session Resource Modify Indication*, the master node asks the AMF to change the downlink data path for some or all of its QoS flows, and includes an IP address and TEID for each one (step 1). For each of the mobile's PDU sessions, the AMF identifies the corresponding SMF and asks it to redirect the downlink data (2). The SMF forwards the request to the corresponding UPF, which acts as instructed and responds (3–6). To complete the procedure, the AMF returns an acknowledgement to the master node, which includes a new set of TEIDs for the delivery of uplink data over N3 (7).

16.4.3 Other Dual Connectivity Procedures

Several other procedures are relevant to dual connectivity, but they all follow a similar theme [19]. The simplest involves changes to the secondary cell group, if the secondary node remains the same. Using the measurement configuration procedure from earlier, the secondary node tells the mobile to measure the signals received from the SCG, and from any neighbouring cells that the secondary node controls. Based on the mobile's measurement reports, the secondary node can trigger changes to the SCG, for example a change of primary SCG cell, or the addition, change or release of a secondary cell. There is no need for the master node to be involved.

Changes to the secondary node can be triggered in two ways: either by a measurement report from the mobile to the master node; or by a measurement report from the mobile to the secondary, which in turn triggers an Xn-AP request from the secondary node to the master. In both cases, the master node responds by replacing or releasing the secondary node, and by reconfiguring the corresponding bearers.

Finally, the handover procedure can also involve changes to the mobile's secondary node. The mobile's measurement reports to a master node include measurements of the master cell group, any secondary cell group and all the potential neighbours. The old master node forwards those measurements to the new one as part of its Xn-AP Handover Request (Figure 16.3, step 3). Using that information, the new master node can decide whether to add a new secondary node, or whether to retain, replace or release an existing secondary node. Those changes take place as an integral part of the handover procedure.

16.5 State Transitions out of RRC_CONNECTED

16.5.1 Core Network Assistance Information

If the mobile is in RRC_CONNECTED but is not transmitting or receiving any data, then the network can prolong its battery life by transferring it to RRC_INACTIVE or RRC_IDLE. The transition is typically triggered by the expiry of timers in the mobile's master and secondary nodes, which started running when the latest uplink or downlink data were delivered. The expiry times are an internal matter for the radio access network, so they do not appear in any specifications, but a typical value in the case of LTE has been 10–20 seconds.

When the timers expire, the master node has a choice of two possible target states. It makes its choice with the help of the core network assistance information that it received from the AMF during earlier procedures such as registration and handover [20]. The information includes expected values for the mobile's active and idle periods, the expected time between handovers and the cells that the mobile is expected to visit. As a simple example, the master node might transfer the mobile to RRC_INACTIVE if the expected idle period is short, or request a transfer to RRC_IDLE if the expected idle period is long.

16.5.2 Transition to RRC_IDLE

Figure 16.7 shows the *access network release* procedure, which transfers the mobile to the states of RRC_IDLE and CM-IDLE [21]. The diagram assumes that the mobile is initially in a state of dual connectivity. The procedure tears down the mobile's communications with its master and secondary nodes, and transfers its PDU sessions into an inactive state in which no data can be delivered.

The procedure begins when internal timers expire in the master node in respect of any MN-terminated bearers, and in the secondary node in respect of any SN-terminated bearers. In the latter case, the secondary node informs the master by means of an Xn-AP *Activity Notification*.

The master node reacts by sending an NG-AP *UE Context Release Request* to the AMF, which asks it to send the mobile to the state of CM-IDLE (step 1). The AMF agrees to do so, and returns an NG-AP *UE Context Release Command* (2). If the mobile is in dual connectivity, then the master node releases the secondary. The master node then tells the mobile to move to RRC_IDLE by means of the message *RRC Release*, in which it can redirect the

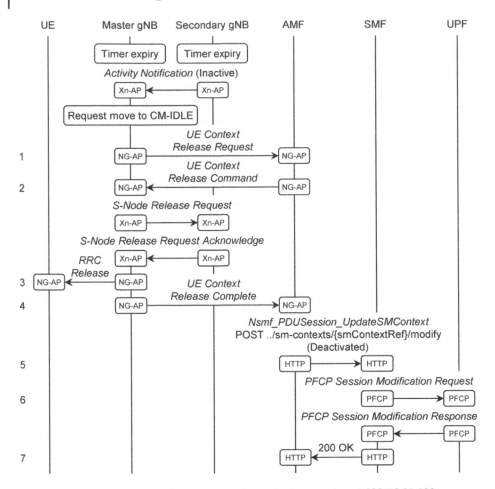

Figure 16.7 Access network release procedure. Source: Adapted from 3GPP TS 23.502.

mobile to a preferred 4G or 5G carrier frequency, and can supply a mobile-specific list of radio frequency priorities for use during cell reselection (3).

The master node then tears down the radio access network's end of the mobile's N3 tunnels, and sends an acknowledgement to the AMF (4). That acknowledgement includes the mobile's radio capabilities, which the AMF stores while the mobile is idle for later delivery to the next master node. The AMF identifies the SMF that is handling each of the mobile's PDU sessions, and instructs it to tear down the core network's end of the N3 tunnels (5). The SMF forwards the relevant information to the UPF (6), which acts as instructed, and responds (7). The mobile is now in the state of RRC_IDLE and behaves in the manner described in Chapter 17.

16.5.3 Transition to RRC_INACTIVE

Alternatively, the master node can suspend the mobile's RRC connection and send it to the state of RRC_INACTIVE. Figure 16.8 shows the resulting procedure, for the case where the

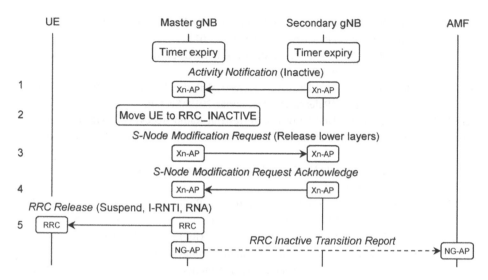

Figure 16.8 Suspension of the RRC connection.

mobile is in dual connectivity and the secondary node is retained [22]. The procedure tears down the lower layers of the mobile's air interface protocol stack, specifically the physical layer, the medium access control (MAC) and radio link control (RLC) protocols, as well as the mobile's Xn and F1 tunnels. If the nodes are split, then the distributed units are released, but the central units are retained.

The procedure is triggered in the same way as before (step 1). This time, however, the master node decides to send the mobile to the state of RRC_INACTIVE (2), and tells the secondary node to release the lower layers of its protocol stack (3). The secondary node acts as instructed, tears down its ends of the corresponding Xn and F1 tunnels, and responds (4).

As before, the master node sends an *RRC Release* message to the mobile, but this time tells the mobile to move to RRC_INACTIVE by including a *suspend configuration* (5). That defines the mobile's RAN-based notification area (RNA) and includes an inactive radio network temporary identifier (I-RNTI) that will identify the mobile. The mobile acts as instructed and releases the lower layers of its air interface protocol stack. At the same time, the master node releases its own lower layers, and tears down its own ends of the corresponding Xn and F1 tunnels.

If the AMF has previously asked it to do so, then the master node sends a report about the state transition by means of an NG-AP *RRC Inactive Transition Report*. The mobile is now in the state of RRC_INACTIVE and behaves in the manner described in Chapter 18.

References

1 3GPP TS 38.300 (2019) NR; NR and NG-RAN overall description; Stage 2 (Release 15), December 2019.
2 3GPP TS 23.502 (2019) Procedures for the 5G system (5GS); Stage 2 (Release 15), December 2019.

3 3GPP TS 37.340 (2019) Evolved Universal Terrestrial Radio Access (E-UTRA) and NR; Multi-connectivity; Stage 2 (Release 15), December 2019.

4 3GPP TS 38.331 (2019) NR; Radio resource control (RRC) protocol specification (Release 15), December 2019, Section 5.5.

5 3GPP TS 37.340 (2019) Evolved Universal Terrestrial Radio Access (E-UTRA) and NR; Multi-connectivity; Stage 2 (Release 15), December 2019, Section 7.

6 3GPP TS 36.331 (2019) Evolved Universal Terrestrial Radio Access (E-UTRA); Radio resource control (RRC); Protocol specification (Release 15), December 2019, Sections 5.3.5, 5.6.2a.

7 3GPP TS 38.423 (2019) NG-RAN; Xn application protocol (XnAP) (Release 15), December 2019, Sections 8.3.4, 8.3.9.

8 3GPP TS 36.423 (2019) Evolved Universal Terrestrial Radio Access Network (E-UTRAN); X2 application protocol (X2AP) (Release 15), December 2019, Sections 8.7.7, 8.7.12.

9 3GPP TS 38.331 (2019) NR; Radio resource control (RRC) protocol specification (Release 15), December 2019, Section 11.2.2 (MeasurementTimingConfiguration).

10 3GPP TS 38.215 (2019) NR; Physical layer measurements (Release 15), December 2019, Sections 5.1.1 to 5.1.6.

11 3GPP TS 38.300 (2019) NR; NR and NG-RAN overall description; Stage 2 (Release 15), December 2019, Section 9.2.4.

12 3GPP TS 37.340 (2019) Evolved Universal Terrestrial Radio Access (E-UTRA) and NR; Multi-connectivity; Stage 2 (Release 15), December 2019, Section 7.2.

13 3GPP TS 38.133 (2019) NR; Requirements for support of radio resource management (Release 15), December 2019, Section 9.1.2.

14 3GPP TS 38.300 (2019) NR; NR and NG-RAN overall description; Stage 2 (Release 15), December 2019, Section 9.2.3.2.

15 3GPP TS 23.502 (2019) Procedures for the 5G system (5GS); Stage 2 (Release 15), December 2019, Section 4.9.1.2.2.

16 3GPP TS 23.502 (2019) Procedures for the 5G system (5GS); Stage 2 (Release 15), December 2019, Section 4.9.1.3.

17 3GPP TS 37.340 (2019) Evolved Universal Terrestrial Radio Access (E-UTRA) and NR; Multi-connectivity; Stage 2 (Release 15), December 2019, Sections 10.2.2, 10.14.3.

18 3GPP TS 23.502 (2019) Procedures for the 5G system (5GS); Stage 2 (Release 15), December 2019, Section 4.14.1.

19 3GPP TS 37.340 (2019) Evolved Universal Terrestrial Radio Access (E-UTRA) and NR; Multi-connectivity; Stage 2 (Release 15), December 2019, Section 10.

20 3GPP TS 38.413 (2019) NG-RAN; NG application protocol (NGAP) (Release 15), December 2019, Section 9.3.1.15.

21 3GPP TS 23.502 (2019) Procedures for the 5G system (5GS); Stage 2 (Release 15), December 2019, Section 4.2.6.

22 3GPP TS 37.340 (2019) Evolved Universal Terrestrial Radio Access (E-UTRA) and NR; Multi-connectivity; Stage 2 (Release 15), December 2019, Section 10.12.2.

17

Mobility Management in RRC_IDLE

This chapter covers the second of 5G's mobility management states, RRC_IDLE. In this state, the mobile is on standby, and is not expected to be transmitting or receiving. The mobile moves from one cell to another by a procedure known as *cell reselection*, and keeps the core network informed about its location by means of registration updates.

As in Chapter 16, we will assume that the mobile is controlled by the next-generation radio access network (NG-RAN) and the 5G core. When discussing the procedures for cell reselection in Section 17.2, we will also assume that the mobile supports standalone operation of the 5G New Radio and is camping on a 5G cell that is controlled by a gNB. We will discuss reselection between 5G and 4G cells within the NG-RAN as part of that section, but we will leave reselection between the NG-RAN and the evolved UMTS terrestrial radio access network (E-UTRAN) until Chapter 19. The main specifications cover the signalling procedures in the NG-RAN and the 5G core, and the mobile's procedures in idle mode [1–3].

17.1 Introduction to RRC_IDLE

17.1.1 Principles

In the state of RRC_IDLE, the mobile is registered with the 5G core network, and may have PDU sessions established with one or more external data networks. However, the mobile is on standby, and is not expected to be transmitting or receiving. The objectives are to maximize the mobile's battery life and minimize any signalling communications with the network.

In this state, it would be inappropriate to control the mobile using the radio resource control (RRC) procedures for measurement reporting and handover, because those procedures require signalling communications every time the mobile changes cell. Instead, the core network only knows the registration area in which the mobile is situated, while the radio access network knows nothing about the mobile at all. The mobile camps on a single serving cell and can move throughout its registration area without the need to inform the network.

To preserve its battery life, the mobile spends most of its time in a low-power state, waking once every discontinuous reception cycle to listen for paging messages from the serving cell and to measure the strength of the received signal. Based on those measurements, the

An Introduction to 5G: The New Radio, 5G Network and Beyond, First Edition. Christopher Cox.
© 2021 John Wiley & Sons Ltd. Published 2021 by John Wiley & Sons Ltd.

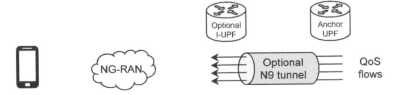

Figure 17.1 Example of an inactive PDU session.

mobile can change its serving cell in a procedure known as *cell reselection*. If the new cell lies outside its registration area, then the mobile informs the network by means of a *registration update*.

17.1.2   Inactive PDU Sessions

If the mobile is in the state of RRC_IDLE, then its PDU sessions are all in an *inactive* state that is illustrated in Figure 17.1. If the mobile is controlled by the NG-RAN and the 5G core, then individual PDU sessions can also be inactive in the states of RRC_CONNECTED and RRC_INACTIVE.

In an inactive PDU session, the data radio bearers are all torn down, and the N3, Xn and F1 tunnels are torn down as well [4]. Optionally, the network can also release any N9 tunnels that the PDU sessions are using, and the corresponding intermediate user plane functions (UPFs). The mobile can then move throughout its registration area without the need for its bearers and tunnels to be redirected.

The mobile retains its network address, so a server can still send packets towards the mobile by way of the PDU session anchor. If it does so, then the network contacts the mobile by sending a paging message through all the possible serving cells within its registration area. On receiving the mobile's reply, the network returns the mobile to the state of RRC_CONNECTED and reactivates the corresponding PDU session. It can then deliver the packets to the mobile.

17.2   Cell Reselection Procedures

17.2.1   Introduction

In the state of RRC_IDLE, the serving cell configures the mobile's measurements by means of cell-specific system information. In the case of a 5G serving cell, the relevant informa-tion is in system information blocks (SIBs) 2, 3, 4 and 5. The mobile can always identify neighbouring cells by itself, so the serving cell does not have to provide any neighbour list as part of its system information. However, it does have the option to provide cell-specific reselection parameters as part of SIBS 3, 4 and 5.

To carry out its measurements, the mobile inspects the synchronization signal/physical broadcast channel (SS/PBCH) blocks of serving and neighbouring 5G cells, and the cell-specific reference signals of serving and neighbouring 4G cells. The channel state information reference signals from 5G are not used. As before, the mobile can carry out two types of measurement, namely intra-frequency and inter-frequency [5, 6]. In the case of a 5G serving cell, a measurement of a 5G neighbour is intra-frequency if the SS/PBCH blocks of the serving and neighbouring cells have the same centre frequency and subcarrier spacing [7]. All other measurements are inter-frequency.

17.2.2 Intra-frequency Measurement Triggering

To begin the discussion, let us assume that the mobile is only carrying out intra-frequency measurements. The mobile wakes once every discontinuous reception cycle, listens for paging messages and measures the received signal power on the secondary synchronization signal from its serving cell. If the signal is strong enough, and no paging message arrives, then the mobile goes back to sleep. That cycle continues until one of the following conditions is met:

$$\text{Either } S_{\text{rxlev}} \leq S_{\text{IntraSearchP}}$$
$$\text{Or } S_{\text{qual}} \leq S_{\text{IntraSearchQ}} \tag{17.1}$$

In this equation, S_{rxlev} and S_{qual} are derived from the synchronization signal reference signal received power (SS-RSRP) and reference signal received quality (SS-RSRQ) for the serving cell, as defined in Equations (13.1) and (13.3). $S_{\text{IntraSearchP}}$ and $S_{\text{IntraSearchQ}}$ are thresholds which the serving cell advertises as part of SIB 2.

As in the state of RRC_CONNECTED, the specifications define how the mobile should convert its beam-specific measurements into a cell-specific quantity. As part of its system information, the serving cell configures the mobile with a threshold value of SS-RSRP. If the highest beam-specific measurement lies below the threshold, then the mobile simply uses that as the cell-specific value. Otherwise, the mobile averages the beam-specific measurements that lie above the threshold, subject to a limit on the number of beams to consider.

17.2.3 Intra-frequency Cell Reselection

If either of the conditions in Equation (17.1) is met, then the mobile starts to measure 5G neighbours whose SS/PBCH blocks have the same radio frequency as those in the serving cell, and the same subcarrier spacing. In the absence of any other information, the mobile has to search for those cells using the acquisition procedure from Chapter 8. However, the base station can help the mobile by advertising three extra parameters as part of SIB 2, namely the period with which the neighbours are transmitting their SS/PBCH blocks (5–160 ms), their timing offset within that period and a bitmap that identifies the precise SS/PBCH blocks being used.

Once it has done so, the mobile computes the following ranking scores for the serving cell and each of its neighbours:

$$R_s = Q_{\text{meas, s}} + Q_{\text{hyst}} - Q_{\text{offset, temp, s}}$$
$$R_n = Q_{\text{meas, n}} - Q_{\text{offset, s, n}} - Q_{\text{offset, temp, n}} \tag{17.2}$$

where R_s and R_n are the ranking scores of the serving cell and one of its neighbours, and $Q_{\text{meas, s}}$ and $Q_{\text{meas, n}}$ are the corresponding measurements of SS-RSRP. Q_{hyst} is a hysteresis parameter with a value of 0 to 24 dB, which the base station advertises in SIB 2 so as to discourage the mobile from bouncing between cells as the signal levels fluctuate. $Q_{\text{offset, s, n}}$ is a cell-specific offset, which the base station can optionally advertise in SIB 3 so as to discourage the mobile from selecting an individual neighbour. $Q_{\text{offset, temp, s}}$ and $Q_{\text{offset, temp, n}}$ are usually zero, and are only set after repeated failures to establish communications with the corresponding cell.

The mobile now chooses the best neighbouring cell in a three-step process. The mobile first identifies the neighbouring cells whose ranking score R_n exceeds a threshold that lies 0 to 24 dB below R_s. From those cells, it identifies the ones in which the largest number of beams have values of SS-RSRP that are above the beam-specific threshold from Section 17.2.2. From the cells that remain, the mobile chooses the one with the highest ranking score.

The mobile then reselects to the best neighbouring cell, provided that three constraints are met. Firstly, the new cell must have a higher ranking score than the serving cell for a time of at least $T_{\text{reselection, NR}}$, which has a value of 0 to 7 seconds. Secondly, the new cell must be suitable according to the criteria laid out in Chapter 13. (In particular, the new cell must belong to the registered public land mobile network [PLMN], and must not be barred.) Finally, the mobile must have been camped on the serving cell for at least one second.

17.2.4 Inter-frequency Measurement Triggering

If the network is using multiple radio frequencies, then it advertises those frequencies as part of its system information. SIB 4 contains the SS/PBCH frequencies and subcarrier spacings of the serving cell's 5G neighbours. Optionally, it can also contain the extra parameters from Section 17.2.3, which are the same for all the cells on a particular frequency. SIB 5 contains the centre frequencies of the serving cell's 4G neighbours.

Each frequency is associated with an integer priority in the range 0 to 7, where 7 is the highest, and with an optional priority offset in units of 0.2. Two frequencies can have the same priority if they are using the same radio access technologies, but must have different priorities if the radio access technologies are different. The network supplies default frequency priorities to the mobile as part of SIBs 2, 4 and 5, but can override those during the RRC Release procedure from Chapter 16.

The aim now is to drive the mobile to the highest-priority frequency that it can find. To help achieve this, the mobile always makes inter-frequency measurements of higher-priority frequencies, no matter how strong the signal from the serving cell. It does so at different times from the rest of the discontinuous reception cycle, as it cannot listen for paging messages on one carrier and measure cells on another at the same time. However, the mobile only has to measure one other carrier frequency every minute, so the load on the mobile is small.

The mobile starts to measure inter-frequency neighbours on equal or lower-priority carriers if one of the following conditions is met:

$$\text{Either } S_{\text{rxlev}} \leq S_{\text{InterSearchP}}$$
$$\text{Or } S_{\text{qual}} \leq S_{\text{InterSearchQ}} \tag{17.3}$$

where $S_{\text{InterSearchP}}$ and $S_{\text{InterSearchQ}}$ are thresholds, which the base station advertises as part of SIB 2. They are typically lower than the corresponding intra-frequency thresholds to ensure that the measurements only begin if intra-frequency reselection is not succeeding.

17.2.5 Inter-frequency Cell Reselection

Now let us move to the cell reselection process itself. The mobile can move to a cell on a higher-priority carrier if one or other of the following conditions is met, depending on a choice that is configured in the serving cell's system information:

$$\text{Either } S_{\text{rxlev, x, n}} > \text{Thresh}_{\text{x, HighP}}$$

$$\text{Or } S_{\text{qual, x, n}} > \text{Thresh}_{\text{x, HighQ}} \tag{17.4}$$

where $\text{Thresh}_{\text{x, HighP}}$ and $\text{Thresh}_{\text{x, HighQ}}$ are thresholds for frequency x that the serving cell advertises in SIBs 4 and 5. In the case of a 5G neighbour, the mobile derives $S_{\text{rxlev, x, n}}$ and $S_{\text{qual, x, n}}$ from the neighbour's values of SS-RSRP and SS-RSRQ, in the manner defined in Equations (13.1) and (13.3). In the case of a 4G neighbour, it derives them from the equivalent measurements for LTE [8]. The comparison does not depend on the signal received from the serving cell, so the mobile moves to a higher-priority frequency whenever it finds a neighbour that is good enough.

The mobile can move to an equal-priority carrier using nearly the same criteria that it did for the same carrier frequency. The only difference is in Equation (17.2), where the serving cell can apply an additional frequency-specific offset, denoted $Q_{\text{offset, frequency}}$, to the neighbour's ranking score R_n.

The mobile can move to a lower-priority carrier if one or other of the following conditions is met, depending once again on the choice configured by the serving cell:

$$\text{Either } S_{\text{rxlev}} < \text{Thresh}_{\text{Serving, LowP}} \quad \text{and} \quad S_{\text{rxlev, x, n}} > \text{Thresh}_{\text{x, LowP}}$$

$$\text{Or } S_{\text{qual}} < \text{Thresh}_{\text{Serving, LowQ}} \quad \text{and} \quad S_{\text{qual, x, n}} > \text{Thresh}_{\text{x, LowQ}} \tag{17.5}$$

As before, the values of S_{rxlev}, S_{qual}, $S_{\text{rxlev, x, n}}$ and $S_{\text{qual, x, n}}$ are derived from measurements of the serving and neighbouring cells, while the other parameters are thresholds that the serving cell advertises in SIBs 4 and 5. If the thresholds are properly configured, then the mobile only moves to a lower-priority carrier if it cannot find a satisfactory cell on the original frequency, or on another frequency with an equal or higher priority.

In each of these cases, the mobile only reselects to the new cell if three additional constraints are met. Firstly, the above conditions must apply for a time of at least $T_{\text{reselection, RAT}}$, where RAT denotes the radio access technology of the neighbouring cell. Secondly, the new cell must be suitable. Finally, the mobile must have been camped on the serving cell for at least one second.

17.2.6 Fast-moving Mobiles

As designed, the cell reselection algorithm includes time delays and hysteresis parameters that encourage the mobile to remain in the serving cell, so as to reduce the number of unnecessary reselections if signal levels fluctuate. However, the usual values of those parameters are too large for fast-moving mobiles.

To deal with the problem, a mobile measures the number of cell reselections that it makes during a time period that lies between 30 and 240 seconds. Based on that number, the mobile places itself into a normal, medium or high-mobility state. In the normal-mobility state, the basic cell reselection algorithm is unchanged. In the medium and high-mobility states, the mobile reduces the hysteresis parameter Q_{hyst} by up to 6 dB, and divides the reselection times $T_{reselection, RAT}$ by a factor of up to 4. The serving cell advertises all the required parameters as part of SIB 2.

17.3 Registration Updating

17.3.1 Registration Update Procedure

After the mobile reselects to a new cell, it reads the new cell's version of SIB 1, which includes its tracking area. If that tracking area lies outside the mobile's tracking area list, then the mobile informs its serving access and mobility management function (AMF) by means of a registration update [9]. The mobile runs the same procedure if its periodic registration update timer expires, to assure the serving AMF that it is still switched on. The timer has a default of 54 minutes and a range of 2 seconds to 310 hours, with the longer values intended for machine-type devices that require low power consumption.

Figure 17.2 shows the simplest version of the procedure. The figure assumes that the mobile begins in RRC_IDLE and that the AMF remains unchanged, and is a simplified version of the initial registration procedure from Chapter 13. The mobile begins with the usual procedures for random access and RRC connection establishment, and then contacts the AMF by means of a Registration Request (steps 1–3). The mobile requests either a mobility

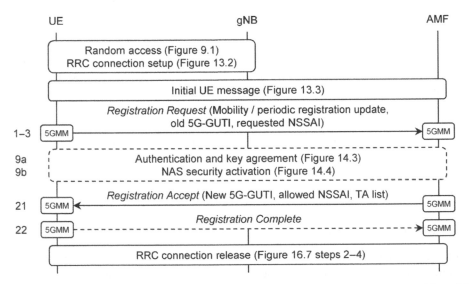

Figure 17.2 Registration update procedure if the AMF is unchanged. Source: Adapted from 3GPP TS 23.501.

or a periodic registration update by means of an information element, but the message is otherwise the same as for an initial registration.

Optionally, the AMF authenticates the mobile and updates the non-access stratum security keys (9a and 9b). It then accepts the mobile's registration request (21) and provides it with a new tracking area list and, usually, a new temporary identity. The new identity is mandatory in the case of a mobility registration update, and is acknowledged by the mobile (22). At this point, the mobile is still in the state of RRC_CONNECTED, so the AMF instructs the serving node to return the mobile to RRC_IDLE.

If the mobile moves to a tracking area that is outside its previous AMF set, then the registration update procedure triggers a change of AMF, by analogy with Figure 13.5. If the selected AMF does not support the mobile's requested network slices, then the procedure can also trigger AMF re-allocation, by analogy with Figure 13.6.

17.3.2 Network Reselection

If a mobile is configured for automatic network selection and is roaming in a visited network, then it periodically runs a procedure for *network reselection* whenever it is in RRC_IDLE [10]. During that procedure, the mobile searches for networks with the same mobile country code as the registered PLMN, which the Universal Subscriber Identity Module (USIM) identifies either as home networks, or as networks in the user or operator-defined lists that have a higher priority than the current one. The search period is stored on the USIM as well, and has a default value of 60 minutes [11].

If the mobile does find a new network, then the resulting procedure is the same as for initial network and cell selection, except that the parameters $Q_{rxlevminoffset}$ and $Q_{qualminoffset}$ from Equations (13.1) and (13.3) are now non-zero. They act as hysteresis parameters, which encourage the mobile to remain in the registered PLMN and discourage it from bouncing back and forth as the signal levels fluctuate.

17.4 State Transitions out of RRC_IDLE

17.4.1 Mobile-triggered Service Request

The mobile invokes the *Service Request* procedure if it wishes to transmit data on an inactive PDU session. Figure 17.3 shows what happens if the mobile is initially in the state of RRC_IDLE [12]. The procedure moves the mobile into RRC_CONNECTED and re-activates one or more of its PDU sessions.

The mobile begins with the usual procedures for random access and RRC connection establishment, and then contacts its serving AMF by means of a *Service Request* (steps 1–2). The message includes the reason for the request, in this case mobile-originated data, and lists the PDU session(s) for which uplink data are pending. Optionally, the AMF can authenticate the mobile and update its non-access stratum security keys (3).

For each PDU session, the AMF identifies the relevant session management function (SMF) and asks for the PDU session to be re-activated (4). In its response (11), the SMF supplies a number of NG application protocol (NG-AP) information elements for delivery to

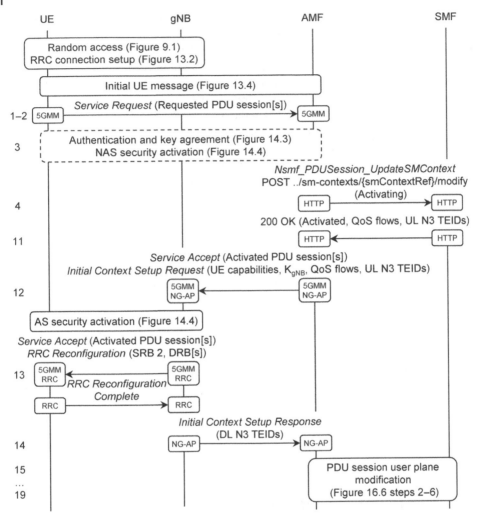

Figure 17.3 Mobile-triggered service request procedure, for a mobile initially in RRC_IDLE. Source: Adapted from 3GPP TS 23.501.

the master node. These include the parameters of the PDU session's QoS flows (for example the 5G QoS identifier), and the IP address and uplink tunnel endpoint identifier (TEID) of the UPF that terminates N3.

The AMF collects all of the replies, and instructs the mobile's master node to establish an NG signalling connection for the mobile (12). The message includes the mobile's radio access capabilities, access stratum security key and core network assistance information, as well as the information elements that it received from the mobile's SMFs. It also includes an embedded *Service Accept* message, which identifies the re-activated PDU sessions. The master node initializes the mobile's access stratum security, reconfigures the mobile's air interface to establish its data radio bearers and signalling radio bearer 2 (SRB 2), and forwards the Service Accept message to the mobile (13).

On receiving the mobile's acknowledgement, the master node responds to the AMF, and includes an IP address and a downlink TEID for each of the mobile's PDU sessions (14). The AMF sends that information to the relevant SMF, which forwards it to the UPF for use during downlink data delivery (15–19). The PDU sessions have now been re-activated, and data can be delivered over the uplink and downlink.

The procedure is also used in the state of RRC_CONNECTED if the mobile wishes to transmit data but the relevant PDU session(s) are inactive. In that procedure, the mobile is already in contact with its serving AMF, so it can send its Service Request right away. Message (12) is replaced by an NG-AP *PDU Session Resource Setup Request*, while message (13) omits the configuration of SRB 2. In addition, the access stratum security keys are only updated if the network authenticated the mobile in step 3. Otherwise, the procedure is unchanged.

17.4.2 Network-triggered Service Request

The Service Request procedure can also be invoked by the network if downlink data arrive on an inactive PDU session. Figure 17.4 shows the procedure [13], once again for a mobile that is initially in RRC_IDLE.

The procedure begins when downlink data arrive on an inactive PDU session at the UPF that is furthest downstream (step 1). The UPF cannot deliver the data any further, so it alerts its SMF by means of a *PFCP Session Report Request* (2). The message includes a *packet detection rule* (PDR) identifier, which identifies the corresponding mobile, the PDU session

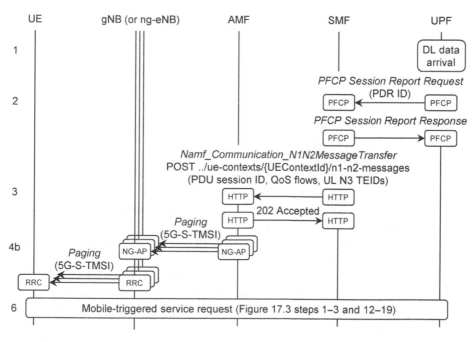

Figure 17.4 Network-triggered service request procedure, for a mobile initially in RRC_IDLE. Source: Adapted from 3GPP TS 23.501.

and the QoS flow. The SMF's acknowledgement suppresses the delivery of further messages, if any more downlink data arrive. In turn, the SMF alerts the AMF (3), identifies the PDU session and includes the same NG-AP information elements that it used in step 11 of Figure 17.3's mobile-triggered procedure.

If the mobile is in RRC_IDLE, then the AMF cannot forward the NG-AP information right away, so its response indicates that the information was accepted but not yet acted upon. Instead, the AMF instructs the potential master nodes in the mobile's registration area to page for the mobile, identifies the mobile by means of its 5G S temporary mobile subscriber identity (5G-S-TMSI) and includes all the information needed to calculate the paging frame and paging occasion (4b). In turn, the nodes page for the mobile.

The mobile receives the paging message and responds by running the mobile-triggered Service Request procedure from earlier (6). There are a few modifications. The mobile does not know which PDU session triggered the procedure, so it omits that information from its service request, and states that the reason is a response to mobile-terminated paging. The AMF already has the information that it needs from the SMF, so it can immediately accept the mobile's request and configure the radio access network.

The procedure is also used for a mobile in RRC_CONNECTED if data arrive on an inactive PDU session. On receiving the SMF's alert, the AMF does not have to page for the mobile, so it can configure the radio access network right away. The Service Request and Service Accept messages are omitted as well.

References

1 3GPP TS 38.300 (2019) NR; NR and NG-RAN overall description; Stage 2 (Release 15), December 2019.

2 3GPP TS 23.502 (2019) Procedures for the 5G system (5GS); Stage 2 (Release 15), December 2019.

3 3GPP TS 38.304 (2019) NR; User equipment (UE) procedures in Idle mode and RRC Inactive state (Release 15), December 2019.

4 3GPP TS 23.501 (2019) System architecture for the 5G system (5GS); Stage 2 (Release 15), December 2019, Sections 5.6.8, 5.8.2.10.

5 3GPP TS 38.304 (2019) NR; User equipment (UE) procedures in idle mode and RRC inactive state (Release 15), December 2019, Sections 5.2.1, 5.2.4.

6 3GPP TS 38.133 (2019) NR; Requirements for support of radio resource management (Release 15), December 2019, Section 4.

7 3GPP TS 38.300 (2019) NR; NR and NG-RAN overall description; Stage 2 (Release 15), December 2019, Section 9.2.4.

8 3GPP TS 36.304 (2019) Evolved Universal Terrestrial Radio Access (E-UTRA); User equipment (UE) procedures in idle mode (Release 15), December 2019, Section 5.2.4.

9 3GPP TS 23.502 (2019) Procedures for the 5G system (5GS); Stage 2 (Release 15), December 2019, Section 4.2.2.2.2.

10 3GPP TS 23.122 (2019) Non-access-stratum (NAS) functions related to mobile station (MS) in idle mode (Release 15), March 2019, Section 4.4.3.2.

11 3GPP TS 31.102 (2019) Characteristics of the Universal Subscriber Identity Module (USIM) application (Release 15), December 2019, Section 4.2.6.

12 3GPP TS 23.502 (2019) Procedures for the 5G system (5GS); Stage 2 (Release 15), December 2019, Section 4.2.3.2.

13 3GPP TS 23.502 (2019) Procedures for the 5G system (5GS); Stage 2 (Release 15), December 2019, Section 4.2.3.3.

18

Mobility Management in RRC_INACTIVE

This chapter addresses the last of 5G's mobility management states, RRC_INACTIVE. This state is designed for mobiles that are communicating with a low data rate. It is only available if the mobile is controlled by the 5G core network and the next-generation radio access network (NG-RAN), although the mobile can be served by either a gNB or an ng-eNB.

It is easiest to understand the state of RRC_INACTIVE by thinking about what happens if the nodes are split into central and distributed units. As part of this chapter, we will take the opportunity to investigate the signalling procedures inside a split gNB, both on the F1 reference point between the gNB central unit (gNB-CU) and the gNB distributed unit (gNB-DU), and on the E1 reference point between the gNB central unit control plane (gNB-CU-CP) and the gNB central unit user plane (gNB-CU-UP).

The main specifications for this chapter overlap with the ones used in the states of RRC_CONNECTED and RRC_IDLE [1–4]. In addition, TS 38.401 [5] contains some valuable examples of the signalling flows inside a split gNB.

18.1 Introduction to RRC_INACTIVE

18.1.1 Principles

The state of RRC_INACTIVE is an optimization for mobiles that are communicating with a low data rate, such as machine-type devices. It uses the most appropriate features of RRC_CONNECTED and RRC_IDLE, so as to limit the mobile's power consumption in between bursts of data, and to limit the signalling overhead once data eventually arrive.

When a mobile enters the state of RRC_INACTIVE, its distributed units tear down their relationships with the mobile, but its central units retain them. In particular, the mobile can still be in a state of dual connectivity towards the central units of a master node (MN) and a secondary node (SN). Consistent with that picture, the mobile's physical layer, medium access control (MAC) and radio link control (RLC) protocols are all released, but its packet data convergence protocol (PDCP) and service data adaptation protocol (SDAP) are both retained. The same principles apply in the case of an integrated gNB or ng-eNB, except that the distributed units are no longer separate physical devices.

Once the mobile is in the state of RRC_INACTIVE, the mobile and the central units store some of the parameters that they used for communications over the air interface, but

An Introduction to 5G: The New Radio, 5G Network and Beyond, First Edition. Christopher Cox.
© 2021 John Wiley & Sons Ltd. Published 2021 by John Wiley & Sons Ltd.

not all of them. For example, they store the air interface's initial security key K_{gNB} and the key K_{RRCInt} that protects the integrity of radio resource control (RRC) signalling, but they delete their other security keys. The stored information is just enough for the mobile to return to RRC_CONNECTED without the need for any signalling communications with the core, but no more.

From the viewpoint of the core network, the mobile is in the state of CM-CONNECTED. If it wishes to do so, then the access and mobility management function (AMF) can ask the master node to keep it informed about the mobile's transitions between RRC_CONNECTED and RRC_INACTIVE, but it does not distinguish the two states otherwise.

The procedures for mobility management follow the same principles as in RRC_IDLE. The mobile maximizes its battery life by entering a state of discontinuous reception, in which it wakes once every discontinuous reception cycle to listen for paging messages and measure the strength of the signal from the serving cell. Using the cell reselection procedures from Chapter 17, the mobile can move throughout its RAN-based notification area (RNA) without the need to inform the network.

18.1.2 Suspended PDU Sessions

If the mobile is in the state of RRC_INACTIVE, then each individual PDU session is either inactive, as described in Chapter 17, or *suspended*. In a suspended PDU session (Figure 18.1), the N9 and N3 tunnels remain in place, but the Xn and F1 tunnels are torn down. The data radio bearers are also in a suspended state, in which they retain their identities as MN or SN-terminated bearers, but lose their identities as master cell group (MCG), secondary cell group (SCG) or split bearers.

If a downlink data packet arrives, then the core network can deliver it to the central unit of the master or secondary node. In response, the NG-RAN sends a paging message through all the possible serving cells in the mobile's RAN-based notification area, before returning the mobile to RRC_CONNECTED and reactivating the corresponding PDU session. The

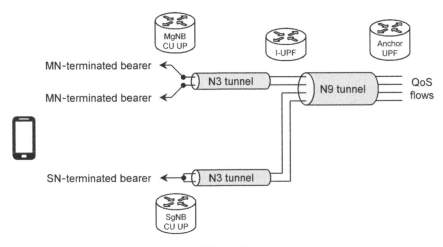

Figure 18.1 Example of a suspended PDU session.

resulting procedure requires little or no signalling between the NG-RAN and the core, so the data can be delivered with little overhead.

18.2 Mobility Management

18.2.1 RAN-based Notification Area Update

When the mobile enters RRC_INACTIVE, the master node assigns it an *inactive radio network temporary identifier* (I-RNTI). This is a 40-bit quantity that identifies both the mobile and the master node, but its exact format is a matter for the network operator [6]. Depending on a choice that is signalled within system information block 1 (SIB 1), the mobile subsequently identifies itself to the network using either the full I-RNTI or a shorter 24-bit version. The latter choice might be suitable if a single central unit controls the whole of the mobile's RAN-based notification area, removing the need to identify the master node.

If a mobile moves outside its RNA, then it alerts its master node by means of the *RAN-based notification area update* procedure [7]. The mobile also uses the procedure on the expiry of a periodic RNA update timer, whose duration is between 5 and 720 minutes. That has two purposes: it assures the master node that the mobile is still switched on and within coverage, and it can trigger a state transition to RRC_IDLE. Figure 18.2 shows the procedure for the case where the mobile is served by a gNB that is split into central and distributed units, there is no dual connectivity, but the central unit changes.

The mobile begins by asking the NG-RAN to resume its RRC connection (step 1). It does so by sending one of two messages, depending on the choice signalled within SIB 1: either *RRC Resume Request*, which includes the short I-RNTI, or *RRC Resume Request 1*, which includes the full version. The message also includes the reason for the request, in this case an RNA update.

The distributed unit processes the incoming message using its physical layer, MAC and RLC protocols. If the distributed unit is split, then it forwards the partly processed message to the central unit, by embedding it into an *Initial UL RRC Message Transfer* that is written using the F1 application protocol (F1-AP). The message requests the establishment of an F1 signalling connection for the mobile, and supplies a 32-bit field that the distributed unit will use to identify the mobile over F1.

If the central unit has changed, then the new central unit extracts the identity of the old one from the mobile's I-RNTI, and asks it to return the details of the mobile (2). In this example, the old central unit decides to hand over control of the mobile, and responds (3). The response includes definitions of the mobile's PDU sessions, security keys and serving AMF.

We will now assume that the new central unit decides to keep the mobile in the state of RRC_INACTIVE (4). Optionally, it sends the old one an IP address and tunnel endpoint identifier (TEID) for each of the mobile's PDU sessions, so that the old central unit can forward any downlink data that arrive (5). As a long-term solution, the new central unit tells the AMF that it is now acting as the mobile's master node, and includes IP addresses and TEIDs for the delivery of any future downlink data over N3. The AMF sends that information to the mobile's session management functions (SMFs) and user plane functions (UPFs), and responds (6–7).

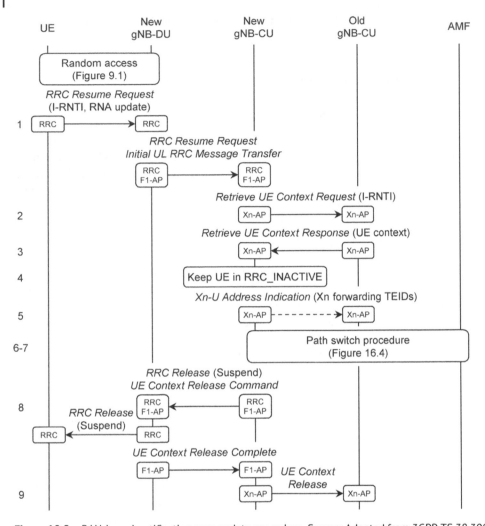

Figure 18.2 RAN-based notification area update procedure. Source: Adapted from 3GPP TS 38.300.

The new central unit now composes an *RRC Release* message for the mobile (8). As part of the message, a suspend configuration tells the mobile to return to RRC_INACTIVE rather than RRC_IDLE, and includes a new I-RNTI, RNA and periodic update timer. If the node is split, then the central unit delivers the message by embedding it into an F1-AP *UE Context Release Command*, which tells the distributed unit to release the F1 identity that it allocated earlier. The distributed unit does so, and responds. To conclude the procedure, the new central unit tells the old one to release its communications with the mobile (9).

As an alternative to this procedure, the old central unit might decide to retain control of the mobile at the end of step 2. That might, for example, be appropriate in the case of a periodic RNA update, if the mobile has remained within the notification area. In that situation, the old central unit sends an error response to the new one with an embedded RRC Release message, which is forwarded to the mobile.

18.2.2 Registration Update

If the mobile moves outside its registration area, then the request for a registration update takes priority over the RNA update [8]. The mobile sends an RRC Resume Request to the NG-RAN, as before, but this time indicates that the reason is mobile-originated signalling. The same thing happens if the periodic registration update timer expires.

The new central unit asks the old one to return the details of the mobile, as before (step 2). There are now two possibilities. If the AMF can remain unchanged, then step 3 succeeds, and the central unit resumes the mobile's old RRC connection in the manner described in Section 18.3.2. If the AMF has to change, then step 3 fails, and the central unit establishes a new RRC connection by means of an RRC Setup message [9].

In both cases, the mobile is now in the state of RRC_CONNECTED and can continue with the registration update procedure from Chapter 17. At the end of the procedure, the core network can either move the mobile to CM-IDLE or leave it in CM-CONNECTED. In the latter case, the master node can either leave the mobile in RRC_CONNECTED or return it to RRC_INACTIVE.

18.2.3 Mobility between a gNB and an ng-eNB

The 3GPP specifications do not support transitions between the 4G and 5G states of RRC_INACTIVE. If a mobile is in RRC_INACTIVE and reselects to a cell belonging to the other radio access technology, then it moves to RRC_IDLE, and discards its air interface configuration parameters and its I-RNTI [10, 11]. It then follows the procedures for RRC_IDLE from Chapter 17.

However, the mobile does not inform the network, which still believes that the mobile is in RRC_INACTIVE. If downlink data arrive at the mobile's central unit, then the paging procedure introduced in Section 18.3.3 will fail because the mobile has lost its I-RNTI. Optionally, depending upon its internal configuration, the master node can now trigger the access network release procedure from Chapter 16 [12]. In response, the core network transfers the mobile to CM-IDLE and can page for the mobile throughout its registration area.

18.3 State Transitions

18.3.1 Transition to RRC_IDLE

If the mobile has been in RRC_INACTIVE for a while without sending or receiving any data, then the network can transfer it to RRC_IDLE. The procedure is typically triggered when the mobile sends its next RNA update request after the expiry of an internal timer in the master node. Figure 18.3 shows the signalling messages that result. The diagram assumes that there is no dual connectivity, as before, but this time assumes that the gNB central units are themselves split into separate control and user planes.

The procedure begins in the same way as for a normal RNA update (step 1). This time, however, the central unit asks the AMF to transfer the mobile to RRC_IDLE by requesting the access network release procedure from Chapter 16 (2). On receiving the AMF's

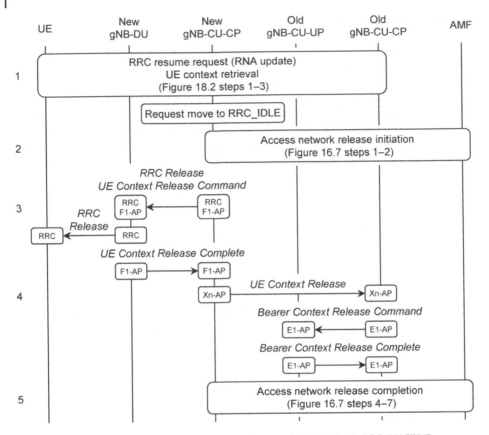

Figure 18.3 Access network release procedure for a mobile initially in RRC_INACTIVE.

response, the master node composes an RRC Release message that omits the suspend configuration, which tells the mobile to move to RRC_IDLE. The message is delivered by way of the distributed unit, as before (3).

The new central unit now tells the old one to release its communications with the mobile (4). If the old central unit is split, then the control plane tells the user plane to release its end of the mobile's N3 tunnels by means of a *Bearer Context Release Command* that is written using the E1 application protocol (E1-AP). (If the initial state were RRC_CONNECTED, then that command would also release the user plane's end of the F1 tunnels.) To complete the procedure, the new central unit sends an acknowledgement to the AMF, which initiates release of the core network's end of the mobile's N3 tunnels (5).

18.3.2 Mobile-triggered Resumption of the RRC Connection

If a mobile is in RRC_INACTIVE and wishes to send data on a suspended PDU session, then it asks the NG-RAN to resume its RRC connection and return it to the state of RRC_CONNECTED. The procedure is shown in Figure 18.4, assuming once again that the central units are split into separate control and user planes, there is no dual connectivity, but the central unit changes [9, 13].

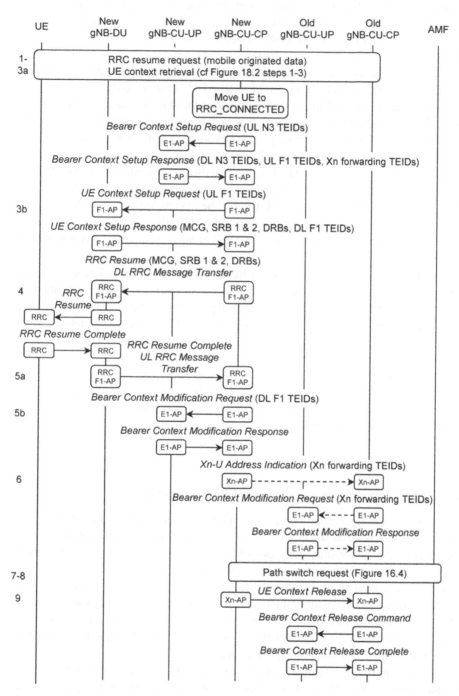

Figure 18.4 Mobile-triggered resumption of the RRC connection. Source: Adapted from 3GPP TS 38.300 and TS 38.401.

To begin the procedure, the mobile sends an RRC Resume Request as before, but this time states that the reason is mobile-originated data (steps 1–3a). If the central unit has changed, then the new central unit retrieves the mobile's details from the old one.

Within the new central unit, the control plane initializes the user plane by means of an E1-AP *Bearer Context Setup Request*. It then tells the distributed unit to establish an F1 signalling connection for the mobile, by means of an F1-AP *UE Context Setup Request* (3b). As part of that message, the central unit tells the distributed unit to establish signalling radio bearers 1 and 2. It also supplies the information needed to resume the suspended data radio bearers, including their quality-of-service parameters, and the TEIDs for uplink data delivery over F1. The distributed unit's response includes the air interface configuration for the master cell group and the requested radio bearers, and the corresponding downlink TEIDs.

The central unit now sends the air interface configuration to the mobile as part of the signalling message *RRC Resume* (4), which the mobile acknowledges using the message *RRC Resume Complete*. On receiving the mobile's reply (5a), the central unit instructs its user plane to establish its own end of the mobile's F1 tunnels, using the downlink TEIDs that it received earlier (5b).

The procedure concludes in a similar way to the one for an RNA update. The new central unit sends IP addresses and TEIDs to the old one, for use in downlink data forwarding over Xn (6). If the old central unit is split, then the control plane passes the relevant information to the user plane. The new central unit tells the AMF that it is acting as the mobile's master node and redirects the mobile's N3 tunnels (7–8), and then tells the old central unit to release its communications with the mobile (9).

If the mobile is in RRC_INACTIVE and wishes to send data on an inactive PDU session, then a different procedure is required. As part of the message RRC Resume Complete, the mobile embeds a 5G mobility management (5GMM) Service Request, which the central unit forwards to the AMF. The procedure then continues in a similar way to the Service Request from Chapter 17.

18.3.3 Network-triggered Resumption of the RRC Connection

If downlink data arrive for a mobile on a suspended PDU session, then the core network can send the packets as far as the central unit's user plane. However, that device cannot send the packets any further, due to the lack of any F1 tunnels for the mobile. Instead, it initiates the paging procedure shown in Figure 18.5 [13, 14].

Triggered by the arrival of downlink data, the central unit's user plane sends an E1-AP *DL Data Notification* to its control plane (step 1). The mobile might be anywhere inside the RAN-based notification area, so the control plane asks all the other central units in the RNA to page for the mobile, and includes the mobile's I-RNTI and the information needed to calculate the paging frame and paging occasion (2).

In turn, each central unit instructs its distributed units to page for the mobile (3), and the distributed units do so. The mobile receives the paging message and responds using the mobile-triggered procedure from Section 18.3.2, stating this time that the cause is mobile-terminated paging (4). Once the F1 tunnels are re-established, the network can deliver the downlink data to the mobile.

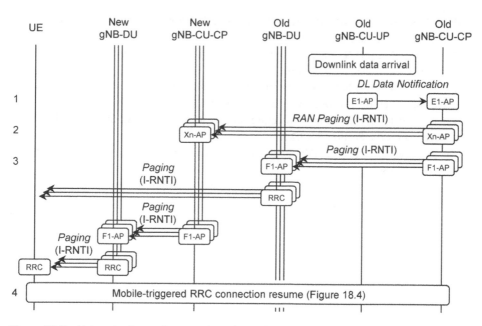

Figure 18.5 Network-triggered resumption of the RRC connection. Source: Adapted from 3GPP TS 38.300 and TS 38.401.

If the mobile is in dual connectivity, then downlink data can also arrive at the user plane of the secondary node's central unit. The user plane alerts the control plane in step 1, as before. In turn, the control plane alerts the master node's central unit by means of an Xn-AP Activity Notification, and the procedure continues from step 2.

References

1 3GPP TS 38.300 (2019) NR; NR and NG-RAN overall description; Stage 2 (Release 15), December 2019.

2 3GPP TS 23.502 (2019) Procedures for the 5G system (5GS); Stage 2 (Release 15), December 2019.

3 3GPP TS 37.340 (2019) Evolved Universal Terrestrial Radio Access (E-UTRA) and NR; Multi-connectivity; Stage 2 (Release 15), December 2019.

4 3GPP TS 38.304 (2019) NR; User equipment (UE) procedures in idle mode and RRC inactive state (Release 15), December 2019.

5 3GPP TS 38.401 (2019) NG-RAN; Architecture description (Release 15), December 2019.

6 3GPP TS 38.300 (2019) NR; NR and NG-RAN overall description; Stage 2 (Release 15), December 2019, Annex C.

7 3GPP TS 38.300 (2019) NR; NR and NG-RAN overall description; Stage 2 (Release 15), December 2019, Section 9.2.2.5.

8 3GPP TS 38.331 (2019) NR; Radio resource control (RRC) protocol specification (Release 15), December 2019, Section 5.3.13.2.

9 3GPP TS 38.300 (2019) NR; NR and NG-RAN overall description; Stage 2 (Release 15), December 2019, Section 9.2.2.1, 9.2.2.4.1.

10 3GPP TS 24.501 (2019) Non-access-stratum (NAS) protocol for 5G system (5GS); Stage 3 (Release 15), December 2019, Section 5.3.1.4.

11 3GPP TS 38.304 (2019) NR; User equipment (UE) procedures in idle mode and RRC inactive state (Release 15), December 2019, Section 5.2.4.8.

12 3GPP TS 23.501 (2019) System architecture for the 5G system (5GS); Stage 2 (Release 15), December 2019, Section 5.3.3.2.5.

13 3GPP TS 38.401 (2019) NG-RAN; Architecture description (Release 15), December 2019, Sections 8.6.2, 8.9.6.2.

14 3GPP TS 38.300 (2019) NR; NR and NG-RAN overall description; Stage 2 (Release 15), December 2019, Section 9.2.2.4.2.

19

Inter-operation with the Evolved Packet Core

The evolved packet core (EPC) will operate alongside the 5G core network for several years, so that network operators can migrate from one system to the other, and so that LTE devices can continue using the older system. Such migration will be gradual: for example, a network operator is unlikely to add the NG reference point to all of its LTE base stations at the same time. That can lead to a situation in which a mobile is registered with the 5G core network, but moves to a geographical area in which the base stations can only communicate with the EPC. At that point, the system has to transfer the mobile from one core network to the other.

In this chapter, we will discuss these issues by looking at the inter-operation architectures that the 5G system supports and the signalling procedures that result. We will assume some previous knowledge of the evolved packet core, but we will review its most important features as we go along.

19.1 Inter-operation Architectures

19.1.1 Migration Architecture

The specifications define two inter-operation architectures, the first of which is illustrated in Figure 19.1 [1]. In this architecture, the two core networks share a single subscriber database, which acts as unified data management (UDM) towards the 5G core network, and as a home subscriber server (HSS) towards the EPC. The radio access networks can overlap in the usual way, with a base station lying in the next-generation radio access network (NG-RAN) if it supports the NG backhaul towards the 5G core, or the evolved UMTS terrestrial radio access network (E-UTRAN) if it supports the S1 backhaul towards the EPC, or both. The diagram only shows base stations which support the signalling backhaul and which can act as master nodes.

Using this architecture, a mobile can register with the network using the three mechanisms that we discussed in Chapter 13: with a master gNB, the NG-RAN and the 5G core (architectural options 2 and 4); with a master ng-eNB, the NG-RAN and the 5G core (options 5 and 7); and with a master eNB, the E-UTRAN and the EPC (options 1 and 3). By implication, the mobile can also de-register from one core network and re-register through the other, for example if it leaves the coverage area of the NG-RAN. Furthermore, the database might also act as a *home location register* (HLR)

An Introduction to 5G: The New Radio, 5G Network and Beyond, First Edition. Christopher Cox.
© 2021 John Wiley & Sons Ltd. Published 2021 by John Wiley & Sons Ltd.

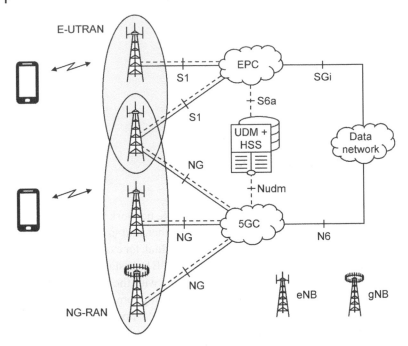

Figure 19.1 Architecture for migration from the evolved packet core. Source: Adapted from 3GPP TS 23.501.

towards the core networks of 2G and 3G. If it does, then the mobile can re-register through those networks as well.

When using this architecture, however, the network cannot transfer any of the mobile's PDU sessions from one network to another. If a mobile does re-register with a new core network, then its PDU sessions are all torn down, and the mobile loses its data network addresses and its communications with external servers. The mobile then has to re-establish its PDU sessions after re-registration.

19.1.2 Interworking Architecture

For more effective inter-operation, we require the architecture shown in Figure 19.2 [2]. As before, a single network function acts as a UDM and as an HSS. In addition, there are three other combined network functions: one for the session management function (SMF) and the PDN gateway control plane; one for the user plane function (UPF) and the PDN gateway user plane; and one for the policy control function (PCF) and the policy and charging rules function (PCRF). The access and mobility management function (AMF) and mobility management entity (MME) remain separate, but they can sometimes communicate over an optional reference point denoted N26.

Using this architecture, the system can transfer a mobile between the 5G core network and the EPC, and can also transfer any or all of its PDU sessions. When that happens, the mobile can retain the corresponding data network addresses and can maintain its communications with external servers. If the system implements the N26 reference point,

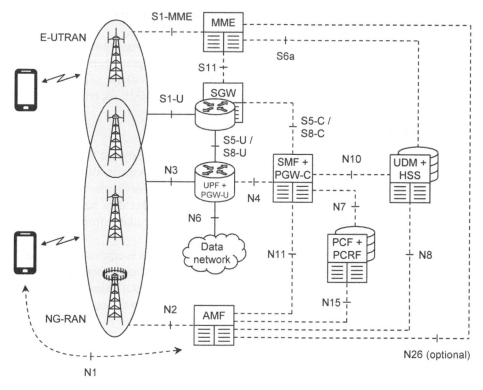

Figure 19.2 Architecture for interworking with the evolved packet core. Source: Adapted from 3GPP TS 23.501.

then it can also forward undelivered data from the old network to the new one to prevent packet loss.

The interworking architecture is especially valuable for mobiles that are roaming with home-routed traffic. In that scenario, the only components in the visited network are the E-UTRAN, MME and serving gateway from the evolved packet system (EPS): the 5G network functions lie in the home network alone. The N26 reference point is optional, so the two networks only need to communicate over the legacy S6a and S8 reference points from LTE. The architecture therefore allows a 5G subscriber to roam in a visited network that has no support for 5G at all.

Once the mobile is in the EPC, it can be transferred to the packet switched core network of 2G and 3G, and to a 2G or 3G cell. However, there is no guarantee that the combined 4G/5G network functions will support the equivalent roles from 2G and 3G, so there is no guarantee that the mobile's PDU sessions will be retained.

19.1.3 Signalling Protocols

The signalling protocols in the EPC are different from those of 5G. The MME communicates with the eNB using the *S1 application protocol* (S1-AP) [3]. It also communicates with the mobile using two non-access stratum protocols, known as *EPS mobility management*

(EMM) and *EPS session management* (ESM) [4]. On the S6a reference point, the MME communicates with the HSS using a *Diameter* application [5].

The other reference points in the EPC use the *GPRS tunnelling protocol version 2 control part* (GTPv2-C) [6]. That protocol is also used on the N26 reference point, between the AMF and the MME. The protocol was enhanced as part of Release 15 in order to support N26 but is otherwise little changed, so only limited enhancements to the MME are required.

19.1.4 State Diagrams

The state diagrams in the EPC are similar to those in 5G [7, 8]. Firstly, the EMM state diagram is equivalent to the 5G state diagram for registration management, although the 5G registration procedure is known as the *attach procedure* in LTE. More significantly, a mobile in Release 8 was always connected to one or more external packet data networks while attached to the EPC. Attach without PDN connectivity is only available from Release 13, and only if the MME supports it.

Secondly, the *EPS connection management* (ECM) state diagram is equivalent to the 5G state diagram for connection management, with no significant differences between the two. Thirdly, the E-UTRAN's state diagram for radio resource control (RRC) is similar to the state diagram used by an ng-eNB within the NG-RAN, but it only supports two states, namely RRC_CONNECTED and RRC_IDLE.

19.2 Registration Modes

19.2.1 Single Registration Mode

When using these inter-operation architectures, the mobile can operate in one of two registration modes. The usual choice is *single registration mode*, in which the mobile can be registered with either the 5G core network in a state known as *N1 mode*, or the EPC in a state known as *S1 mode*, but not both. If a mobile supports both types of core network, then support for single registration mode is mandatory. The mobile can then use that mode in conjunction with any of the inter-operation architectures discussed in this chapter.

If the mobile is in single registration mode, then the network usually maintains a single registration for the mobile as well. However, an exception arises if the AMF and MME both support the interworking architecture from Figure 19.2, but are not connected by way of N26. In that situation, the network maintains two independent registrations for the mobile: one with an AMF in the 5G core network, and one with an MME in the EPC. That possibility is included to support the use of dual registration mode in Section 19.2.2.

19.2.2 Dual Registration Mode

In *dual registration mode*, the mobile maintains two independent registrations, so it can be registered with either the 5G core network, or the EPC, or both. The network behaves in the same way. It is optional for the mobile to support dual registration mode, and it is only available when using the architecture for inter-operation without N26.

If the mobile is registered with both core networks, then it can also maintain two separate RRC connections, one with a master gNB in the NG-RAN, and one with a master eNB in the E-UTRAN. The mobile can then communicate through both radio access networks at the same time. There is a restriction, however: a dual-registered mobile is never served by a master ng-eNB in the NG-RAN, to ensure that it never has two separate RRC connections to the same node.

In dual registration mode, each individual PDU session is routed through either the 5G core network and the NG-RAN, or the EPC and the E-UTRAN, but not both. As a result, the mobile can simultaneously use the 5G core network and NG-RAN for some of its PDU sessions, and use the EPC and E-UTRAN for the remainder.

19.2.3   Temporary Identities

The EPC identifies a mobile using the LTE *globally unique temporary identity* (GUTI). In single registration mode, the mobile has either an LTE GUTI or a 5G-GUTI, but not both. It can then convert between the two using mapping rules that are defined in the specifications [9]. In dual registration mode, the mobile can have both types of GUTI.

19.3   Use of the Migration Architecture

19.3.1   Configuration Procedures

We will now look at the signalling procedures for inter-operation. We will address each of the architectures in turn, focussing mainly on mobility from the 5G core network to the EPC. We will begin with the migration architecture from Figure 19.1.

During the registration procedure from Chapter 13, the mobile exchanges information elements with the 5G core network that refer to inter-operation with the EPC. To start this process, the mobile indicates whether it supports the EPC as part of its registration request (Figure 13.4, step 1). If it does, then the AMF supplies an information element in its registration accept, to indicate whether it supports the interworking architecture from Figure 19.2 without the use of N26 (Figure 13.4, step 21). If the AMF is only using the migration architecture from Figure 19.1, then it denies such support.

In its registration accept, the AMF also includes a list of *equivalent public land mobile networks*, in other words networks that the mobile should treat as equivalent to the registered PLMN for the purposes of cell selection, reselection and handover. If the EPC has a different PLMN identity from the 5G core, then the network operator can include it in the list of equivalent PLMNs, so that the mobile can reselect between the two.

During the PDU session establishment procedure from Chapter 15, the SMF tells the mobile whether it can transfer each individual QoS flow to the EPC by means of N26 signalling. To do so, the SMF supplies a *mapped EPS bearer context* for each QoS flow that it can transfer, as part of its PDU session establishment accept (Figure 15.11, steps 11–13). If the mobile has any mapped EPS bearer contexts, then it knows that the network is using the N26-based interworking architecture from Figure 19.2. If it does not, then there are three possible causes: either the network is not using N26; or the network is using N26 but cannot transfer any of the mobile's PDU sessions; or the mobile did not set up any PDU sessions in the first place. In the discussion that follows, we will assume that the network is not using N26, so the mobile does not have any mapped EPS bearer contexts. The mobile knows that it cannot transfer any PDU sessions to the evolved packet core, but it does not yet know whether N26 is supported.

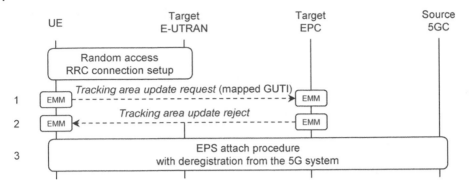

Figure 19.3 Mobility to the evolved packet core, using the migration architecture in RRC_IDLE. Source: Adapted from 3GPP TS 23.502.

19.3.2 Mobility in RRC_IDLE

Now consider what happens if the mobile carries out a reselection to a 4G cell in the manner described in Chapter 17. The mobile reads the new cell's system information, as before. However, we will now assume that the system information only contains a list of EPC network identities, together with their associated 16-bit tracking area codes. The mobile concludes that it has to move to the EPC. From the system information, the mobile can also discover whether the network supports attach without PDN connectivity. The signalling then continues as summarized in Figure 19.3 [10].

If the network supports attach without PDN connectivity, then the mobile begins by assuming that the N26 reference point is supported. Using the equivalent to a 5G mobility registration update, it sends a *tracking area update request* to the EPC, indicates that it is currently registered with the 5G core and identifies itself by mapping its 5G temporary identity to the LTE equivalent. In this example, the MME does not support N26, so it cannot use that identity, and it rejects the mobile's request (step 2). (If the MME does support N26, then the procedure continues in the manner described in Section 19.5.2.)

The mobile reacts by running the LTE attach procedure, which is equivalent to a 5G initial registration (3). (If the network does not support attach without PDN connectivity, then the mobile omits steps 1 and 2, and runs the attach procedure right away.) The resulting signalling is the same as a normal LTE attach procedure, and leads to the mobile being de-registered from the 5G core. The mobile's old PDU sessions are torn down.

If the mobile is initially in the NG-RAN state of RRC_INACTIVE, and carries out a reselection to a 4G cell, then it immediately moves to the state of RRC_IDLE in the manner described in Chapter 18 [11]. If the new cell is only connected to the EPC, then the mobile continues with the procedure for RRC_IDLE. This same behaviour applies to all of the other architectures that we will discuss in this chapter.

Similar procedures are used when the mobile moves from the EPC to the 5G core, but there are two main differences. Firstly, the 5G core network always supports registration without a PDU session, so the equivalent steps to (1) and (2) are always used. Secondly, there is no equivalent to the mapped EPS bearer context, because the 5G core network can accept any of the mobile's EPS bearers without the need for special mapping rules.

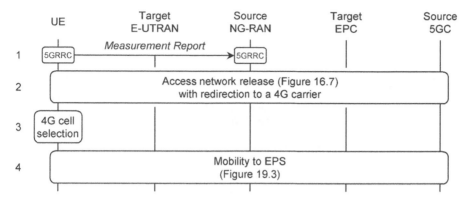

Figure 19.4 RRC release with redirection, using the migration architecture for a mobile initially in RRC_CONNECTED.

19.3.3 RRC Release with Redirection from RRC_CONNECTED

If the mobile is in the state of RRC_CONNECTED, then the 5G core network can still transfer it to the EPC using the procedure for *RRC release with redirection* in Figure 19.4. The procedure can be triggered by 5G measurement event A2, in which the mobile states that the signal received from the primary cell has fallen below a threshold (step 1).

In the absence of any measurement reports about neighbouring cells, the master node runs the access network release procedure from Chapter 16 (2). As part of the RRC Release message, the master node sends the mobile details of a 4G carrier frequency on which to look for a suitable cell (3). The mobile moves to the requested frequency, and selects a cell as instructed. If the new cell is only connected to the EPC, then the mobile continues in the same way as it did earlier in the state of RRC_IDLE (4). The old PDU sessions are torn down as before, but the mobile can re-establish them once it arrives in the EPC.

19.4 Interworking Without N26

19.4.1 Configuration Procedures

The configuration procedures for interworking without N26 begin in the same way that we described earlier in this chapter. This time, however, the AMF indicates in its registration accept that it does support interworking without N26. If the mobile supports dual registration mode, then it can now choose to use it. Otherwise, the mobile enters single registration mode [12].

19.4.2 Mobility in Single Registration Mode

Now consider what happens if the mobile is in single registration mode in the state of RRC_IDLE, and reselects to a 4G cell that is only connected to the EPC. This time, the mobile knows that the network is not using the N26 reference point, so it knows that a tracking area update will not be suitable. Instead, it runs the LTE attach procedure right away, in a procedure that is summarized in Figure 19.5.

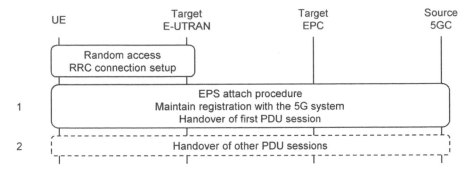

Figure 19.5 Mobility to the evolved packet core, using the interworking architecture in RRC_IDLE but without the use of N26.

In its attach request, the mobile indicates that it is currently registered with the 5G core (step 1). In the equivalent of step 14a of Figure 13.5, the MME asks the HSS to maintain that previous registration in case the mobile is using dual registration mode. As a result, the mobile is not de-registered from its previous AMF.

If the mobile has any PDU sessions, then it embeds an ESM *PDN Connectivity Request* into its attach request. That message identifies the mobile's first PDU session, and asks the MME to hand it over to the EPC and to maintain its data network address. In response, the network converts the PDU session's QoS flows to the corresponding EPS bearers, and redirects their user planes through the EPC. If the mobile has any other PDU sessions, then it sends a separate PDN connectivity request after the end of the attach procedure for each one (2).

If the mobile does not have any PDU sessions, but the EPC supports attach without PDN connectivity, then the mobile omits the PDN connectivity request. Otherwise, the mobile includes the request as before, but asks for connectivity to a new PDN.

If the mobile begins in the state of RRC_CONNECTED, then the network can use the procedure of RRC release with redirection in the same way as before. Once the mobile arrives in the EPC, it can re-activate its PDU sessions by means of a 4G service request, at the expense of a short break in communications.

19.4.3 Mobility in Dual Registration Mode

If the mobile has chosen to use dual registration mode, then it can remain served by a 5G cell within the NG-RAN and can independently select a 4G cell within the E-UTRAN. Once it is registered with both core networks, the mobile does not have to transfer all of its PDU sessions to the EPC. Instead, it can transfer none, some or all of them.

19.5 Interworking with N26

19.5.1 Configuration Procedures

The configuration procedures for interworking with N26 begin in the same way as they did before. As in the case of the migration architecture, the AMF denies support for interworking without N26, and the mobile enters single registration mode.

During the procedure for PDU session establishment, the SMF supplies a mapped EPS bearer context for each QoS flow that it can transfer to the EPC by means of N26 signalling.

The information maps the QoS flow onto an EPS bearer and defines the corresponding LTE QoS parameters such as the QoS class identifier [13]. If the mobile receives any mapped EPS bearer contexts, then it knows that the network is using N26.

19.5.2 Mobility in RRC_IDLE

As before, let us assume that the mobile is in the state of RRC_IDLE and reselects to a 4G cell that is only connected to the EPC. Let us also assume that the mobile has at least one mapped EPS bearer context. The resulting procedure begins as shown in Figure 19.6 [14].

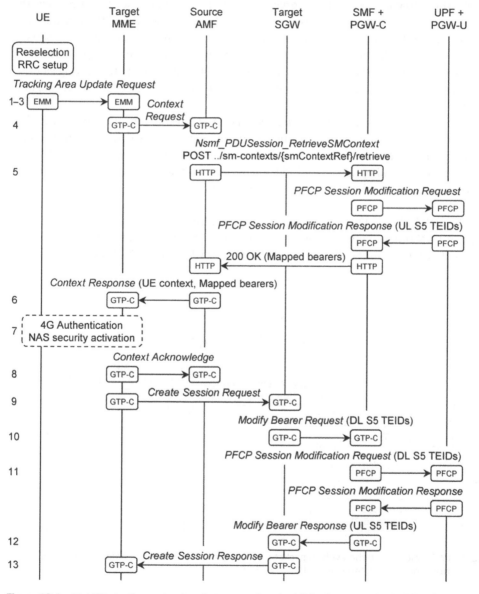

Figure 19.6 Mobility to the evolved packet core, using the N26 reference point in RRC_IDLE. (1) Initiation. Source: Adapted from 3GPP TS 23.502.

As in step 1 of Figure 19.3, the mobile sends an LTE tracking area update request to the EPC, in which it indicates that it is currently registered with the 5G core, and identifies itself by mapping its 5G temporary identity to the LTE equivalent (steps 1–3). This time, however, the MME uses that information to identify the mobile's AMF, and asks it to return the UE context (4). In turn, the AMF asks each of the mobile's SMFs for the mapped EPS bearer contexts that they sent to the mobile earlier (5). It then collects their responses, and sends all the information to the MME (6). After the optional 4G procedures for authentication and security activation (7), the MME confirms that it has accepted the mobile (8).

The MME now selects a suitable serving gateway. For each PDU session, it contacts the serving gateway, identifies the network function that combines the SMF with the PDN gateway control plane and includes the mapped EPS bearer contexts (9). The serving gateway forwards the information to the PDN gateway, which maps the QoS flows onto EPS bearers, redirects their data paths and responds (10–12). In turn, the serving gateway responds to the MME and indicates which bearers have been accepted: any others are torn down (13).

Figure 19.7 shows how the procedure ends. The MME records the mobile's registration in the HSS, which cancels the previous registration with the AMF and returns the LTE subscription data (14–16). The MME accepts the mobile's original request, sends it a new tracking area list and optionally includes a new temporary identity for LTE (17). If it does so, then the mobile acknowledges receipt (18). Finally, the MME returns the mobile to RRC_IDLE.

If the mobile does not have any mapped EPS bearer contexts, but the EPC supports attach without PDN connectivity, then the mobile begins in the same way, by sending a tracking area update request to the MME (steps 1–3). If the MME supports N26 then it accepts the

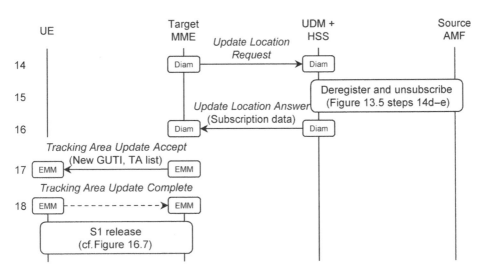

Figure 19.7 Mobility to the evolved packet core, using the N26 reference point in RRC_IDLE. (2) Completion. Source: Adapted from 3GPP TS 23.502.

mobile's request. However, the network does not establish any EPS bearers, so steps 5 and 9–13 are omitted.

19.5.3 Handovers in RRC_CONNECTED

To close this chapter, let us consider what happens if the mobile is in the state of RRC_CONNECTED and leaves the coverage area of the NG-RAN. The resulting procedure is very like a normal NG-based handover, and begins as shown in Figure 19.8. To keep the diagram simple, we have assumed that the mobile is not in dual connectivity, and we have not shown the network functions that combine the SMF and PDN gateway control plane, or the UPF and PDN gateway user plane.

As part of the procedure, the 5G core network tears down the tunnels that delivered data on the N3 reference point, and any other tunnels that were used on N9, Xn and F1. The EPC replaces them with tunnels on the S5 and S1 reference points, and with any other tunnels that are required on X2. The networks also set up temporary tunnels on the N3, S5 and S1 reference points, and tear them down at the end of the procedure. Those tunnels are respectively used to forward undelivered data packets from the NG-RAN to the UPF, from there to the serving gateway, and from there to the E-UTRAN.

On the basis of a measurement report, for example measurement event B2, the source node decides to hand the mobile over to a target 4G cell. It has no Xn communications with the target eNB, so it tells the AMF that a handover is required and includes the information from the mobile's measurement report (step 1). The AMF retrieves the mapped EPS bearer contexts from each of the mobile's SMFs (2). It then selects a suitable MME, asks it to take control of the mobile and forwards all the information there (3).

The MME selects a suitable serving gateway, initializes it and retrieves a set of uplink S1 tunnel endpoint identifiers (TEIDs) (4, 5). The MME forwards those to the target eNB, includes the measurement information from earlier and asks it to take control of the mobile (6). In its reply, the eNB identifies the EPS bearers that it has accepted, includes a downlink S1 TEID for each one and includes an embedded 4G RRC Connection Reconfiguration message for the mobile (7). The MME now contacts the serving gateway a second time, so as to set up a temporary forwarding tunnel from the serving gateway towards the eNB (8). It then responds to the AMF and includes the information that it received from the target eNB (9).

The AMF tells the mobile's master node to hand it over, states which QoS flows are to be transferred and includes the 4G RRC message that the eNB composed earlier (11a). The node extracts the embedded message, and forwards it to the mobile as a handover command (11b). The mobile reconfigures itself as instructed, contacts the target cell using the 4G random access procedure and acknowledges (12a). At the same time, the master node starts to return any undelivered downlink packets to the UPF, which forwards them to the serving gateway and eventually to the eNB.

The procedure ends as shown in Figure 19.9. The eNB tells the MME that the mobile has arrived, and the MME tells the AMF (12b–12d). The MME also sends the S1 downlink TEIDs to the serving gateway (13). In turn, the serving gateway contacts the PDN gateways, which map the QoS flows onto EPS bearers and redirect the corresponding data paths (14–17).

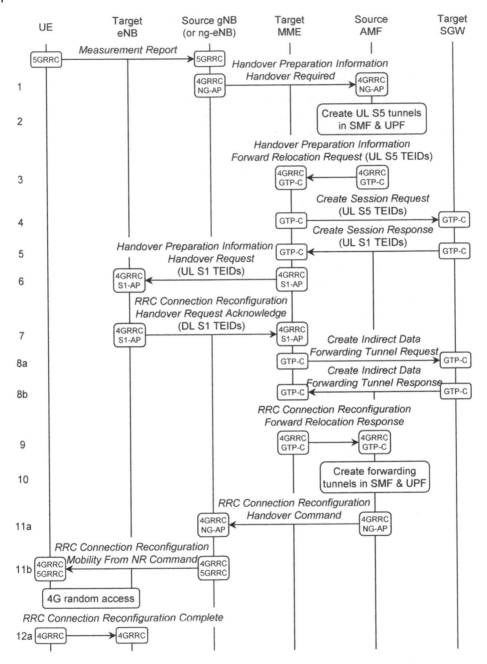

Figure 19.8 Handover to the evolved packet core, using the N26 reference point in RRC_CONNECTED. (1) Initiation. Source: Adapted from 3GPP TS 23.502.

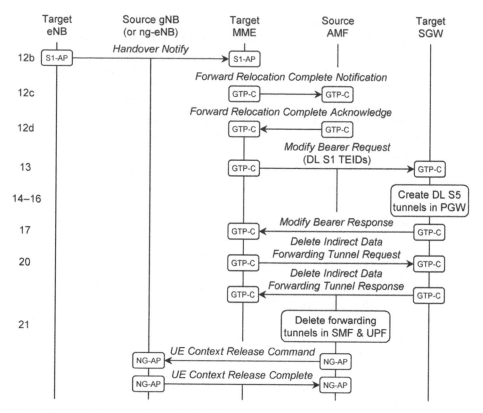

Figure 19.9 Handover to the evolved packet core, using the N26 reference point in RRC_CONNECTED. (2) Completion. Source: Adapted from 3GPP TS 23.502.

On the expiry of a timer, the MME tells the serving gateway to tear down the temporary forwarding tunnels that it created earlier, and the AMF tells the SMF in the same way (20–21). Finally, the AMF tells the original master node to release the resources that it had allocated for the mobile.

References

1 3GPP TS 23.501 (2019) System architecture for the 5G system (5GS); Stage 2 (Release 15), December 2019, Section 5.17.1.

2 3GPP TS 23.501 (2019) System architecture for the 5G system (5GS); Stage 2 (Release 15), December 2019, Sections 4.3, 5.17.2.

3 3GPP TS 36.413 (2019) Evolved Universal Terrestrial Radio Access Network (E-UTRAN); S1 Application Protocol (S1AP) (Release 15), December 2019.

4 3GPP TS 24.301 (2019) Non-access-stratum (NAS) protocol for evolved packet system (EPS); Stage 3 (Release 15), December 2019.

5 3GPP TS 29.272 (2019) Evolved packet system (EPS); Mobility management entity (MME) and serving GPRS support node (SGSN) related interfaces based on diameter protocol (Release 15), December 2019.

6 3GPP TS 29.274 (2019) 3GPP evolved packet system (EPS); Evolved General Packet Radio Service (GPRS) tunnelling protocol for control plane (GTPv2-C); Stage 3 (Release 15), September 2019.

7 3GPP TS 23.401 (2019) General Packet Radio Service (GPRS) enhancements for Evolved Universal Terrestrial Radio Access Network (E-UTRAN) access (Release 15), December 2019, Section 4.6.

8 3GPP TS 36.331 (2019) Evolved Universal Terrestrial Radio Access (E-UTRA); Radio resource control (RRC); Protocol specification (Release 15), December 2019, Section 4.2.1.

9 3GPP TS 23.003 (2019) Numbering, addressing and identification (Release 15), September 2019, Section 2.10.2.

10 3GPP TS 23.502 (2019) Procedures for the 5G system (5GS); Stage 2 (Release 15), December 2019, Section 4.11.2.

11 3GPP TS 24.501 (2019) Non-access-stratum (NAS) protocol for 5G system (5GS); Stage 3 (Release 15), December 2019, Section 5.3.1.4.

12 3GPP TS 24.501 (2019) Non-access-stratum (NAS) protocol for 5G system (5GS); Stage 3 (Release 15), December 2019, Section 5.5.1.2.4.

13 3GPP TS 24.501 (2019) Non-access-stratum (NAS) protocol for 5G system (5GS); Stage 3 (Release 15), December 2019, Sections 6.1.4, 9.11.4.8.

14 3GPP TS 23.502 (2019) Procedures for the 5G system (5GS); Stage 2 (Release 15), December 2019, Section 4.11.1.3.2.

20

Release 16 and Beyond

In this final chapter, we will address the enhancements that are being made to the 5G system as part of Releases 16 and 17. The main topics include vehicle communications, location services, integrated access and backhaul, non-terrestrial networks and massive machine-type communications. We will also address a number of other enhancements towards the end of the chapter, and proposals for the longer term.

20.1 Vehicle-to-everything (V2X) Communications

20.1.1 Introduction

Vehicle-to-everything (V2X) communication refers to the exchange of information to and from a road vehicle. Such communication is part of an intelligent transport system (ITS), in other words a system in which information and communication technologies are applied in the field of road transport; it is often used to support the operation of driverless cars and other autonomous vehicles. The term *everything* covers communications with other vehicles (V2V), pedestrians (V2P), servers based in the network (V2N) and infrastructure elements (V2I) that are known as *roadside units* (RSUs).

LTE was enhanced to offer basic support for vehicle communication as part of 3GPP Release 14 [1, 2]. However, that was limited by LTE's capabilities for data rate, latency and reliability, and simply allows vehicles to exchange information such as their position, speed and direction of travel, and to disseminate urgent messages such as a warning of a possible collision.

5G is enhanced to support vehicle communications as part of 3GPP Release 16 [3, 4], in which the greater capabilities of 5G allow it to support a wider range of applications. There are four new categories. In an application known as *extended sensors*, vehicles can exchange information that they have gathered from sensors such as video cameras. That extends their knowledge of the surrounding environment and improves the driving experience, for example by allowing a vehicle to brake more gently if a queue appears in front of another vehicle further ahead. In a second application known as *advanced driving*, nearby vehicles use that information to share their driving intentions and coordinate their movements, so as to support semi-automated or fully automated driving. *Vehicle platooning* allows a lead vehicle to control a convoy of other vehicles that are behind it. The separation between

An Introduction to 5G: The New Radio, 5G Network and Beyond, First Edition. Christopher Cox.
© 2021 John Wiley & Sons Ltd. Published 2021 by John Wiley & Sons Ltd.

successive vehicles is limited by computational delays and by the latencies within the communication system, and can be much shorter than it is in traditional driving. Finally, *remote driving* allows a vehicle to be controlled from a distance, either by a remote driver or by a V2X application.

To help support the development of autonomous vehicles, SAE International has defined six levels of driving automation, which range from no automation (level 0) to full automation (level 5) [5]. The 5G system is intended to support all of these automation levels.

20.1.2 Architectural Enhancements

As part of Release 14, the architecture of LTE was enhanced to allow external V2X applications to communicate with the mobile and with the evolved packet core [6, 7]. The architecture supports device-to-device communications over an air interface known as the sidelink, which had previously been introduced to LTE in Release 12.

5G is enhanced in a similar way as part of Release 16 [8, 9]. Figure 20.1 shows the high-level architecture, on the assumption that the mobile is not roaming. The *V2X application server* runs vehicle-related applications and acts as an application function towards the 5G core network. PC5 is the air interface's sidelink, which is discussed in more detail in Section 20.1.3. V1 and V5 are reference points within the application layer, which the air interface transports using the Uu and PC5 interfaces respectively. The lower parts of V1 and V5 are specified in the case of LTE [10, 11], but they are currently unspecified in the case of 5G.

Figure 20.2 shows more details of the application server's interactions with the 5G core. The application server exchanges information about the mobile's V2X communications

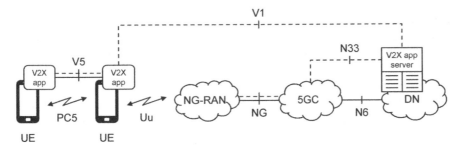

Figure 20.1 Architecture for vehicle-to-everything (V2X) communications. Source: Adapted from 3GPP TS 23.287.

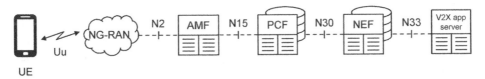

Figure 20.2 Representation of the V2X application server's interactions with the 5G core, using reference points. Source: Adapted from 3GPP TS 23.287.

with the network exposure function (NEF), in a similar way to the interactions that we introduced in Chapter 15. The policy control function (PCF) determines the mobile's V2X configuration parameters, taking on the role of the *V2X control function* from LTE. The access and mobility management function (AMF) conveys the relevant parameters to the mobile, using the procedures for the delivery of UE policy data after registration.

20.1.3 Device-to-device Communications

Originally, the LTE sidelink was introduced to support *mission-critical communications* (MCCs) in a public safety network. Using the sidelink, two devices can continue communicating if they are outside the coverage area of the network, or if the base stations have been taken down by the very situation that triggered the emergency. Later, the LTE sidelink was enhanced to support the more stringent requirements of vehicle communications, such as the high Doppler shifts that result from fast-moving vehicles. In the case of the 5G sidelink, vehicle communications is the initial motivation.

The 5G sidelink is closely based on the designs for the uplink and downlink [12–14]. It supports both of the 5G frequency ranges, although the only frequency bands defined at the time of writing are bands n38 and n47, the second of which encompasses the bands defined for intelligent transport systems in Europe and the USA. The sidelink supports the same subcarrier spacings that the uplink and downlink use for data, namely 15, 30 and 60 kHz in frequency range 1, and 60 and 120 kHz in FR2.

A mobile is configured with a single bandwidth part for each sidelink frequency band. That is the same for transmission and reception, but can be different from the bandwidth parts that are used by the uplink and downlink.

There are then two *resource allocation modes*. In mode 1, the base station schedules a mobile's use of the sidelink explicitly, either semi-statically by setting up a configured grant using radio resource control (RRC) signalling, or dynamically by transmitting downlink control information (DCI) using two new DCI formats that are denoted 3_0 and 3_1. In mode 2, the mobile is pre-configured with one or more *resource pools*, each of which is a set of time-frequency resources that can be used for sidelink transmission and/or reception. A mobile indicates its intended use of those resource pools to mobiles that are nearby, detects the corresponding indications from other mobiles and schedules its transmissions in the resources that are available. In both modes, the specifications support three types of transmission: unicast transmissions to a single mobile, groupcast transmissions to mobiles that are in a predefined communication group and broadcast transmissions to all of the mobiles that are nearby.

The mobiles exchange traffic using *PC5 QoS flows*, which are mapped onto *sidelink radio bearers*. Each PC5 QoS flow is associated with a *PC5 5G QoS identifier* (PQI). Like the 5G QoS identifier from Chapter 15, this field acts as a pointer into a look-up table, which defines parameters of the QoS flow such as its target error rate and target delay. However, the PQI defines a completely different set of QoS targets from the 5QI, which are listed in Table 20.1.

Table 20.1 Standardized values of the PC5 5QI.

PQI	Type	Default priority level	Packet delay budget (ms)	Packet error rate	Default maximum data burst	Default averaging window (ms)	Examples
1	GBR	3	20	10^{-4}	—	2000	Platooning[a]
2		4	50	10^{-2}	—	2000	Sensor sharing[a]
3		3	100	10^{-4}	—	2000	Automated driving[a]
55	Non-GBR	3	10	10^{-4}	—	—	Changing lanes[a]
56		6	20	10^{-1}	—	—	Platooning[b]
57		5	25	10^{-1}	—	—	Changing lanes[b]
58		4	100	10^{-2}	—	—	Sensor sharing[b]
59		6	500	10^{-1}	—	—	Platooning reports
82	Delay-critical GBR	3	10	10^{-4}	2000 bytes	2000	Collision avoidance[a]
83		2	3	10^{-5}	2000 bytes	2000	Emergency trajectory alignment[a]

a) High degree of automation.
b) Low degree of automation.
Source: Adapted from 3GPP TS 23.287.

20.2 Location Services

20.2.1 Introduction

Location services (LCS) allow an application to find out the geographical location of a mobile. LTE supported location services from 3GPP Release 9, for regulatory services such as emergency calls and lawful interception, and also for mapping. In LTE, location measurement is most often carried out using a *global navigation satellite system* (GNSS), for example the Global Positioning System (GPS). However, a mobile may be unable to detect the satellites if it is surrounded by tall buildings or is indoors, while mobiles used for machine-type communications may not be equipped with a suitable receiver at all. To handle those situations, the mobile's location can also be determined using the LTE air interface itself.

In Release 15, 5G supports location services using the architecture that we will discuss in Section 20.2.2. As in LTE, the mobile's location can be estimated either by means of GNSS, or by using measurements over the 4G air interface between the mobile and nearby eNBs. That allows the network to support the regulatory services identified above, with a horizontal positioning accuracy of 50 m for 80% of mobiles, a vertical positioning accuracy of 5 m and an end-to-end latency of 30 seconds [15].

Release 16 extends the location measurement capabilities of 5G to support a wider variety of commercial positioning applications, for example augmented reality, unmanned aerial vehicles and traffic control [16]. To do so, the architecture is improved so as to reduce the end-to-end location measurement latency, and the 5G New Radio is enhanced to support location measurements between the mobile and nearby gNBs. The resulting system should deliver a horizontal positioning accuracy of 3 m indoors and 10 m outdoors, a vertical positioning accuracy of 3 m in both environments and an end-to-end latency of one second. These capabilities are similar to those of GPS, for which the US government guarantees a mobile-to-satellite distance that is accurate to 7.8 m for 95% of devices [17].

20.2.2 System Architecture

Figure 20.3 is a reference point representation of the architecture for 5G location services, for the case where the mobile is not roaming [18, 19]. Figure 20.4 represents the same architecture by means of service-based interfaces, and also shows the signalling protocols that are used towards the next-generation radio access network (NG-RAN) and the mobile.

Using this architecture, an external application can request a mobile's location in two ways. In Release 15, the application can act as a legacy location service client from LTE, by sending a location request to a *gateway mobile location centre* (GMLC) across the Le reference point. That reference point is not defined by the 3GPP specifications but typically uses the Open Mobile Alliance's *Mobile Location Protocol* [20]. From Release 16, the application can also act as a 5G application function by sending a location request to the network exposure function. The NEF delivers that request to the GMLC by means of its service-based interface [21].

In both cases, the GMLC looks up the identity of the mobile's serving AMF and asks it to return the mobile's location. In turn, the AMF delegates that request to the location management function (LMF) [22]. To determine the mobile's location, the LMF communicates with the NG-RAN using the *New Radio positioning protocol A* (NRPPa) [23], with the messages relayed by way of the AMF. It also communicates with the mobile using the *LTE positioning protocol* (LPP) [24], with the messages relayed by the AMF and the NG-RAN.

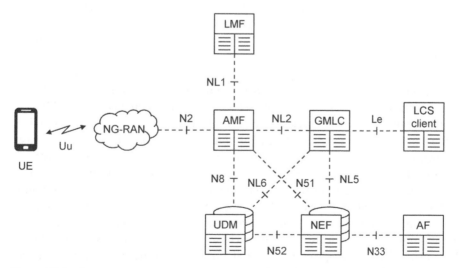

Figure 20.3 Representation of the location service architecture, using reference points. Source: Adapted from 3GPP TS 23.273.

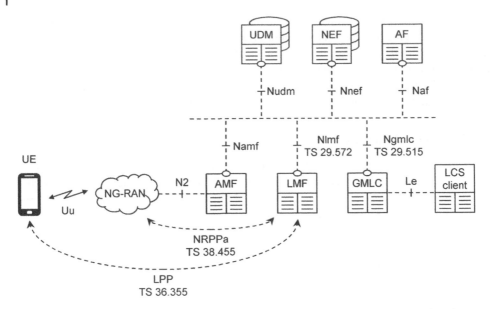

Figure 20.4 Representation of the location service architecture, using service-based interfaces. Source: Adapted from 3GPP TS 23.273.

In Release 15, the GMLC and LMF are part of the 5G core network. From Release 16, a network operator can reduce the location measurement latency by deploying those functions as part of the NG-RAN [25].

20.2.3 Enhancements to the Air Interface

From Release 16, the LMF can determine a mobile's location over the 5G New Radio using six different techniques [26]. The simplest is *enhanced cell identity*, which only uses measurements that the gNB is making anyway. These include the identities of the mobile's serving cells, the corresponding values of the uplink timing advance and the base stations' choices of spatially filtered beams.

The positioning accuracy can be improved by more sophisticated timing measurements. When using *downlink time difference of arrival*, the mobile measures the times at which downlink reference signals arrive from its serving cells and their nearest neighbours, and reports the time differences back to the network. The network can then determine the mobile's position by triangulation. *Uplink time difference of arrival* uses the base stations' measurements of the mobile's uplink reference signals, while measurements of the *round-trip time* use both. At least in line-of-sight conditions, the resulting position estimates can be far more accurate than in LTE, because the larger bandwidth of 5G results in a shorter sample duration. However, the positioning accuracy degrades in a multipath environment, in which the measurements are corrupted by additional time delays.

The positioning accuracy can also be improved by a more sophisticated use of beamforming. When using *downlink angle of departure*, the mobile identifies the spatially filtered beams that arrive from its serving cells and their nearest neighbours, and reports the reference signal received power from each one. By interpolating between the individual beams, the network can determine the mobile's direction from each individual base station more accurately than before, and can combine the results to determine the mobile's location. *Uplink angle of arrival* is based on similar measurements by the base stations themselves.

When using these techniques, the mobile's measurements are made using a new downlink reference signal, which is known as the *positioning reference signal* (PRS). The base stations' measurements are made using the sounding reference signals from Release 15.

20.3 Integrated Access and Backhaul

20.3.1 Introduction

In the long term, the best way to increase the capacity of a 5G network will be through the use of millimetre wave communications. There are two reasons: each cell occupies a large amount of radio spectrum; and the individual cells are small, so their spectrum can be re-used many times. However, the cost of deploying those cells could be large. One particular problem lies in the backhaul, where the cost of equipping each node with a fixed fibre-optic connection or a dedicated point-to-point wireless link could be prohibitive.

One solution is *integrated access and backhaul* (IAB), in which the backhaul is implemented using the 5G New Radio itself. IAB is especially attractive for the case of millimetre wave communications, for two reasons. Firstly, the wide radio bandwidth implies that the air interface and backhaul can often use different radio frequencies without limiting the system capacity unduly. Secondly, even if their radio frequencies are the same, the network operator can minimize interference between them by the use of beamforming.

IAB is introduced into 5G as part of 3GPP Release 16 [27, 28], with the work treated as a priority because of its potential impact on the roll-out of 5G. The design is similar to that of relays in LTE, but those had a slightly different motivation, namely that of increasing the coverage of an LTE base station without the need to install a new small cell.

20.3.2 High-level Architecture

The IAB architecture supports a number of topologies, two of which are shown in Figure 20.5. In these architectures, an *IAB node* is a base station with an integrated access and backhaul, which is controlled by a normal base station known as an *IAB donor*. The IAB node communicates with one or more mobiles by means of the 5G New Radio, and with the IAB donor by using the New Radio as a wireless backhaul. The simplest

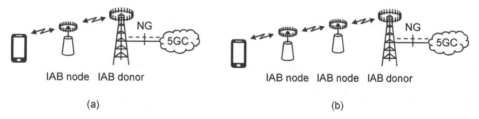

(a)

(b)

Figure 20.5 Example topologies for the integrated access and backhaul. (a) Single-hop backhaul. (b) Multi-hop backhaul.

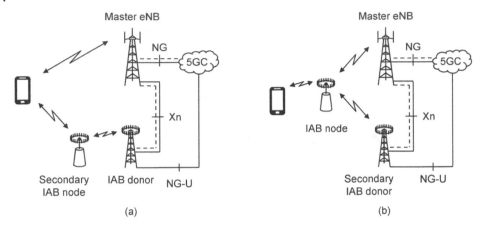

Figure 20.6 Example dual connectivity topologies for the integrated access and backhaul. (a) Mobile in dual connectivity with a master ng-eNB and a secondary IAB node. (b) IAB node in dual connectivity with a master ng-eNB and a secondary IAB donor. Source: Adapted from 3GPP TR 38.874.

architecture uses a single-hop backhaul, but a multi-hop backhaul is a valuable alternative because of the short range of millimetre wave communications.

The architecture also supports multi-radio dual connectivity for both the mobile and the IAB node. There is one restriction: any communications between an IAB node and a master eNB are limited to configuration and control of the IAB node itself, while wireless backhauling is carried out using the 5G New Radio alone. Figure 20.6 shows some examples that involve the 5G core network alone, but the mobile and/or IAB node can also be controlled by the evolved packet core.

IAB nodes are intended to be physically fixed. Even so, the network should be able to reconfigure the backhaul automatically, because individual millimetre wave links can be blocked by obstructions such as vehicles, seasonal foliage or new buildings. The resulting procedure is known as *topology adaptation*.

20.3.3 Architectural Details

Figure 20.7 shows some more detail of the architecture, for the case of the multi-hop backhaul from Figure 20.5. The IAB donor contains gNB central and distributed units,

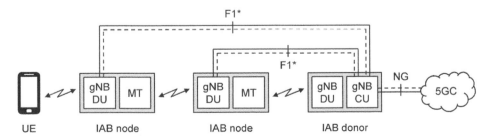

Figure 20.7 Integrated access and backhaul architecture. Source: Adapted from 3GPP TR 38.874.

in the usual way. Each IAB node contains a mobile termination (MT) for its upstream communications and a gNB distributed unit for its downstream communications, but no central unit. It therefore implements the air interface's physical layer, medium access control (MAC) and radio link control (RLC) protocols, but not the packet data convergence protocol (PDCP) or service data adaptation protocol (SDAP). Adding some more detail, an IAB node uses the MAC protocol to schedule its own downstream communications, and the RLC protocol to implement hop-by-hop re-transmissions.

The IAB donor's central unit communicates with the IAB nodes' distributed units over a modified version of the F1 reference point, denoted F1*. That reference point transports information using the wireless backhaul instead of IP, so the new *backhaul adaptation protocol* (BAP) provides an interface between the overlying F1 protocols and the underlying RLC [29]. In the case of a multi-hop backhaul, the adaptation layer also routes traffic and signalling messages between the IAB donor and the appropriate IAB node.

There are two possible relationships between the wireless backhaul and the air interface. In the case of an *out-of-band backhaul*, the air interface and wireless backhaul span different radio frequencies, so they can operate independently. In the case of an *in-band backhaul*, their radio frequencies overlap or coincide, so the interference between them has to be minimized by means of time division multiplexing, frequency division multiplexing or beamforming.

20.4   Non-terrestrial Networks

20.4.1   Introduction

A *non-terrestrial network* (NTN) provides some or all of its coverage from platforms such as satellites or unmanned aircraft systems. 3GPP defined the requirements for NTNs as part of Release 15 [30, 31], and investigated the system architecture and the resulting modifications to the 5G New Radio in Release 16 [32, 33]. Two IEEE frequency bands have been of particular interest: S band transmissions around 2 GHz, and transmissions in the satellite Ka band in which the downlink and uplink are around 20 and 30 GHz respectively.

A satellite might be in low, medium or geostationary orbit, while the possible aircraft systems include balloons, airships and drones. As illustrated in Figure 20.8, the satellite or aircraft can be deployed transparently as part of the air interface or backhaul, or non-transparently as a distributed unit or a gNB. There is also the possibility of relaying the transmission, for example by deploying an additional satellite or aircraft as an IAB node.

The main use of non-terrestrial networks is to improve the system's coverage in unserved or under-served areas of the globe. For example, a network operator might offer coverage in remote land areas or over the oceans by deploying a satellite in the air interface or as a gNB, or in remote villages by deploying a satellite in the backhaul. NTNs can also improve the system's reliability and resilience for applications such as public safety, and can address market sectors such as shipping and aircraft that terrestrial networks are unable to handle. Furthermore, integrating the non-terrestrial component into 5G offers the extra benefit of seamless coverage and roaming. For example, a shipping company might track the delivery of containers over the ocean using satellite communications, and then follow their subsequent route on land using the cellular network [34].

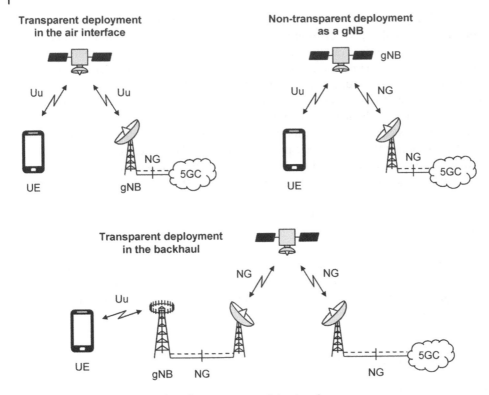

Figure 20.8 Deployment options for a non-terrestrial network.

20.4.2 Design Challenges

NTNs have a number of design challenges [35], and the specifications are expected to form part of Release 17. The first problem is the large propagation delay. A geostationary satellite is at an altitude of 35 786 km above sea level. If the satellite is deployed transparently as part of the air interface, then the distance between the mobile and the base station is at least twice that amount, 71 572 km, which is over 200 times greater than the distance of 300 km that 5G was designed to handle. In turn, the one-way propagation delay is at least 238 ms, or even larger if the satellite is not directly overhead. As partial compensation, the mobiles are each at similar distances from the satellite so the differential delay between them is much smaller, no more than about 16 ms.

There are several impacts. Firstly, the re-transmission procedures have to be modified to handle the full propagation delay, both in the physical layer and in the RLC protocol. Secondly, the guard period in TDD mode has to be increased to handle the full propagation delay, which makes TDD mode unsuitable for the geostationary case. Thirdly, the air interface's control loops operate more slowly, so they are unable to compensate for issues such as fading. Finally, the timing advance procedure has to be modified to handle the differential delay.

The second problem is the platform's speed. A satellite in low earth orbit travels at about 28 000 km h^{-1}, which is over 50 times greater than the speed of 500 km h^{-1} that 5G was designed to handle. The resulting Doppler shift can be as great as 50 kHz at a carrier frequency of 2 GHz, or 500 kHz at a carrier frequency of 20 GHz, and can change very quickly as the satellite passes overhead. These Doppler shifts are too large for the Release 15 acquisition procedures to handle, and also lead to inter-carrier interference.

A third problem is mobility management. The beams of a satellite in low earth orbit move quickly over the ground, which requires fast, reliable handovers for mobiles in RRC_CONNECTED, and careful management of tracking areas for mobiles in RRC_IDLE. If the tracking areas are fixed with respect to the satellite's beams, then they move over the ground, and the number of registration updates is large. If they are fixed with respect to the ground, then the mobile has to discover its tracking area from a separate location service, and the network has to maintain a dynamically changing conversion between tracking areas and satellite beams. 3GPP has adopted the second approach.

20.5 Massive Machine-type Communications

20.5.1 Introduction

When writing the specifications for the 5G New Radio, 3GPP focussed on the use cases for enhanced mobile broadband and for ultra-reliable low-latency communication. However, the New Radio does not support all of the stated requirements for massive machine-type communications, notably those for enhanced coverage and a long battery life.

Instead, 3GPP decided to support these applications through the earlier air interface technologies of enhanced machine-type communications (eMTC) and the narrowband internet of things (NB-IoT) [36–38]. From Release 16, a machine-type device can access the 5G core network by means of eMTC and NB-IoT. The resulting system addresses all the requirements for machine-type communications, while allowing machine-type devices to benefit from the new capabilities of the 5G core. It also supports mobility between the 5G core network and the evolved packet core to handle situations where the mobile moves away from a base station that supports the NG reference point, and reselects or hands over to a base station that only supports S1.

20.5.2 Enhancements to the 5G Core Network

Release 16's support for machine-type communications involves several enhancements to the 5G core network, which replicate features that were introduced to the evolved packet core under the name of the *cellular internet of things* (CIoT). The first helps the delivery of small data packets, while minimizing the signalling overhead and maximizing the mobile's battery life. From Release 16, the system can maintain a mobile's access stratum parameters, for example its access stratum security keys, while the mobile is idle. That replicates the mobile's behaviour in the 5G state of RRC_INACTIVE, and allows it to run the service request procedure and re-activate its radio bearers with minimal signalling overhead.

In another technique, a mobile can send and receive small data packets by embedding them into the 5G session management messages that it exchanges with the session management function (SMF). From idle mode, the mobile can do so after running the procedure of RRC connection setup, but without any need to run the service request procedure or to re-activate its data radio bearers. The SMF can exchange those packets with an external server in two ways: either on the user plane by way of the user plane function (UPF), or on the control plane by way of the NEF.

A second set of enhancements reduces the mobile's power consumption. Release 15 already allowed a mobile to request a mode known as *mobile-initiated connection only* (MICO), in which it does not listen for paging messages in the state of CM-IDLE, and only contacts the network to carry out a periodic registration update or some other mobile-initiated transmission. Release 16 adds an *active time*. After a transition from CM-CONNECTED to CM-IDLE, a MICO-enabled mobile stays in the normal idle mode for a time that is limited by the interval between its periodic registration updates, and only then enters MICO mode.

Another power-saving technique is *extended discontinuous reception* (eDRX). This increases the mobile's discontinuous reception cycle length to a maximum of 10 485.76 seconds, which is nearly 3 hours. For both of these power-saving techniques, the SMF, NEF and UPF can be enhanced to support *extended buffering*, in which they agree to buffer incoming downlink data for longer than they usually do.

20.5.3 NR Light

A study is taking place as part of Release 17 into a modified version of the 5G New Radio, which is provisionally known as *NR Light* [39]. Once implemented, NR Light will offer a solution for machine-type devices which is fully integrated into the 5G New Radio, but which addresses a slightly different set of performance requirements from before. In particular, the data rate, reliability and latency of NR Light are likely to be better than those of eMTC and NB-IoT, but the coverage is likely to be worse. Applications include higher-end machine-type devices such as security cameras and wearable devices, which are not so well addressed by the earlier technologies.

20.6 Other New Features and Studies

20.6.1 Enhancements to the Service-based Architecture

We introduced the Release 15 procedures for network function registration and discovery in Chapter 2. In Release 15, a network function consumer explicitly discovers individual instances of a network function producer by sending a request to the network repository function (NRF). It then selects a single producer from the list that the NRF returns and contacts the producer directly by means of its service-based interface.

Release 16 builds upon that architecture by introducing the *service communication proxy* (SCP) [40, 41]. The SCP acts as an application-layer router for HTTP/2 signalling on the core network's service-based interfaces, by receiving HTTP/2 requests from a network function consumer, selecting a suitable producer and forwarding the signalling on. As part of that task, the SCP carries out network function service discovery implicitly, without the need for the consumer to contact the NRF at all. It also carries out centralized load balancing by selecting a suitable producer using its overall knowledge of the load within the network.

20.6.2 Support for Vertical and LAN Services

Release 16 extends 5G's support for ultra-reliable low-latency communication to *cyber-physical control applications*, in other words industrial control systems containing a mix of physical components such as sensors and actuators, and computational components such as a centralized control system [42–45].

Two enhancements are especially important [46]. The first is time-sensitive networking. In some control systems, the sensors and actuators have to be time-synchronized, with a tolerance that can be as small as one microsecond. Release 16 handles this issue by distributing an external timing signal over the 5G network, treating the 5G system as a *time-aware system* in accordance with IEEE specification 802.1AS [47].

The second is support for a *non-public network* (NPN), in other words part or all of a 5G network that is private to an industrial client. A NPN can be implemented in three ways: as a stand-alone network, identified by combining the public land mobile network identity (PLMN-ID) with a separate network identifier; as a slice of a public network; or by reserving cells in a public network for private use by assigning them to a *closed access group*.

20.6.3 Self-optimizing Networks

LTE supports a number of techniques for automatic configuration and optimization of the radio access network, under the name of *self-optimizing* or *self-organizing networks* (SONs). 5G replicated two of these techniques in Release 15. The first is the configuration of *automatic neighbour relations* and the Xn interface. If a node receives a measurement report about a neighbouring physical cell identity that it wasn't previously aware of, then it can ask the mobile to read system information block 1 from the neighbouring cell, and to return the corresponding New Radio cell identity, tracking area code and PLMN list [48]. It can then initiate an NG-based handover. It can also ask the AMF to retrieve an IP address from the neighbouring node, and can use that IP address to establish signalling communications with the neighbour over Xn [49]. The second technique is *energy saving*, which allows a network to switch its small cells on and off automatically in response to changes in the network load, while maintaining the macrocells that are required for coverage purposes [50].

Other features are replicated in Release 16 [51, 52]. Using *automatic PCI configuration*, a distributed unit can assign physical cell identities to its cells without intervention from the network operator, avoiding identities that are being used by other cells nearby by means of a downlink receiver. *Mobility robustness optimization* allows the radio access network to detect handover failures that arise due to poor choices of measurement reporting thresholds, and to correct them. *Mobility load balancing* allows nearby base stations to exchange information about their load levels by means of Xn signalling, so that a congested base station can hand over some of its mobiles to less congested neighbours. Finally, *RACH optimization* allows nearby cells to exchange information about the parameters used for the random access channel, so as to minimize the interference between their random access transmissions.

20.6.4 Use of Unlicensed Spectrum

5G networks mainly use licensed spectrum, in which the network is centrally planned for low interference and seamless coverage, which in turn delivers high data rates and high reliability. However, the capacity of a 5G network can also be increased by the use of unlicensed spectrum.

LTE was enhanced to operate in unlicensed spectrum from Release 13, and 5G is enhanced in a similar way as part of Release 16 [53]. An important aspect is the need to avoid causing interference to WiFi receivers that are operating in the same frequency band. That is implemented by adopting a WiFi mechanism known as *listen before talk* (LBT), in which a 3GPP device pauses before transmission to ensure that no other devices are attempting to do so.

20.6.5 Reduction of Cross-link Interference

In TDD mode, there is a risk of interference from the downlink transmitter of one base station to the uplink receiver of another. Usually, that interference is only a problem over short distances, and can be controlled by co-ordinating the TDD configurations of base stations that are nearby. Sometimes, however, it can also be a problem over longer distances, as much as 300 km or so, due to scattering and reflection from the upper atmosphere. It is then known as *cross-link interference* (CLI).

Release 16 introduces measures to detect cross-link interference and to reduce it [54–56]. If a victim TDD cell detects excess levels of uplink interference that are consistent with the expected behaviour of CLI, then it can start transmitting a *remote interference management reference signal* (RIM-RS) on the downlink. In the presence of channel reciprocity, the aggressor detects the reference signal, and takes action by adjusting its TDD configuration. It can also send an indication back to the victim, either by a reference signal of its own or by signalling messages over the backhaul, to help the victim establish whether the aggressor's actions have been enough.

20.6.6 Further Enhancements to the 5G New Radio

Release 16 includes a number of other enhancements to the 5G New Radio. There are several measures to reduce the mobile's power consumption, for example by dynamically reducing the number of physical downlink control channel (PDCCH) monitoring occasions if the data rate is low [57]. The usage of multiple antennas is enhanced, to increase the system capacity, and to improve its robustness when using multiple spatially filtered beams [58]. The handover procedure is also enhanced, to improve the procedure's reliability, and to reduce the interruption time for mobiles with a dual protocol stack that can simultaneously communicate with the source and target nodes [59, 60]. Another new feature is the introduction of a random access procedure that contains only two steps rather than four, with a view to reducing the procedure's latency [61].

Other proposed enhancements include the operation of 5G in new frequency ranges. 3GPP studied the ranges from 7 to 24 GHz [62] and beyond 52.6 GHz [63] as part of Release 16, and plans to implement the range from 52.6 to 71 GHz in Release 17 [64]. Issues being addressed include national regulatory requirements, suitable architectures for the base station and mobile, the precise dividing lines between the different frequency ranges, and the additional problems at high radio frequencies due to atmospheric absorption and phase noise. Another study concerns *non-orthogonal multiple access* (NOMA) [65]. 3GPP investigated several other multiple access schemes for 5G as part of Release 14, but abandoned them in favour of the continued use of OFDMA. However, those other schemes can offer a number of benefits, for example signalling and latency reductions, by skipping OFDMA's explicit allocation of mobiles to subcarriers.

References

1 3GPP TR 22.885 (2015) Study on LTE support for vehicle to everything (V2X) services (Release 14), December 2015.

2 3GPP TS 22.185 (2018) Service requirements for V2X services; Stage 1 (Release 15), June 2018.

3 3GPP TR 22.886 (2018) Study on enhancement of 3GPP support for 5G V2X services (Release 16), December 2018.

4 3GPP TS 22.186 (2019) Enhancement of 3GPP support for V2X scenarios; Stage 1 (Release 16), June 2019.

5 SAE International J3016_201806 (2018) Taxonomy and definitions for terms related to driving automation systems for on-road motor vehicles, June 2018.

6 3GPP TR 23.785 (2016) Study on architecture enhancements for LTE support of V2X services (Release 14), September 2016.

7 3GPP TS 23.285 (2019) Architecture enhancements for V2X services (Release 16), December 2019.

8 3GPP TR 23.786 (2019) Study on architecture enhancements for the evolved packet system (EPS) and the 5G system (5GS) to support advanced V2X services (Release 16), June 2019.

9 3GPP TS 23.287 (2019) Architecture enhancements for 5G system (5GS) to support vehicle-to-everything (V2X) services (Release 16), December 2019.

10 3GPP TR 23.795 (2018) Study on application layer support for V2X services (Release 16), December 2018.

11 3GPP TS 23.286 (2019) Application layer support for vehicle-to-everything (V2X) services; Functional architecture and information flows (Release 16), December 2019.

12 3GPP TR 38.885 (2019) NR; Study on NR vehicle-to-everything (V2X) (Release 16), March 2019.

13 3GPP TR 37.985 (2019) Overall description of radio access network (RAN) aspects for vehicle-to-everything (V2X) based on LTE and NR (Release 16), November 2019.

14 3GPP TR 38.886 (2019) V2X Services based on NR; User equipment (UE) radio transmission and reception (Release 16), February 2020.

15 3GPP TR 38.855 (2019) Study on NR positioning support (Release 16), March 2019, Section 4.

16 3GPP TR 22.872 (2018) Study on positioning use cases; Stage 1 (Release 16), September 2018.

17 US Department of Defense (2008) GPS Standard Positioning Service (SPS) Performance Standard, 4th ed., US Department of Defense, September 2008.

18 3GPP TR 23.731 (2018) Study on enhancement to the 5GC location services (Release 16), December 2018.

19 3GPP TS 23.273 (2019) 5G system (5GS) location services (LCS); Stage 2 (Release 16), December 2019.

20 Open Mobile Alliance (2011) Mobile Location Protocol Version 3.1. http://www .openmobilealliance.org/release/MLP/V3_1-20110920-A/OMA-LIF-MLP-V3_1-20110920-A.pdf (accessed 18 January 2020).

21 3GPP TS 29.515 (2019) 5G system; Gateway mobile location services; Stage 3 (Release 16), December 2019.

22 3GPP TS 29.572 (2019) 5G system; Location management services; Stage 3 (Release 16), December 2019.

23 3GPP TS 38.455 (2019) NG-RAN; NR positioning protocol A (NRPPa) (Release 15), January 2019.

24 3GPP TS 37.355 (2019) LTE positioning protocol (LPP) (Release 15), December 2019.

25 3GPP TR 38.856 (2019) Study on local NR positioning in NG-RAN (Release 16), December 2019.

26 3GPP TR 38.855 (2019) Study on NR positioning support (Release 16), March 2019.

27 3GPP TR 38.874 (2018) NR; Study on integrated access and backhaul (Release 16), December 2018.

28 3GPP TS 38.174 (2019) NR; Integrated access and backhaul radio transmission and reception (Release 16), September 2019.

29 3GPP TS 38.340 (2019) NR; Backhaul Adaptation Protocol (BAP) specification (Release 16), November 2019.

30 3GPP TR 22.822 (2018) Study on using satellite access in 5G; Stage 1 (Release 16), June 2018.

31 3GPP TR 38.811 (2019) Study on New Radio (NR) to support non-terrestrial networks (Release 15), September 2019.

32 3GPP TR 23.737 (2019) Study on architecture aspects for using satellite access in 5G (Release 17), December 2019.

33 3GPP TR 38.821 (2019) Solutions for NR to support non-terrestrial networks (NTN) (Release 16), January 2020.

34 3GPP TR 22.822 (2018) Study on using satellite access in 5G; Stage 1 (Release 16), June 2018, Section 5.1.

35 3GPP TR 38.811 (2019) Study on New Radio (NR) to support non-terrestrial networks (Release 15), September 2019, Sections 5, 7.

36 3GPP RP-180581 (2018) Interim conclusions on IoT for Rel-16, March 2018.

37 3GPP TR 23.724 (2019) Study on cellular internet of things (CIoT) support and evolution for the 5G system (5GS) (Release 16), June 2019.

38 3GPP TS 23.501 (2019) System architecture for the 5G system (5GS); Stage 2 (Release 16), December 2019, Section 5.31.

39 3GPP RP-193238 (2019) New SID on support of reduced capability NR devices, December 2019.

40 3GPP TR 23.742 (2018) Study on enhancements to the service-based architecture (Release 16), December 2018.

41 3GPP TS 23.501 (2019) System architecture for the 5G system (5GS); Stage 2 (Release 16), December 2019, Sections 6.2.19, 6.3.1, 7.1.1, E, G.

42 3GPP TR 22.804 (2018) Study on communication for automation in vertical domains (Release 16), December 2018.

43 3GPP TR 22.821 (2018) Feasibility study on LAN support in 5G (Release 16), June 2018.

44 3GPP TS 22.104 (2019) Service requirements for cyber-physical control applications in vertical domains; Stage 1 (Release 16), December 2019.

45 3GPP TR 23.734 (2019) Study on enhancement of 5G system (5GS) for vertical and local area network (LAN) services (Release 16), June 2019.

46 3GPP TS 23.501 (2019) System architecture for the 5G system (5GS); Stage 2 (Release 16), December 2019, Sections 4.4.8, 5.27, 5.28, 5.30.

47 IEEE P802.1AS (2019) IEEE draft standard for local and metropolitan area networks – Timing and synchronization for time-sensitive applications, Draft 8.3, October 2019.

48 3GPP TS 38.331 (2019) NR; Radio resource control (RRC) protocol specification (Release 15), December 2019, Section 6.3.2 (CGI-InfoEUTRA, CGI-InfoNR).

49 3GPP TS 38.423 (2019) NG-RAN; Xn application protocol (XnAP) (Release 15), December 2019, Section 8.4.1.

50 3GPP TS 38.423 (2019) NG-RAN; Xn application protocol (XnAP) (Release 15), December 2019, Sections 8.4.2, 8.4.3.

51 3GPP TR 28.861 (2019) Telecommunication management; Study on the self-organizing networks (SON) for 5G networks (Release 16), December 2019.

52 3GPP TR 37.816 (2019) Study on RAN-centric data collection and utilization for LTE and NR (Release 16), July 2019.

53 3GPP TR 38.889 (2018) Study on NR-based access to unlicensed spectrum (Release 16), December 2018.

54 3GPP TR 38.828 (2019) Cross link interference (CLI) handling and remote interference management (RIM) for NR (Release 16), September 2019.

55 3GPP TR 38.866 (2019) Study on remote interference management for NR (Release 16), March 2019.

56 3GPP TS 38.300 (2019) NR; NR and NG-RAN overall description; Stage 2 (Release 16), December 2019, Section 17.

57 3GPP TR 38.840 (2019) NR; Study on user equipment (UE) power saving in NR (Release 16), June 2019.

58 3GPP RP-192271 (2019) Revised WID: Enhancements on MIMO for NR, September 2019.

59 3GPP RP-192534 (2019) New WID: NR mobility enhancements, December 2019.

60 3GPP TS 38.213 (2019) NR; Physical layer procedures for control (Release 16), December 2019, Section 15.

61 3GPP RP-190711 (2019) Revised work item proposal: 2-step RACH for NR, March 2019.

62 3GPP TR 38.820 (2019) NR; 7–24 GHz frequency range (Release 16), December 2019.

63 3GPP TR 38.807 (2019) Study on requirements for NR beyond 52.6 GHz (Release 16), December 2019.

64 3GPP RP-193229 (2019) Extending current NR operation to 71GHz, December 2019.

65 3GPP TR 38.812 (2018) Study on non-orthogonal multiple access (NOMA) for NR (Release 16), December 2018.

Further Reading

Long-term Evolution (LTE)

Cox, C.I. (2014). *An Introduction to LTE: LTE, LTE-Advanced, SAE, VoLTE and 4G Mobile Communications*, 2e. Wiley.

Dahlman, E., Parkvall, S., and Sköld, J. (2013). *4G: LTE/LTE-Advanced for Mobile Broadband*, 2e. Academic Press.

Ghosh, A. and Ratasuk, R. (2011). *Essentials of LTE and LTE-A*. Cambridge University Press.

Ghosh, A., Zhang, J., Andrews, J.G., and Muhamed, R. (2010). *Fundamentals of LTE*. Prentice Hall.

Holma, H. and Toskala, A. (2011). *LTE for UMTS: Evolution to LTE-Advanced*, 2e. Wiley.

Holma, H. and Toskala, A. (2012). *LTE-Advanced: 3GPP Solution for IMT-Advanced*. Wiley.

Johnson, C. (2012). *Long Term Evolution in Bullets*, 2e. Createspace.

Khan, F. (2009). *LTE for 4G Mobile Broadband: Air Interface Technologies and Performance*. Cambridge University Press.

Kreher, R. and Gaenger, K. (2015). *LTE Signaling: Troubleshooting and Performance Measurement*, 2e. Wiley.

Olsson, M., Sultana, S., Rommer, S. et al. (2012). *EPC and 4G Packet Networks: Driving the Mobile Broadband Revolution*, 2e. Academic Press.

Rumney, M. (2013). *LTE and the Evolution to 4G Wireless: Design and Measurement Challenges*, 2e. Wiley.

Sesia, S., Toufik, I., and Baker, M. (2011). *LTE: The UMTS Long Term Evolution: From Theory to Practice*, 2e. Wiley.

Voice over LTE (VoLTE) and the IP Multimedia Subsystem

Camarillo, G. and Garcia-Martin, M.-A. (2008). *The 3G IP Multimedia Subsystem (IMS): Merging the Internet and the Cellular Worlds*, 3e. Wiley.

Noldus, R., Olsson, U., Mulligan, C. et al. (2011). *IMS Application Developer's Handbook: Creating and Deploying Innovative IMS Applications*. Academic Press.

Poikselkä, M., Holma, H., Hongisto, J. et al. (2012). *Voice over LTE (VoLTE)*. Wiley.

An Introduction to 5G: The New Radio, 5G Network and Beyond, First Edition. Christopher Cox.
© 2021 John Wiley & Sons Ltd. Published 2021 by John Wiley & Sons Ltd.

Spectrum, Antennas and Propagation

Balanis, C.A. (2009). *Antenna Theory: Analysis and Design*, 3e. Wiley.

Bertoni, H.L. (1999). *Radio Propagation for Modern Wireless Systems*. Prentice Hall.

Parsons, J.D. (2000). *The Mobile Radio Propagation Channel*, 2e. Wiley.

Saunders, S. and Aragón-Zavala, A. (2007). *Antennas and Propagation for Wireless Communication Systems*, 2e. Wiley.

Wireless Communications

Du, K.-L. and Swamy, M.N.S. (2010). *Wireless Communication Systems*. Cambridge University Press.

Goldsmith, A. (2005). *Wireless Communications*. Cambridge University Press.

Molisch, A.F. (2010). *Wireless Communications*, 2e. Wiley.

Proakis, J.G. and Salehi, M. (2007). *Digital Communications*, 5e. McGraw-Hill.

Rappaport, T.S. (2001). *Wireless Communications: Principles and Practice*, 2e. Prentice Hall.

Rappaport, T.S., Heath, R.W., Daniels, R.C., and Murdock, J.N. (2014). *Millimeter Wave Wireless Communications*. Prentice Hall.

Sauter, M. (2017). *From GSM to LTE-Advanced Pro and 5G: An Introduction to Mobile Networks and Mobile Broadband*, 3e. Wiley.

Tse, D. and Viswanath, P. (2005). *Fundamentals of Wireless Communication*. Cambridge University Press.

Multiple Antennas

Björnson, E., Hoydis, J., and Sanguinetti, L. (2018). *Massive MIMO Networks: Spectral, Energy, and Hardware Efficiency*. Now Publishers.

Clerckx, B. and Oestges, C. (2012). *MIMO Wireless Networks: Channels, Techniques and Standards for Multi-Antenna, Multi-User and Multi-Cell Systems*, 2e. Academic Press.

Hampton, J.R. (2013). *Introduction to MIMO Communications*. Cambridge University Press.

Heath, R.W. and Lozano, A. (2018). *Foundations of MIMO Communication*. Cambridge University Press.

Marzetta, T.L., Larsson, E.G., Yang, H., and Ngo, H.Q. (2016). *Fundamentals of Massive MIMO*. Cambridge University Press.

Digital Signal Processing

Lyons, R.G. (2010). *Understanding Digital Signal Processing*, 3e. Prentice Hall.

Proakis, J.G. and Manolakis, D.K. (2013). *Digital Signal Processing*, 4e. Pearson.

Smith, S.W. (1998). *The Scientist and Engineer's Guide to Digital Signal Processing*. California Technical Publishing.

Mathematics

Bird, J. (2017). *Higher Engineering Mathematics*, 8e. Routledge.
Singh, K. (2011). *Engineering Mathematics Through Applications*, 2e. Palgrave.
Stroud, K.A. and Booth, D.J. (2013). *Engineering Mathematics*, 7e. Palgrave.

5G System

Chandramouli, D., Liebhart, R., and Pirskanen, J. (2019). *5G for the Connected World*. Wiley.
Marsch, P., Bulakci, Ö., Queseth, O., and Boldi, M. (2018). *5G System Design: Architectural and Functional Considerations and Long Term Research*. Wiley.
Penttinen, J.T.J. (2019). *5G Explained: Security and Deployment of Advanced Mobile Communications*. Wiley.
Rommer, S., Hedman, P., Olsson, M. et al. (2019). *5G Core Networks: Powering Digitalization*. Academic Press.
Webb, W. (2018). *The 5G Myth: When Vision Decoupled from Reality*. De G Press.

5G Air Interface

Ahmadi, S. (2019). *5G NR: Architecture, Technology, Implementation, and Operation of 3GPP New Radio Standards*. Academic Press.
Dahlman, E., Parkvall, S., and Sköld, J. (2018). *5G Physical Layer: Principles, Models and Technology Components*. Academic Press.
Holma, H., Toskala, A., and Nakamura, T. (2019). *5G Technology: 3GPP New Radio*. Wiley.
Johnson, C. (2019). *5G New Radio in Bullets*, 2e. Createspace.
Zaidi, A., Athley, F., Medbo, J. et al. (2018). *5G Physical Layer: Principles, Models and Technology Components*. Academic Press.

Index

a

access and mobility management function
(AMF)   31–32
  change   255, 258, 317
  inter-operation with EPC   348
  network slicing   38
  non-3GPP access   37
  protocols   41, 42, 147
  roaming   34
  security   262
  selection   248, 251
  serving   31
  state diagrams   40, 41
  see also AMF
access and mobility management function
(AMF) services
  Namf_Communication   255, 258, 297,
    302, 334
access network release procedure   41, 259,
  321–322, 341–342
access point name (APN)   38
access stratum (AS)   42, 55, 145
account balance management function
(ABMF)   295
acknowledgement   113–115, 220–221, 240
acquisition procedure   173–175, 249, 310,
  327
active antenna system (AAS)   122
additional maximum power reduction   250
Advanced Encryption Standard (AES)   270
Advanced Mobile Phone System (AMPS)   6
aggregate maximum bit rate (AMBR)
  282–283, 295

aggregation level
  PDCCH   209
  PDSCH   212
  PUSCH   212
air interface   1, 22, 55, 58, 145–146
allocation and retention priority (ARP)
  244–245, 295, 315
AMF identifier (AMFI)   39
AMF pointer   39
AMF region   39
AMF region ID   39
AMF set   39, 251, 258
AMF set ID   39
analogue processing   87, 94–95, 146,
  168–169
analogue-to-digital (A/D) converter   95, 110,
  121, 122, 138
angular spread   85, 124
antenna   73, 75–76
antenna array   78–79, 117
antenna gain   75–77
antenna panel   138, 141, 196, 203
antenna port   164–166, 168
  CSI-RS   165, 191, 193, 208, 217
  DM-RS   165, 208, 217–219
  PDCCH   165, 208
  PDSCH   165, 217, 219
  PT-RS   165, 217–219
  PUSCH/SRS   165, 203, 218
antenna spacing   119, 121, 124, 127, 138
apiName   47–49
apiRoot   47, 49
apiSpecificResourceUriPart   48, 49

An Introduction to 5G: The New Radio, 5G Network and Beyond, First Edition. Christopher Cox.
© 2021 John Wiley & Sons Ltd. Published 2021 by John Wiley & Sons Ltd.

Printed and bound by CPI Group (UK) Ltd, Croydon, CR0 4YY

27/10/2024